TESLA
Wizard at War

BOOKS BY MARC J. SEIFER

Nonfiction

Wizard: The Life and Times of Nikola Tesla

Transcending the Speed of Light

Where Does Mind End?

The Definitive Book of Handwriting Analysis

Framed! Murder, Corruption, and a Death Sentence in Florida

Ozone Therapy for the Treatment of Viruses

The Art & Life of Robert J. Adsit, with Marta Waterman and Terrance Neuage

Fiction

Rasputin's Nephew

Doppelgänger

Crystal Night

Fate Line

TESLA
Wizard at War

The Genius, the Particle Beam Weapon, and the Pursuit of Power

Marc J. Seifer

CITADEL PRESS
Kensington Publishing Corp.
www.kensingtonbooks.com

CITADEL PRESS BOOKS are published by

Kensington Publishing Corp.
119 West 40th Street
New York, NY 10018

Copyright © 2022 Marc Seifer

All Kensington titles, imprints, and distributed lines are available at special quantity discounts for bulk purchases for sales promotions, premiums, fund-raising, educational, or institutional use. Special book excerpts or customized printings can also be created to fit specific needs. For details, write or phone the office of the Kensington sales manager: Kensington Publishing Corp., 119 West 40th Street, New York, NY 10018, attn: Sales Department; phone 1-800-221-2647.

CITADEL PRESS and the Citadel logo are Reg. U.S. Pat. & TM Off.

ISBN: 978-0-8065-4096-2

First Citadel hardcover printing: September 2022

10 9 8 7 6 5 4 3 2 1

Printed in the United States of America

Library of Congress Control Number: 2020945319

First electronic edition: September 2022

ISBN: 978-0-8065-4098-6

CONTENTS

In knowing George Lucas all these years,
I've experienced the Force. . . . He's a pathfinder,
a pioneer like Edison, Bell, and Tesla and Jobs
in the way that everything they touched changed a paradigm.

—STEVEN SPIELBERG
at the Kennedy Center,
December 30, 2015

For my mother, Thelma Imber Seifer,
and father, Stanley Seifer

NOTE TO THE READER

Our planet has been capitalized as "the Earth" herein whenever it is referred to, even if an original quote has kept it uncapitalized. The reason for this is to help understand Tesla's idea of attempting to send wireless waves through the ground. Conceptually, it makes a difference if he is sending them through the earth or the Earth, and it is the latter idea that he means to convey.

PREAMBLE

In 1896, Tesla receives Swami Vivekananda in his laboratory. And what is the laboratory for Tesla?. . . It is not just [a location], some place where he sees some experiments to sell them, as many scientists do nowadays. No, it is his temple where he explores, where he is in connection with the universe and higher beings and where he works.

When Vivekananda came, he was completely shocked because Tesla first attached himself to various currents and glowed in the dark, and these were emanations . . . emanations of Indian gods. Indra is the one in Indian mythology who throws lightning . . . and the other divinities also appear for certain visions that are very similar to the electrical sights Vivekananda saw in Tesla's laboratory. These were strong electrical discharges to which Tesla was attached. He would calculate the capacitance of his body, I mean the electricity, so that it doesn't damage him and he can glow for half an hour without a problem. Vivekananda was completely puzzled. Tesla is the only Western man who was granted the status of a saint in India, for example.

If you go to Kolkata, [you will find] Vivekananda's Center there. . . . And if you go inside, you will not see either Ramakrishna's or Vivekananda's portrait. You will find Tesla's portrait in the first place. Because Vivekananda says in his letters, "I met a man who is completely different from all the other men here in the west . . . his origin is above the origin of some Indian divinities." He [discusses] Shiva, Vishnu, and Krishna. . . .

"In Tesla's electrical discharges," Vivekananda says, "I even felt the presence of Brahma himself."

—VELIMIR ABRAMOVIĆ[1]

INTRODUCTION

WIZARD AT WAR

On Saturday, January 9, 1943, I was on the staff of the Office of the Alien Property [as a] Custodian (Senior Examiner Attorney) stationed in New York Office (120 Broadway). At about noon ... we were alerted by telephone from the Washington Office ... that a Nikola Tesla had just died [and] that he was reputed to have invented, and [was] in possession of a "death ray," a significant military device capable of destroying incoming war planes (presumably Japanese on the West Coast) by "projecting" a beam into the skies creating a "field of energy" which would cause the planes to "disintegrate." Moreover, it was suspected that Germans (enemy agents) were in "hot pursuit" to locate the device or the plans for its production. I was instructed to proceed to the hotel where Tesla had died and to impound all of his belongings.[1]

—IRVING JUROW,
personal letter to Marc J. Seifer,
July 5, 1993

I rving Jurow, who was a young attorney during the height of World War II, recalled that Tesla's nephew, Sava Kosanović, had already been to Tesla's room and had removed at least one item, presumably a photograph. Jurow was told that Kosanović was an ambassador from "Rumania," although he was actually the ambassador from Yugoslavia, but at this point, he had no idea who Tesla was.

Here it was, fifty years later, and Jurow recounted that his first impression of the inventor was that the deceased man, whose body had been found by a maid, naked except for a pair of leg warmers, was a "deadbeat" who didn't pay his rent to other hotels and that his trunks were filled with "birdseed and newspapers."

1

Jurow had received working orders derived from a bevy of top officials who were concerned that the Germans were trying to obtain Tesla's death ray, a number of whom were in direct contact with Franklin Delano Roosevelt, president of the United States. The initial order to impound Tesla's effects stemmed from Vannevar Bush, founder of the Raytheon Corporation, former dean of the Engineering Department at the Massachusetts Institute of Technology (MIT), and then head of the Office of Scientific Research and Development (OSRD), who was also at that time in charge of the top secret Manhattan Project. Bush sent his directive to L. M. C. Smith, heir to the Smith-Corona fortune and head of the Special War Policies Unit in Washington, DC, who alerted J. Edgar Hoover of the FBI that the Office of Alien Property (OAP) had jurisdiction and that they would take charge.[2] Walter Gorsuch, head of the OAP in New York, was given the directive. He gave Jurow the order to seize Tesla's estate, and the young attorney complied.

Jurow stated that there was the fear that the Russians had offered Tesla $50 million for his death ray, and because of that, he was accompanied by military advisors from Army Intelligence, the Office of Naval Intelligence, and the Office of Strategic Services (OSS) to the Hotel New Yorker, where Tesla had lived, to impound his effects. In short order, Jurow verified that indeed Tesla was a "deadbeat" who still owed rent to the Waldorf-Astoria, the Hotel Pennsylvania, the Hotel St. Regis, and the Hotel Governor Clinton, where he had in some or all of these places various trunks being held as collateral against his various debts. One can only imagine Jurow's state of mind when he was told by the people at the Westinghouse Corporation that "there would be no Westinghouse if it weren't for Tesla."

I first became aware of Nikola Tesla in 1976, which was thirty-three years after his death. Clearly, Tesla's name had disappeared from the mainstream, his legacy kept alive in the underground by John O'Neill's 1944 watershed biography *Prodigal Genius*, which was being distributed by a UFO organization. What this meant was that very few individuals in the country knew who Tesla was and that many of those who did know Tesla's name connected him to occult circles and extraterrestrial beings.

The 1970s were an exciting time for me. In the early part of the decade I left a job in New York City to begin graduate work at the University of Chicago. I had no idea that the university was such a prestigious place. Already by that time there were something like fifty Nobel Prize winners coming from that school. To add to the university's cachet, it sat at the site of the 1893

Chicago World's Fair, with the science building from the fair just several hundred feet from my dorm. When I had a class in a building along the midway, that referred to the midway from the fair, where the great Ferris wheel had been situated.

I had never heard of Tesla at that time (ca. 1972–1973), so I had no way of knowing that this location would later loom so large in my research and my psyche because the fair marked the introduction of Nikola Tesla onto the world stage as the inventor of our AC power system, the global electrical network, often called "the grid," which is essentially unchanged today, well over a hundred years later.

I decided to make the study of the unconscious the subject of my master's thesis, and this took me beyond the theories of Freud and Jung to the realm of parapsychology and the topic of telepathy. To my astonishment, I found that J. B. Rhine had conducted many successful telepathy experiments at Duke University stemming back to the 1930s. I also became aware that in Great Britain, such famous scientists and electrical engineers as Sir William Crookes, Lord Rayleigh, and Sir Oliver Lodge (who I would later find out were all close associates of Tesla), as members of the British Society of Psychical Research, also studied telepathy and were thus able to verify this ability in human beings.

This work enabled me to design a course on parapsychology, which I taught at the night schools of the University of Rhode Island and Providence College. And that endeavor led me to such books as *Psychic Discoveries Behind the Iron Curtain* by Sheila Ostrander and Lynn Schroeder and Andrija Puharich's masterwork *Beyond Telepathy*, wherein Puharich discussed his experiments with various psychics, in particular, Eileen Garrett, who was able to obtain telepathic information when seated inside a Faraday cage, which was a mesh enclosure that blocked out electromagnetic waves.

TALKING WITH THE PLANETS

COLLIERS MAGAZINE, 1901: *I can never forget the first sensations I experienced when it dawned upon me that I had observed something possibly of incalculable consequences to mankind. I felt as though I were present at the birth of a new knowledge or the revelation of a great truth. Even now, at times, I can vividly recall the incident, and see my apparatus as though it were actually before me. My first observations positively terrified me, as there was present in them something mysterious, not to say supernatural, [It was shortly after a*

massive thunder storm in the summer of 1899] and I was alone in my
[Colorado Springs] laboratory at night.[3]

LETTER TO THE RED CROSS, 1900: *I have observed elec-*
trical actions, which have appeared inexplicable. Faint and uncertain
though they were, they have given me a deep conviction and foreknowl-
edge, that ere long all human beings on this globe, as one, will turn
their eyes to the firmament above, with feelings of love and reverence,
thrilled by the glad news: "Brethren! We have a message from another
world, unknown and remote. It reads: one . . . two . . . three . . ."[4]

—Nikola Tesla

In a massive article adorned with H.G. Wellsian drawings of what Martians
might look like and astronomers like Percival Lowell "photographs" of their
supposed canals, the Italian upstart Guglielmo Marconi boldly proclaimed,
"Ten years from now we may be talking to the inhabitants of the planet
Mars." Not to be outdone, having already decreed that he might actually
have already achieved that feat, the Serbian wonder-maker got a bit bolder:

NEW YORK AMERICAN, 1906: *"With a billion horsepower*
drawn from Niagara," Tesla said, he was not only   going to send a
signal back to Mars, but "one hundred times further . . . to Neptune.
. . . Whether we can get an answer or not depends on who is there.
More than likely, the first answer [from] my neighbor will be: 'Well—
well—at last. We have been calling you the last ten thousand years,'"[5]

—Nikola Tesla

In 1975, I met Howard Smukler, who was teaching a course on UFOs at the
University of Rhode Island and also the New School in New York, and so,
together, we created an academic journal focused on the topic of parapsychology. Soon after, he became editor of two national magazines, *Ancient
Astronauts* and *ESP Magazine*, and I became one of the main writers. This
work led to my discovery of an odd book on avatars that suggested that
a man with the strange name of Nikola Tesla had arrived to the Earth
from the planet Venus to give us such inventions as the induction motor,
the hydroelectric power system, fluorescent and neon lights, remote-control
capabilities, and robotics. I thought the story absurd; however, since I was
in the New York Public Library, I looked up the name and found to my
astonishment an article by Tesla on high-frequency phenomena.

When I returned to Rhode Island, Howard handed me two books, *Prodigal Genius*, an exceptional biography of Tesla by science writer John O'Neill, and a nutty book called *Nikola Tesla and the Venusian Spaceship* by Arthur Matthews, which contained Tesla's entire autobiography (unavailable in any other form at that time) and a ridiculous story that Tesla was still alive at the age of 114, flying around in a UFO, occasionally landing in Matthews' back yard, which was some place north of Quebec.

By that time, I had met Andrija Puharich at a conference on consciousness held at Harvard University, and thus became aware that Andrija had visited Arthur Matthews to find out more about his so-called Teslascope, which supposedly interacted with extraterrestrials. The rumor was that Puharich was funded for his ESP work by the CIA, so I suggested we set up an interview and Howard agreed. Andrija's living near Sing Sing Prison in Ossining, New York, added to the allure of the meeting.

Andrija cordially invited us into his living room. Among other things, Andrija stated that everything in his room had levitated except for the piano. I asked him what it was like working for the CIA, and he said he felt like he was sitting in the dentist's chair with his hands holding the dentist's testicles. With regards to Tesla, he said he had met with Matthews at his home north of Quebec and accepted at face value the possibility that the Teslascope was real, but was disappointed that the odd fellow was unable to demonstrate the equipment. Fearful of secret agents, Matthews told Andrija that he had hidden the various components in different places. We also discussed Puharich's work with the Israeli superpsychic Uri Geller, whom, by this time, we had also met, and we tried to find out more information as to why their relationship fell apart.

While we were walking the property, Andrija showed me his watch, which he said was controlled by extraterrestrials. When they wanted to contact him, the watch moved. Clearly, spending time with Andrija Puharich strained one's credulity. The field of parapsychology was a minefield, and it became near impossible to determine where reality ended and fantasy began.[6] However, Andrija also had written a well-constructed book on technical aspects of Tesla's wireless tower at Wardenclyffe,[7] and he also had a great deal of information about one of Tesla's engineering associates, whom Puharich had worked with.

Many of Puharich's telepathy experiments were conducted with John Hays Hammond Jr. at the Hammond castle located in Gloucester, Massachusetts. Jack, as he was also called, a wunderkind inventor himself with over seven hundred patents, as I would later find out, was a pre–World War I business partner of Tesla. Hammond's father, who had made his fortune in gold and diamond mines in South Africa, had helped fund Tesla's remote-controlled robot, according to O'Neill, giving Tesla $40,000 towards the creation of this invention, so Hammond had known Tesla ever since he was a boy. Having studied this device, which Tesla called the telautomaton, Hammond duplicated and improved on what became known as the radio-guidance system, which the younger inventor eventually sold to the military. The basis of this invention, which Hammond hailed as Tesla's "prophetic genius patent," also lay at the heart of protected privacy, wi-fi and drone warfare devices. Tesla's ability to combine frequencies and alter their sequencing as far back as 1898 enabled him to create a virtually unlimited number of separate wireless channels, and that set the stage for what became cell phone technology, radio-guidance systems, encryption, and even remote-controlled robotics.

TESLA'S PARTICLE BEAM WEAPON

WESTERN UNION TELEGRAM:

March 1, 1941
To: Sava Kasanovic

One will need nine stations: four for Serbia, three for Croatia and two for Slovenia, and everyone needs 200 KV which can defend our dear homeland against any type of attack.

—Nikola Tesla[8]

In 1940, at the very outset of what came to be called the Second World War, the actor and later president Ronald Reagan starred in a movie that capitalized on the public's great interest in secret weapons, and in particular, Tesla's so-called death ray.

MURDER IN THE AIR!

1940: A Major Motion Picture
Starring Ronald Reagan

This Is The Most Terrifying Weapon Ever Invented!
The Death Ray Projector!

[The death ray] not only makes the United States invincible in war, but in so doing promises to become the greatest force for World Peace ever discovered. Join Ronald Reagan, Operative 207 battling 20,000 unseen enemies to protect its amazing secret! . . .

"Well, it seems the spy ring has designed one of the greatest war weapons ever invented." Ronald Reagan, as secret agent Brass Bancroft warns, "Which, by the way, is the exclusive property of Uncle Sam!"

The most exciting story of the Secret Service!

—Murder in the Air! 1940

This theme of portraying Tesla as a fictionalized madman actually predates the Ronald Reagan movie by a year, with the very first 1939 Superman comic made into a ten minute cartoon two years later under the title *Superman Captures Mad Scientist!* In the premier cartoon appearing at the height of WWII, the diabolical mad scientist not only threatens the modern city of Metropolis, high on a mountaintop with his laser-like "electrothanasia ray," the crazed genius actually kills a bunch of motorists as he zaps out a bridge resembling the Golden Gate and then, after capturing Lois Lane, starts to take down a major sky scraper before Superman comes to the rescue!

JANUARY 1943

It was the height of World War II. Ignoring yet another treaty, the Germans had invaded Vichy France. In retaliation, D-Day was imminent, U.S. troops were impinging, and Berlin and much of Germany were being bombed with impunity; on the Eastern Front, the Battle of Stalingrad was in full swing. Having already perfected their death-dealing V-2 rockets, Germany was getting closer to manufacturing jet planes, the actor Ronald Reagan was protecting America in the movies with his death ray, and way off in Illinois at the University of Chicago, below the bleachers at Stagg Field, a team led by Enrico Fermi initiated the first nuclear chain reaction. President Roosevelt was now considering two separate kinds of superweapons, a nuclear bomb suggested by Albert Einstein and a death ray suggested by Nikola Tesla. Had he listened just to Tesla, the president would have never considered the bomb, because Tesla had no belief whatever in Einstein's theories. It would be, by far, the inventor's greatest theoretical blunder. Concerning the properties of long-distance wireless messages, Tesla said:

> This . . . bending of the beam . . . is not so much due to refraction
> and reflection [off the ionosphere] as to the properties of a gas-
> eous medium and certain peculiar actions *which I shall explain
> some time in the future.* . . . (emphasis added) [However,] . . . the
> downward deflection always occurs, irrespective of wave length
> . . . all the more pronounced, the bigger the planet. . . . It might
> be inferred that I am alluding to the curvature of space sup-
> posed to exist . . . in the presence of large bodies . . . according to
> the teachings of relativity, but nothing could be further from my
> mind. . . . I, for one, refuse to subscribe to such a view.[9]

Although Tesla was wrong about Einstein, having completely misunderstood
the potential for the atom bomb, hidden in this revealing passage is the hint
of one of Tesla's greatest secrets, his so-called dynamic theory of gravity.
Replacing a curved space with a force field surrounding stars and planets,
Tesla hypothesized that this view accounts for the propagation of electromagnetic surface waves that bend around large bodies. This is a far-reaching idea
directly linked to what today is called "the God particle," the particle that gives
matter its mass. This idea will be explored in the last section of this treatise.

Embedded in Tesla's FBI files was the highly classified Trump Report,
which was an official dossier drafted shortly after Tesla's death in January
of 1943 by MIT physics professor John G. Trump. Assigned by Vannevar
Bush, head of the National Defense Research Committee (NDRC), and his
emissaries, Irvin Stewart and Karl Compton, president of MIT, Trump was
tasked to study Tesla's notes and determine if they contained information
that should be prevented from getting into the hands of a foreign government. Trump, who would turn out to be the uncle of a future U.S. president,
was a close associate of both Robert van de Graaff, inventor of the famous
Van de Graaff generator, who also worked at MIT, and Karl Compton's
brother, Nobel Prize winner Arthur Compton, who cabled New York to
express his condolences regarding Tesla's death. Where Trump's specialty
included the ability to generate one million volts for X-ray machines for hospitals, Arthur Compton would use information derived from this research
for top secret participation in the Manhattan Project.

An expert in the construction of high-voltage particle accelerators,
Trump was a superb choice to study Tesla's papers. Spending just a long
weekend, Trump concluded in his report that the last fifteen years of
Tesla's life were "primarily of a speculative, philosophical and somewhat

promotional character," and thus his papers did not present any lines of research for the military to pursue.[10]

This was not at all the view of the people of Serbia, who expected the elderly wizard to return to his native land and protect them from the Nazi invasion, recalled Mike Markovitch, an economics professor from New York University who had escaped war-torn Yugoslavia and emigrated to America. Attending one of my lectures, Markovitch told me that the Nazis and their allies located in Croatia had murdered ninety thousand Serbs. "I saw the bodies floating down the river," he revealed.

"How did you survive?" I asked.

"By sheer luck," he stated matter of factly.

WESTERN UNION TELEGRAM:

from Nikola Tesla to Sava Kosanović

New York, N.Y.
March 4, 1941

Although I am poor with words, I still didn't explain . . . what would be necessary to increase up to twelve stations: eight in Croatia, each of the same construction like at Wardenclyffe and only 20 meters high—a ball five meters in diameter—the station would be using diesel oil for energy with mechanical action—my air turbines, steam powered, electrically or other manners of transforming into non-alternating electrical current with sixty billion volts pressure without danger. I am waiting for Governor Subasic to select one station on top of Mt. Lovcen. There will not be any light. Electrical energy will deliver particles through space with the speed of 118,837,370,000 centimeters per second . . . 394579 part of the speed of light. . . . The particles can be larger than that of the diameter of a Hydrogen atom with metals of all kinds of materials and sent to all distances [with] good results in war and bring about peace. . . . In my attempts with an effective 20 million volts, electrons carried 40 times more electricity than normally and penetrated two meters in depth and terrible damage in a moment each. I have to finish [so] that I [can later] give you a fresh view.

Warm Greetings, I remain your uncle,
Nikola[11]

Throughout the 1930s, Tesla made no secret of the fact that he had invented a particle beam weapon as he revealed numerous details of the device to a plethora of newspaper reporters. However, it wasn't until he alerted William Laurence of the *New York Times* in 1940 that his prognostications gained the attention of the FBI, as well as various members of military intelligence and individuals associated with the White House. As World War II approached and Roosevelt began to investigate seriously the possibility of creating an atomic bomb, a number of scientists associated with the Manhattan Project also became privy to details of Tesla's secret weapon. These individuals included James Conant, a chemist and president of Harvard University; Richard Tolman, a physicist, theoretical cosmologist, and associate of Albert Einstein who speculated that the origins of the universe were oscillatory in nature; Frank Jewett, first president of Bell Labs; Leo Crowley, head of the OAP who shortly after Tesla's death joined Roosevelt's cabinet; and Vannevar Bush, an early administrator of the Manhattan Project and head of the OSRD, who, like Crowley, had a direct line to the president.[12]

When Tesla's body was discovered on January 7, 1943, management from the Hotel New Yorker alerted Tesla's nephew, Sava Kosanović, ambassador from Yugoslavia, and he in turn notified Tesla acolyte Kenneth Swezey; George Clark, who ran a research lab at RCA; and Hugo Gernsback, editor-in-chief of a variety of science fiction publications. As Gernsback ran off to get the materials to create a death mask, Kosanović, Swezey, and Clark entered Tesla's apartment, supposedly in association with two managers from the hotel who in fact were from military intelligence: Colonel Ralph Doty and Army Private Bloyce Fitzgerald.

Noticing that Tesla owned a sizeable safe, Kosanović contacted a locksmith, who successfully picked the lock and then changed the combination, giving Kosanović the key. Looking for a will, which turned out not to exist, Kosanović removed one or two photographs, and Swezey took a book he had put together in 1931 containing letters from numerous dignitaries such as Nobel Prize winners Wilhelm Roentgen, William Bragg, Ernest Rutherford, Robert Millikan, Arthur Compton, and Albert Einstein, wishing Tesla a happy seventy-fifth birthday. They also noticed that Tesla's Edison Medal was there, along with a set of keys, one of which was to a safety deposit box at the Hotel Governor Clinton, where Tesla had claimed a prototype of his particle beam weapon was housed. Other keys presumably were to other safety deposit boxes throughout the city, probably at the Hotel Pennsylvania, where Tesla owed the hotel $2,000 in back rent, and the Hotel New

Yorker, which had a massive bank of safety deposit boxes on its main floor. Along with other papers, these keys and the Edison Medal were returned to the safe.

Although the FBI had been following Tesla for several years, it was determined that Tesla's estate fell under the auspices of the OAP. Kosanović was a foreigner, that was the rationale, and so the FBI yielded jurisdiction.

At the time of Tesla's death, the world was embroiled in the very height of World War II. As OAP attorney Irving Jurow intimated, there was indeed a fifth column of German sympathizers centered in New York City, and that was probably one key reason why "Wild Bill" Donovan, head of the OSS, was interested in the Tesla death ray. In August of 1942, two German spies who were American citizens and six infiltrators from Germany were dropped off by a German U-boat and were later captured before they carried out plans to sabotage railroad lines, bridges, tunnels, hydroelectric plants, and dams (when several of them defected). And a year after Tesla's death, with the outcome of the war still uncertain, Hitler tried again to directly attack America, sending out two more submarines to "surface on the banks of Amagansett, past the Hamptons, at the far end of Long Island. Four Nazi saboteurs emerged." Although they were spotted, it was dark and they slipped their tail and somehow made their way, most likely by rail, 107 miles to the heart of New York City. Their targets were the great Tesla turbines that ran the subway system, hidden in a colossal subbasement beneath Grand Central Terminal; their weapon, sand. Had they succeeded in simply infiltrating the great generators with their sand, they would have literally brought the city to a halt. Commuters, supplies, and troops would have no longer been able to be transported. This act, had it succeeded, would have crippled the Big Apple and made it impossible to ship goods to the ports, and thus to the front lines. Fortunately, they were caught; two of the men were executed and FDR commuted the executions of the other two in exchange for their revealing the details of their orders.[13]

The Trump report was explicit in expressing Trump's opinion that, although Tesla sold the details of his particle beam weapon to the Soviets, there really was nothing of military importance in the details of its construction.

"Although the investigation brought to light nothing of value of significance to the war effort," the OAP's Homer Jones wrote to Lawrence M.C. Smith, chief of the Special War Policies Unit, War Department, on February 4, 1943, "it is believed the search was entirely worthwhile and justified as a precautionary measure in the national interest."[14]

[In] a few years hence, it will be possible for nations to fight without armies, ships or guns, by weapons far more terrible to the destructive action and range of which there is virtually no limit. Any city at any distance whatsoever from the enemy can be destroyed by him and no power on Earth can stop him from doing so. If we want to avert an impending calamity and state of things which may transform this globe into an inferno, we should push the development of flying machines and wireless transmission of energy without an instant's delay and with all the power and resources of the nation.

—Nikola Tesla,
circa 1925, as told to John O'Neill[15]

Even though Trump was dismissive of Tesla's invention, Exhibits D, F, and Q of his report refer to the highly technical and heretofore underground Tesla treatise written in 1937 titled "The New Art Of Projecting Concentrated Non-dispersive Energy Through Natural Media." This article, in contradiction to Trump's statement, seemed to offer the promise of describing explicit information on the actual workings of a particle beam weapon for destroying tanks and planes and for igniting explosives. And thus, for most of the next decade, that is, in my case, from 1976 until 1984, part of my task was to track down this elusive and highly sought after paper.

SUMMER 1984:
TESLA CONFERENCE, COLORADO SPRINGS

Toby Grotz, head of the International Tesla Society, and physicist Elizabeth Raucher decided to put together the first of what would turn out to be seven biennial international Tesla conferences, most of them run by Steve Elswick. All were held in Colorado Springs, near the site of Tesla's 1899 experimental wireless station. At this first conference, I was invited to speak, along with war games analyst Colonel Tom Bearden; Ozone Therapist George Freibott, MD; Tesla researchers Dr. James Corum and his brother Ken Corum; Robert Golka, who had built two enormous Tesla coils; Tesla's grandnephew William Terbo; Serbian theoretician Rastko Maglic; researchers Oliver Nichelson, Charles Yost, and Moray King; and the medical doctor and parapsychologist Andrija Puharich, who was scheduled to speak about the problem of using Tesla technology to alter brain wave patterns at great distances.[16]

My talk was centered on Tesla's relationship with J. Pierpont Morgan. I explained, with slides and a discussion, precisely what went wrong with their partnership and at the same time showed that Tesla's handwriting fell apart when their deal finally fell through in 1906. This was the first time any researcher, as far as I knew, had speculated that Tesla suffered a nervous breakdown when he finally realized that he would not have the funds to complete his wireless plans at Wardenclyffe.

My plane landed the night before in Denver, but my bus was late in arriving at Colorado Springs, so I came into Puharich's talk about ten minutes after it started. Puharich had been to the U.S. Psychotronics meeting, and Bob Beck, head of the society, had passed along Tesla's particle beam weapon patent application. This was the long-sought top-secret paper, and Puharich was presenting all of the details right before my eyes! It was history in the making. What most likely happened was that one of the military personnel and a veteran from World War II who had met with Tesla during the last months of his life, Ralph Bergstresser, whom I would later interview, had sat on the paper for literally forty years and then finally turned it over to Beck, who gave it to Puharich.

Soon after, Bearden noticed tremendous similarities between Tesla's schematics and the designs of a Soviet particle beam weapon published seven years earlier, in the May 2, 1977, issue of *Aviation Week and Space Technology*. In comparing the drawings, I agreed and therefore speculated that back in 1943, when Trump issued his report, in which he downplayed Nikola Tesla and his inventions, this top secret paper was not removed from the estate, and thus its details were smugly allowed to be sent behind the Iron Curtain, where they were analyzed by the Soviets. Ironically, the details of this weapon were barred from American public view, yet it was studied, worked on, and developed in the Soviet Union! However, as it turned out, as discussed later herein, Tesla's link to the Soviet Union was a lot more complicated than I had first assumed.

One of the key features of the death ray was the realization that this was not a "ray" at all. What Tesla realized was that if the device remained a ray, the electromagnetic beam would spread out too much and lose its effectiveness. Tesla's genius was in conceiving the idea of shooting out a thin beam of particles, tiny pieces of tungsten that would be propelled out the barrel at a terrific force.

I had learned from Nancy Czito, the granddaughter-in-law of Tesla's right-hand man, Coleman Czito, that around 1918, Tesla and Czito's son Julius Czito were bouncing laser or laser-like beams off the Moon and timing the impulse. I knew that Tesla had designed laser-like effects as far back as the 1890s, but Tesla soon realized that if he wanted to design a weapon that would have a range of several hundred miles, a beam of light or electromagnetic energy would be ineffectual. That is when he came up with the idea of shooting microscopic pellets. Since they would not disperse, they would maintain their destructive force.

Andrija Puharich, for me, had come full circle. He had notified me of important Tesla archives at the Hammond Castle in Gloucester, Massachusetts; he had given me inside information about Tesla's biographer, John O'Neill, including the speculation that Bernard Baruch had played a role in dissuading Morgan from completing his funding of the Wardenclyffe project; he had written an important treatise on Tesla's magnifying transmitter; and now he had brought to the world the long-sought top secret Tesla weaponry paper.

> Tesla's life pattern is so rare as life patterns go, even among extraordinary humans, that it needs to be clearly explicated at the outset. Tesla himself considered that he had only one life which was his work. It is true that he had never developed any emotional relationship (outside his parents and siblings) to any human male or female. I say this with some confidence because I was privileged to know many of the people who knew Tesla as closely as anyone knew him. John Hays Hammond, Jr., whose father was a friend and financial backer of Tesla's, told me many a story about this relationship with Tesla for some 30 years. Agnes Holden, who was the daughter of Robert Underwood Johnson, Tesla's closest friend, knew Tesla both as a family friend and as her escort to some of New York Society's parties. I spent many an evening with John J. "Jack" O'Neill, Tesla's official biographer, going over his intimate personal recollections of his PRODIGAL GENIUS. From these and other intimates of Tesla, I caught glimpses of his presence, his mind, his ideas, and his personality that are nowhere to be found in his dry and dusty archives. How this absence of close human relationship came about can be traced, and it has meaning in his work and ideas.
>
> —ANDRIJA PUHARICH[17]

DISAPPEARANCE FROM HISTORY

Unknown to me, my real introduction to Tesla began when I was a boy in the late 1950s. My father, Stan Seifer, at the time was a TV repairman, and I would go out with him on some of his jobs. At that time, he would go up onto the roofs of the various houses to install TV antennas and also wire their homes for TV reception. My father also built TVs from scratch, and there were a number of them at his workbench in various stages of completion along with testing equipment, many varieties of radio tubes, and other electronic paraphernalia. With my father, I constructed an induction motor using large nails and wires wrapped to create magnetic lines of force, and with his help for a Boy Scout assignment, I constructed a fully working crystal radio set made out of a jar with wires wrapped around it, a dial made from a beer can, a crystal detector, aerial and ground connectors, and earphones.

In 1986, under the tutelage of my mentor, Dr. Stanley Krippner, I completed a 715-page doctoral dissertation, "Nikola Tesla: Psychohistory of a Forgotten Inventor," and ten years later, I published the book *Wizard: The Life and Times of Nikola Tesla, Biography of a Genius.*

When *Wizard* came out in 1996, it was several years before the internet became ubiquitous and what we call "smart phones" really took off. At that time, Tesla was still a very obscure individual. Yet surprisingly, as early as 1992, when cell phones were the size of sixteen-ounce beer cans, Tesla was listed by the *New York Times* as being in the top ten people of interest by scholars using the *Encyclopedia Britannica*'s Instant Research Service. The other nine individuals were, in order: John F. Kennedy, Martin Luther King Jr., Albert Einstein, Adolf Hitler, Edgar Allan Poe, Abraham Lincoln, Jesus Christ, Pablo Picasso, and Al Capone. Tesla came in tenth.[18]

Twenty years later, on April 11, 2013, *GeekWire* announced that Nikola Tesla had won a contest crowning him "Greatest Geek Ever" over such rivals as Bill Gates, Alan Turing, Steve Jobs, Elon Musk, Mark Zuckerberg, Jeff Bezos, Gordon Moore, Linus Torvalds, Albert Einstein, Isaac Newton, Tom Edison, Galileo, Ben Franklin, Archimedes, Guglielmo Marconi, Madame Curie, Gutenberg, Alexander Graham Bell, and Leonardo da Vinci. Clearly, Tesla's name was back on the rise.

When I wrote *Wizard*, I set out to uncover Tesla's actual role in the development of the many inventions listed above. At that time, there was a lot of disinformation about Tesla's role in the development of his most important achievements, especially in regard to the invention of the induction motor, the AC polyphase system (our hydroelectric system), remote-control capabilities, robotics, and wireless communication. Even now there is a myth that Tesla was incapable of understanding James Clerk Maxwell's equations on electromagnetic theory, even though his notebooks are filled with incredibly complex equations.[19] However, since my biography contained a strict chronology and cited all its sources, I found that disinformation about Tesla generally seemed to dissipate, and I'd like to think that my book helped spawn the publication of a number of new books on Tesla that are a welcome addition to the literature on this topic. Excellent new works that stand out include the well-illustrated *Tesla: The Wizard of Electricity* by David J. Kent; *Lightning Strikes: Timeless Lessons in Creativity from the Life and Work of Nikola Tesla*, which contains singular insights by John F. Wasik; and a new definitive, highly original, beautifully illustrated coffee table–sized biography, *Wireless: The Life, Work, and Doctrine of Nikola Tesla* by the director of the Tesla Museum, Branimir Jovanovic.

Other works include the impressive picture book *The Truth About Tesla* by Christopher Cooper, with an introduction by myself; the weighty compendium *Tesla: Inventor of the Electrical Age* by W. Bernard Carlson, which argues that Tesla used illusion to sell his wares;[20] an attractive oversized picture book, *Tesla: The Life and Times of an Electric Messiah* by Nigel Cawthorne; and a well-written compact text *Tesla: Inventor of the Modern* by Richard Munson.

Although most of his creations were generated in the nineteenth century, in actuality, Tesla was a modern thinker. For his entire adult life, the wizard cautioned the world to stop its dependence on fossil fuels and seek alternative sources of energy. In his article titled "Our Future Motive Power," written in the 1930s, Tesla discusses the very topic that Tesla Motors CEO and SpaceX founder Elon Musk propounds as the most pressing problem of our time: how to obtain alternative sources of nonpolluting renewable energy. In this comprehensive piece, Tesla suggests not only waterfalls but also harnessing the tides, wind power, solar energy, and geothermal energy, all avenues that would, in Tesla's words, "harness the wheelwork of nature." It was not by chance that Tesla sought to solve the electrical power distribution problem, filling his notebooks with advanced calculations; he was also always thinking of the best way to help humanity move forward but be respectful of the planet at the same time.[21]

This present book is centered on Tesla's link to war, particularly his connection to Telefunken, the German wireless concern, during the run-up to World War I and the related complicated legal battles in wireless when Marconi sued Telefunken and the U.S. Navy, and Tesla countersued. With regards to World War II, since both the FBI and CIA have recently declassified their Tesla archives, I was able to compare redacted with unredacted passages. Along with new information I have obtained from the Tesla Museum in Belgrade, from the research facilities of Prometheus Films in our production of *The Tesla Files* TV show, and from Tesla aficionados from around the globe, I have been able to greatly expand my understanding of what Tesla's role was during these wars, including which world leaders he interacted with, particularly during the onset of World War II. Further, I was also able to shed a great deal more light on the rather complex story of discovering precisely what happened to his top secret papers after he died.

Tesla is becoming more and more a household name, in part because of his take-your-breath-away photos that grace the internet and because he appears frequently on TV shows such as *American Experience, Ancient Aliens, The Tesla Files*, and, in particular, *The Big Bang Theory* (2007–2019), with a worldwide audience of nearly fifty million. Tesla also appears in motion pictures, such as *The Current War, La La Land, Tesla* starring Ethan Hawke, *Final Frequency*, and also *The Prestige*, wherein the wizard is played by rock star Davie Bowie. In print media, Tesla graces such popular books as James Redfield's *The Tenth Insight*, the sequel to his blockbuster breakout book, *The Celestine Prophecy*; and *The Seventh Plague* by James Rollins; and Tesla's name plays key roles as nonhuman characters: the advanced roadster in *Razor Girl* by the sardonic author Carl Hiaasen and the self-driving electrical vehicle in *Origin* by mega-author Dan Brown.

In 2010, Tesla was mentioned by President Barack Obama when the former was listed along with Sergey Brin, Andrew Carnegie, and Albert Einstein in a talk on immigration. Five years later, Steven Spielberg dropped Tesla's name along with those of Tom Edison, Alexander Graham Bell, and Steve Jobs when he compared these iconoclasts to George Lucas when his close friend became a Kennedy Center honoree. Another key aspect to this is the personality, accomplishments, and genius of Elon Musk, who has not only created an incredibly advanced electric vehicle named after the inventor that he launched literally into space in February of 2018, but he also contributed other accomplishments in perfecting solar cells and his incredible SpaceX rocket ships. Thus, Tesla's name is frequently mentioned when the most advanced technologies are discussed in unrelated stories. For example, in *An Inconvenient Sequel: Truth to Power,* the Al Gore sequel to *An Inconvenient Truth,* as the former vice president seeks to get India to agree to sign onto an international agreement addressing climate change, he calls Elon Musk, mentions the word *Tesla,* and gets Musk to agree to provide solar cells to India as an incentive to move them away from building air-polluting coal-operated power plants. In the case of the recent Ebola and COVID-19 pandemics, ozone therapy has made the news as a possible treatment, and thus Tesla's name arises again because Tesla invented and patented an ozone generator in 1896 (patent #568,177), and this simple fact is therefore mentioned in scientific articles that discuss the possible efficacy of this treatment. In other words, the word *Tesla* has simply reentered the lexicon as a matter of course. As solid evidence of Tesla's newfound worldwide appeal, when Matt Inman of The Oatmeal website worked with Jane Alcorn, head of the Wardenclyffe project, to set up a crowd-funding campaign in 2012 to purchase Tesla's laboratory out on Long Island, thirty-three thousand people from 108 countries raised $1.4 million in ten days to save Wardenclyffe![22]

Tesla was a great writer and an amazing predictor of the future. One of his most brilliant quotes came in 1904 when he was trying desperately to talk his backer, J. Pierpont Morgan, into providing the rest of the funds he needed to complete his great wireless enterprise known as Wardenclyffe:

> The results attained by me have made my scheme of "World Telegraphy" easily realizable. It constitutes a radical and fruitful departure from what has been done heretofore. It involve[s] the employment of a number of plants each of [which] will be preferably located near some important center of civilization and the news it receives through any channel will be flashed to all points of the globe. A cheap and simple [pocket-sized] device, may then be set up somewhere on sea or land, and it will record the world's news or such special messages as may be intended for it. Thus the entire Earth will be converted into a brain as it were, capable of response in every one of its parts. Since a single plant of but one hundred horse-power can operate hundreds of millions of instruments, the system will have a virtually infinite working capacity.

> —NIKOLA TESLA, 1904[23]

Several things about this quote are truly amazing, foremost being the idea of conceiving of the Earth itself as one totality able to feel in all its parts. Compare this prescient idea with the actual development and invention of what came to be called the World Wide Web. The scene is the late 1980s in Geneva, Switzerland, as a group of scientists from all over the world was building on the work of DARPA (Defense Advanced Research Projects Agency) expert Vint Cerf on the development of the internet. Their goal was to augment communication with each other in their attempts to complete the construction by CERN, the European Organization for Nuclear Research, of a huge particle accelerator that came to be known as the Large Hadron Collider, and also to share information with other comparable centers of research around the globe:

> To make communication between the experiments and the particle physicists scattered all over the world easier and more effective, Sir Timothy "Tim" Berners-Lee, a British computer scientist working at CERN, developed a new networking system that eventually became the "World Wide Web" (WWW) in 1989. During the development period, its basic concepts such as URL, HTTP and HTML were defined. They are still in use today. The first browser and server software was also created at this time. Berners-Lee made his ideas freely available with no patent and no royalties due. CERN management decided to give this invention to the "public domain" and make it available to everyone for free. The Internet, a global network of information and easy accessibility that was previously used by only a handful of scientists, military personnel and computer nerds, was made available to the general public in this act of generosity.
>
> —MICHAEL KRAUSE, 2014[24]

Ironically, as we shall come to see, the Higgs boson, often called the God particle, the hypothesized particle that gives matter its mass, which was apparently discovered at CERN, is directly linked to Tesla's little-known dynamic theory of gravity. By exploring Tesla's theory in this light, a new view on cosmogenesis and the ultimate structure of matter may emerge.

Tesla: Wizard at War covers a lot of territory. The book seeks to fill in previous blanks, explore relationships Tesla had with individuals he has never been linked to before, create a more full-bodied description of who the man was and how his wireless tower at Wardenclyffe was supposed to operate, explain precisely his role in providing advanced technology and weapons development for both world wars, and also explain how his more advanced theories on the nature of gravity result in a rather elegant explanation of precisely what the God particle is. Our story begins with Tesla's own recollections about his childhood and later life.

PART I

High Life

There is no doubt about it—your heart does beat faster when you are about to meet a famous electrical scientist, one of the foremost in the world. . . . It didn't beat a second faster when [I met similar] men, but not to any such extent as when I entered the room where the master electrical wizard lives; he who has produced electrical discharges resembling lightning bolts, the largest electrical discharges ever attempted by man. . . . The allotted time grew short . . . the great inventor bid adieu and I left the presence of one of the world's most distinguished scientists of whom it has rightly been said, "He lives a hundred years ahead of his time."

—SAMUEL COHEN,
"An Interview with Nikola Tesla,"
Electrical Experimenter,
June 1915, pp. 39, 45

1

Interview with Nikola Tesla

I have seen men hung, beaten to death, shot, quartered, stuck on a pointed stick, heads chopped off and children on a bayonette like quails "on broche" at Delmonicos.

—Nikola Tesla[1]

In 1895, or thereabouts, Tesla brought several Serbian ballads to his friends Robert and Katharine Johnson. One of them, "Luka Filipov," was written by Zmai Jovanovich, Serbia's renowned poet whom Tesla met when he returned to Serbia in 1892. It is about a recent war, circa 1870s, between the Serbs and the Turks.

> One more hero to be part
> Of the Servians' glory!
> Lute to lute and heart to heart
> Tell the homely story:
> Let the Moslem hide for shame,
> Trembling like the falcon's game,
> Thinking on the falcon's name—
> Luka Filipov!

During the ensuing conflict, Luka is wounded in battle and a Turkish prisoner is caught.

> We have fired, but Luka's hand
> Rose in protestation,
> While his pistol's mute command
> Needed no translation:
> For the Turk retraced his track,
> Knelt, and took upon his back
> (as a peddler lifts his pack)
> Luka Filipov!

How we cheered him as he passed
Through the line, a-swinging
Gun and pistol—bleeding fast—
Grim—but loudly singing:
"Lucky me to find a steed
Fit to give the Prince for speed!
Rein or saddle ne'er shall need
Luka Filipov!"

And so, Luka is brought to meet the prince.

And as couriers came to say
That our friends had won the day,
Who should up and faint away?
Luka Filipov.

One can easily envision Tesla in the Johnsons' living room translating on the fly this classic verse to Robert and Katharine's delight.[2] An unforgettable moment of deep bonding, this would be a watershed evening that would reverberate throughout their lifelong friendship. And so, Robert became "Dear Luka," and Katharine became "Dear Mrs. Filipov."

When Tesla's translation appeared in Johnson's book of poems, Tesla included a discussion of the "sad fate" of the Serbian people. The turning point occurred in 1389 at the Battle of Kossovo, whereby thirty thousand Turks destroyed the Serbian nation. Tesla writes that "Europe can never repay the great debt it owes the Servians for checking, by the sacrifice of its own liberty, that barbarian influx." The Battle of Kossovo, I would learn, is as important to the Serbs as the flight of the Hebrew slaves from Egypt is for Jews and the crucifixion of Jesus is to the Christians. Thus, to understand Tesla, we must keep in mind that he's a Serb and that, as such, he carries this complex heritage. As Professor Mike Markovitch explained to me regarding the Battle of Kossovo, "It follows us always."[3]

Due to deep-rooted enmities among the Serbs, Croats, and Moslems, even today they are unable to get along, with over a hundred thousand perishing when civil war erupted in the mid-1990s, thereby splitting Yugoslavia into seven separate nations. As a young man, when Tesla took to the hills, in part to avoid the draft, he encountered what Joseph Conrad called the "heart of darkness," where villages were sacked and the men murdered and their heads put on stakes, but the women and children suffered similar fates.

I was chokin' mad with thirst,
An' the man that spied me first
Was our good old grinnin', gruntin' Gunga Din.

Both Katharine and Tesla read Rudyard Kipling's *The Jungle Book* and *Rikki-Tikki-Tavi*, which Tesla liked "best."[4] On April 1, 1901, after his return from his experiments in Colorado, Tesla met the world-famous poet for lunch. "What is the matter with inkspiller Kipling?" Tesla exclaimed to Mrs. Filipov. "He dared to invite me to dine in an obscure hotel where I would be sure to get hair and cockroaches in the soup!"[5]

'E carried me away
To where a dooli lay,
An' a bullet come
an' drilled the beggar clean.
'E put me safe inside
An' just before 'e died
"I 'ope you liked your drink," sez Gunga Din

Both Tesla and Kipling shared similar views of the horror of war and praise for great heroes. Tesla writes in Robert's book *Poems* that the Serbian hero Marko Kraljevich, "when vanquishing Musa, the Moslem chief, exclaims, 'Woe unto me, for I have killed a better man than myself!'" Kipling, of course, uses the same observation in his iconic verse.

About the time Tesla was conceived, according to Tesla's father, "a divine phenomena" occurred. A large meteor, perhaps the size of a house, streaked across the sky, "pouring sparks [leaving] a trail of purple stripes behind. And when [it disappeared beyond] the first hill there was a dawning, as though some great tower had collapsed and the echo reverberated on the south side of the Velebit for a long time."[6] The following year, during a terrific lightning storm, at midnight on July 10, 1856, Nikola Tesla was born. When the midwife tried to suggest that being born during a storm was a bad omen, according to legend, Tesla's father instead insisted, "No, this will be a child of light!"

Myths continued to abound, captured most eloquently by Tesla's first biographer, John O'Neill, who wrote:

Even the Gods of old, in the wildest imaginings of their worshipers, never undertook such gigantic tasks of world-wide dimension as those of Tesla . . . himself an invention, . . . a

self-made superman . . . designed specifically to perform wonders; and he achieved them in a volume far beyond the capacity of the world to absorb.[7]

In 1919, at the age of sixty-three, Tesla penned his autobiography. Two
years later, in April of 1921, *American Magazine* published another rendition.
The piece was described as follows: "Great inventor tells the romantic story
of his life." The piece, recaptured in part below, begins with Tesla stating
that he was born "in an old fashioned home . . . two miles away" from
any neighbor. His mother is described as "an indefatigable worker," arising
every morning at four a.m. "while others slumbered," making breakfast,
assisting the servants in feeding the animals, and fashioning various tools
through her own inventive spirit.

Tesla then describes "the fountain of my enjoyment," his "magnificent
Macak," the cat whom Tesla played with as a child, rolling around down
a hill, "indulging in this enchanting sport day by day." Pausing in the
story, he relates an indelible occurrence that made a profound impact on
his later life:

> It happened on the day . . . we had a cold drier than ever
> observed before. People walking in the snow left a luminous
> trail behind them and a snowball thrown against an obstacle
> gave a flare of light like a loaf of sugar hit with a knife. . . . [At]
> dusk . . . I felt compelled to stroke Macak's back . . . [which
> became] a sheet of light and my hand produced a shower of
> sparks loud enough to be heard all over the place.
>
> My father was a very learned man. . . . But this phenom-
> enon was new, even to him. Well he finally remarked, "This
> is nothing but electricity, the same thing you see on the trees
> in a storm."
>
> My mother seemed alarmed. "Stop playing with the cat!"
> she said. "He might start a fire!"
>
> I was thinking abstractedly. Is nature a gigantic cat? If so,
> who strokes its back? It can only be God, I concluded. . . . Here
> I was, only three years old, and already philosophizing!
>
> I cannot exaggerate the effect of this marvelous sight on my
> childish imagination. Day after day I asked myself what is elec-
> tricity and found no answer. Eighty years have gone by since
> and I still ask the same question, unable to answer it.[8]

Tesla goes on to describe his two earliest inventions, a fishhook he used to capture frogs and a motor created by fastening very large May bugs to a propeller.

> The May bugs never knew when to stop; the hotter it was, the harder they worked . . . until one day I saw the son of a retired officer of the Austrian army eating May bugs, and seeming to enjoy them. I never played with the bugs after that, and to this day I shrink from touching any kind of insect.[9]

> My childhood . . . might have passed blissfully if I did not have a powerful enemy, relentless and irreconcilable. This was our gander, a monstrous ugly brute, with the neck of an ostrich, mouth of a crocodile and a pair of cunning eyes radiating intelligence and understanding like that of the human. I aroused his ire by throwing pebbles at him, a most foolish and reckless act which I bitterly regretted afterwards. . . . The moment I entered the poultry yard, he would attack me and as I fled, grab me by the seat of my trousers and shake me viciously. When I finally managed to free myself and run away he would flap his huge wings in glee and raise an unholy chatter in which all the geese joined. . . .

> [I had] two aunts. . . . One was Aunt Veva who had two protruding teeth like the tusks of an elephant. She loved me passionately and buried them deep in my cheek kissing me. I cried out from pain, but she thought it was from pleasure and dug them in still deeper. Nevertheless, I preferred her to the other whose name has slipped from my memory and she used to glue her lips to mine and suck and suck until by frantic efforts I managed to free myself gasping for breath.

> These two aunts amused themselves by asking me all sorts of questions of which I remember a few. . . . They asked me who was the prettiest of the two. After examining their faces intently, I answered thoughtfully, pointing to one of them, "This here is not as ugly as the other."

> They asked me, . . . "Are you afraid of the bad wolf?"

> "No! No!" This was the wolf I met in the woods near the church. . . .

After a number of such questions one of the aunts asked me, "Are you afraid of the gander?"

"Yes! Yes!" I replied emphatically, "I am afraid of the gander!"

I had good reason to be. One summer day my mother had given me a rather cold bath and put me out for a Sun's warming in Adam's attire. When she stepped in the house, the gander espied me and charged. The brute knew where it would hurt most and seized me by the nape almost pulling out the remnants of my umbilical cord.

My mother, who came in time to prevent further injury, said to me, "You must know that you can never make peace with the gander or a cock whom you have taunted. They will fight you as long as they live."

But now and then . . . on certain days our geese led by the gander, rose high in the air and flew down to the meadow and brook and sported like swans in the water and probably found some food. . . . The sight of the flying geese was a joy and inspiration to see.[10]

As a very small boy I was weak and vacillating, and made many childish resolves, only to break them. But when I was eight years old I read *The Son of Aba*, a Serbian translation of a Hungarian writer, Josika, whose lessons are similar to those of Lew Wallace in *Ben-Hur*. This book awakened my will power. I began to practice self-control, subdued many of my wishes, and resolved to keep every promise I ever made, whether to myself or to anyone else. The members of my family were not long in learning that if I promised a thing I would do it.

Long before I was twenty, I was smoking excessively—fifteen or twenty big black cigars every day. My health was threatened, and my family often tried to get me to promise to stop. . . . [And finally, after his sister almost died, he agreed]. . . . I have freed myself [from this and other] passions, and so have preserved my health and my zest for life. The satisfaction derived from demonstrating my own strength of will has always meant more to me in the end than the pleasurable habits I gave up. I believe that a man can and should stop any habit he recognizes to be "foolish."

When I was about twenty, I contracted a mania for gambling. We played for very high stakes; and more than one of my companions gambled away the full value of his home. My luck was generally bad, but on one occasion I won everything in sight. Still I was not satisfied . . . I lent my companions money so that we might continue, and before we left the table I had lost all that I had won and was in debt.

My parents were greatly worried by my gambling habits. My father . . . often expressed his contempt at my wanton waste of time and money. However, I never would promise him to give up gambling, but instead defended myself with a bad philosophy that is very common. I told him that, of course, I could stop whenever I pleased, but that it was not worth while . . . because the pleasure was more to me than the joys of Paradise.

My mother understood . . . that a man cannot be saved from his own foolishness . . . by someone else's efforts or protests, but only by the use of his own will. One afternoon, when I had lost all my money, but still was craving to play, she came to me with a roll of bills in her hand—a large sum of money for those times and conditions—and said, "Here, Niko. Take these. They're all I have. But the sooner you lose everything we own, the better it will be. Then I know you will get over this."

She kissed me.

So blinded was I by my passion that I took the money, gambled the whole night, and lost everything, as usual. It was morning when I emerged from the den, and I went on a long walk through sunlit woods pondering my utter folly. The sight of nature had brought me to my senses, and my mother's act and faith came vividly to mind. Before I left the woods, I had conquered this passion. I went home to my mother and told her I never would gamble again. And there never has been the slightest danger of my breaking the promise.

My father was the son of an officer who served in the army of the Great Napoleon. He himself had received military training, [however] both my father and mother were . . . eager that I should become a preacher; but I had no leaning in that direction. From the age of ten I had been inventing all sorts of things in my mind: flying machines, a submarine tube for carrying letters and packages under the Atlantic, and means of getting power from the rotation of the planets; all fanciful, but even after I had gone to study at the gymnasium at Carlstadt, Croatia, where I became intensely interested in physics and electricity, my parents still wanted me to become a preacher.

Perhaps, if I had not become very ill, I should have given [up] my promise. But because of overstudy, I had my first serious breakdown in health. Physicians absolutely gave me up. It was an American genius who saved my life. During my illness . . . one day I was handed a few volumes unlike anything I had ever read, and . . . I forgot my hopeless state. My recovery seemed miraculous.

The books [included] early works of Mark Twain—among them "Tom Sawyer," and "Huckleberry Finn." Twenty-five years later, when I met Mr. Clemens and we formed a lifelong friendship, I told him of this experience and of my belief that l owed my life to his books. I was deeply moved to see tears come to the eyes of this great man of laughter.

Tesla's memory is faulty here, as *Huckleberry Finn* was first published in 1884. *Tom Sawyer* was a possibility, published in 1876 when Tesla was twenty. And he may have also read "The Celebrated Jumping Frog of Calaveras County," which came out in 1865, when Tesla was nine.

After graduating from the Higher Realschule at Carlstadt . . . on the very day of my arrival [home I] was stricken with cholera, which was then epidemic in those parts. Again I was near death. My father tried to cheer me with hopeful words.

"Perhaps," I said, "I might get well, if you would let me become an engineer instead of a clergyman."

He promised solemnly that I should go to the best technical institution in the world. This, literally, put new life into me; and, owing partly to my improved state of mind and partly to a wonderful medicine, I recovered. My father kept his word by sending me to the Polytechnic School in Graz, Styria, one of the oldest institutions of Europe.

All during my first year there I started work at three A.M. and continued until eleven P.M., neither Sundays nor holidays being excepted. Such leisure as I allowed myself I spent in the library. It was during my second year that something happened that has determined the whole course of my life.

We [had] received a Gramme dynamo from Paris. It had a horseshoe form of field magnet and a wire-wound armature with a commutator—a type of machine that has since become antiquated. While . . . demonstrating . . . this machine, the brushes [of the commutator] sparked badly, and I suggested that it might be possible to operate a motor without such appliances. The professor declared that I could never create such a motor, because the idea was equivalent to a perpetual motion scheme.

This statement from such a high authority caused me to waver in my belief. . . . Then I took courage and began to think intently of the problem, trying to visualize the kind of machine I wanted to build, constructing all its parts in my imagination. . . . I conceived many schemes, changing them daily, but I did not at that time succeed in evolving a workable plan.[11]

> *From time to time, in rare intervals, the Great Spirit of Invention descends to Earth to tell a secret which is to advance humanity. He selects the best fitted . . . and whispers the secret in his ear. Like a flash of light, precious knowledge comes. . . . The marvel that he sees . . . He knows it . . . [and in] his body he feels: It is a Great Idea!*
>
> —NIKOLA TESLA[12]

Four years later, in 1881, I was in Budapest, Hungary, studying the American telephone system, which was just being installed. . . . Never for a day had I given up my attempt to visualize an electric motor without a commutator. In my anxiety to visualize one that would work, my health again broke down, just when I was feeling that the long sought solution was near; but after six months of careful nursing I recovered.

Then, one afternoon I was walking with a friend in the City Park . . . reciting poetry. At that time I knew entire books by heart. . . . One . . . was Goethe's "Faust;" and the setting Sun reminded me of the passage:

> The glow retreats, done is the day of toil;
> It yonder hastes, new fields of life exploring;
> Ah, that no wing can lift me from the soil,
> Upon its track to follow, follow soaring!

Even while I was speaking these glorious words, the vision of my induction motor, complete, perfect, operable, came into my mind like a flash. . . . Pygmalion seeing his statue come to life could not have been more deeply moved. A thousand secrets of nature which I might have stumbled upon . . . , I would have given for that one which I had wrestled from her, against all odds and at the peril of my existence. . . . I drew with a stick on the sand the vision I had seen. They were the same diagrams I was to show six years later before the American Institute of Electrical Engineers. My friend understood the drawings perfectly; [when] . . . suddenly I cried, "Look! Watch me reverse my motor!" And I did it, demonstrating with my stick.[13]

Alternating current . . . [transmission] over longer distances was handicapped in the power field for lack of a motor. There was no satisfactory [one] . . . [until] Nikola Tesla [announced one] in 1888. His polyphase system involved two or three currents from the same generator, each alternating . . . in regular sequence . . . overcom[ing] the dead points of a single crank engine . . . produc[ing] a rotation which the single phase system does not. The unique feature of the new system was the induction motor, with no commutator (an essential part of the direct current motor) and with most ideal simplicity of construction and operation.

—EDWARD DEAN ADAMS,
president, Niagara Falls Power Company[14]

This discovery is known as the "rotating magnetic field." It is the principle on which my induction motor operates. In this invention I produced a sort of magnetic cyclone which grips the rotable part and whirls it—exactly what my professor had said could never be done.

After inventing this motor, I gave myself up more intensely than ever to the enjoyment of picturing in my mind new kinds of machines. It was my great delight to imagine motors constantly running. In less than two months, I had created mentally nearly all the types of motors and modifications of the system which are now identified with my name.

In the course of our investigations into the subject of power transmission we came upon the inventions of Mr. Nikola Tesla, now so well known for his remarkable achievements. . . . Recogniz[ing] the value and originality of his discoveries, and the fundamental character of his patents, [we] at once secured the exclusive right to manufacture and sell apparatus covered by these patents.

—GEORGE WESTINGHOUSE,
January 16, 1893[15]

The inventions I have conceived in this way, have always worked. In thirty years there has not been a single exception. My first electric motor, the vacuum tube wireless light, my turbine engine, and many other devices have all been developed in exactly this way.

From Budapest I went to Paris, and there became associated with Mr. Charles Batchelor, an intimate friend and assistant of Mr. Edison. From Paris I made many trips throughout France and Germany, repairing the disorders of powerhouses; but I had no success in raising money for the development of my invention . . . when Mr. Batchelor urged me to go to America and undertake the design of dynamos and motors for the Edison Company. So I decided to try my fortunes in this Land of Golden Promise.

On arriving here . . . one of the great events in my life was my first meeting with Edison. This wonderful man, who had received no scientific training, yet had accomplished so much, filled me with amazement. I felt that the time I had spent studying languages, literature and art was wasted; though later, of course, I learned this was not so.

It was only a few weeks after first meeting Mr. Edison, that I knew I had won his confidence. The fastest steamship afloat at that time, the Oregon, had disabled both her lighting engines, so that her sailing was delayed. . . . The difficulty annoyed Mr. Edison considerably, because it seemed that the ship would be held in port some length of time.

That evening I took the necessary instruments and went aboard. . . . The dynamos were in bad condition, with short circuits and breaks; but with the aid of the crew I put them in shape. At five that morning, on my way home, I met Mr. Edison on Fifth Avenue, with Mr. Batchelor and their assistants, just going home from their own work. When Mr. Edison saw me, he laughed and said, "Here's our young man just over from Paris running around at all hours of the night." Then I told him I was coming from the "Oregon," and that I had repaired the machines. Without a word he turned away; but as they went on I heard him say, "Batchelor, this is a damn good man!"

Soon after I left Mr. Edison's employment a company was formed to develop my electric arc-light system. This system was adopted for street and factory lighting in 1886, but as yet I got no money—only a beautifully engraved stock certificate. Until April of the following year I had a hard financial struggle. Then a new company was formed, and provided me with a laboratory on Liberty Street, in New York City. Here I set to work to commercialize the inventions I had conceived in Europe [and soon after sold the patents to George Westinghouse].

After returning from Pittsburgh, where I spent a year assisting the Westinghouse Corporation in the design and manufacture of my motors, I resumed work in New York in a little laboratory on Grand Street, where I experienced one of the greatest moments of my life—the first demonstration of the wireless light. . . . Without any connection whatever between me and the machine to be tested . . . in each hand I held a long glass tube from which the air had been exhausted. "If my theory is correct," I said, "when the switch is thrown, these tubes will become swords of fire." I ordered the room darkened and the switch thrown—and instantly the glass tubes became brilliant swords of fire.

Under the influence of great exultation I waved them in circles round and round my head. My men were actually scared, so new and wonderful was the spectacle. They had not known of my wireless light theory, and for a moment they thought I was some kind of a magician or hypnotizer. But the wireless light was a reality, and with that experiment I achieved fame overnight.

Following this success, people of influence began to take an interest in me. I went into "society." And I gave entertainments in return; some at home, some in my laboratory—expensive ones, too. For the one and only time in my life, I tried to roar a little bit like a lion.

But after two years of this, I said to myself, "What have I done in the past twenty-four months?" And the answer was, "Little or nothing." I recognized that accomplishment requires isolation. I learned that the man who wants to achieve must give up many things—society, diversion, even rest—and must find his sole recreation and happiness in work. He will live largely with his conceptions and enterprises; they will be as real to him as worldly possessions and friends.

In recent years I have devoted myself to the problem of the wireless transmission of power. Power can be, and at no distant date will be, transmitted without wires, for all commercial uses, such as the lighting of homes and the driving of aeroplanes. I have discovered the essential principles, and it only remains to develop them commercially. When this is done, you will be able to go anywhere in the world—to the mountain top overlooking your farm, to the arctic, or to the desert—and set up a little equipment that will give you heat to cook with, and light to read by. This equipment will be carried in a satchel not as big as the ordinary suitcase. In years to come wireless lights will be as common on the farms as ordinary electric lights are nowadays in our cities.

The matter of transmitting power by wireless is so well in hand that I can say I am ready now to transmit 100,000 horse-power by wireless without a loss of more than five percent in transmission. The plant required to transmit this amount will be much smaller than some of the wireless telegraph plants now existing, and will cost only $10,000,000, including water development and electrical apparatus. The effect will be the same whether the distance is one mile or ten thousand miles, and the power can be collected high in the air, underground, or on the ground.[16]

2

The Wizard's Lab

The following letter, written in 1894, captures the zeitgeist of Tesla's impact on his era, the Gay Nineties, from Marion Crawford:

The daily life of this man has been the same practically ever since he has been in New York. . . . He starts for his laboratory before 9 o'clock in the morning, when he has not passed the night there. All day long he lives in his weird, uncanny world, reaching forth to capture new power to glean fresh knowledge. . . .

Usually he works until 6 o'clock but he might stay later. The absence of natural light does not trouble him. Tesla makes sunlight in his workshop.

At exactly 8 o'clock he enters the Waldorf . . . attired in irre- proachable evening clothes. . . . He walks directly to a table in the farthest corner of the Palm Garden . . . reserved for him. He carries two or three newspapers in his hands and one of them he perches before him. No gourmet who dines at the Waldorf can order a better dinner than Tesla. It is an elaborate dinner of many courses. He drinks a fine burgundy . . . it is practically the only meal he eats during his day. He never tips . . . less than a dollar. He finishes at exactly 10 o'clock and leaves the hotel, either to go to his rooms to study or to return to his laboratory to work through the night. . . . Tesla is above all . . . undoubtedly the most serious man in New York. Yet he has a keen sense of humor and the most beautiful manner. He is the most genuinely modest of men. He knows no jealousy. . . . When he talks you listen.[1]

Due to the "War of the Currents," with Edison electrocuting animals with AC and appearing at the trial of axe murderer William Kemmler to drive home the point that the best way to electrocute a man on death row was to use AC, Tesla countered in several ways. He wrote to Edward Dean

Adams, head of the Niagara Falls Power Company, that "the whirl of the wheels of Niagara will soon drown out the foolish talk."[2] And then he allowed himself to be interviewed by journalist Arthur Brisbane, helping to choreograph a front page spread in the *World* of this younger wizard sending five hundred thousand volts through his body to show the public how wrong Edison could be.

When discussing his inventions, Tesla put a positive spin on his audacious experiments, but they were, in fact, fraught with danger. In June of 1896, the wizard made a mistake that almost cost him his life, admitting, "I got a shock of about three and half million volts from one of my machines. The spark jumped three feet through the air and struck me . . . on the right shoulder. If my assistant had not turned off the current instantly, it might have been the end of me. As it was, I have to show for it a queer mark on my right breast where the current struck and a burned heel in one of my socks where it left my body."[3] To the public, however, his experiments were generally presented in a different light.

> To illustrate, let me mention here [an] experiment of mine . . . in [which] the body of a person is subjected to the rapidly-alternating pressure of an electrical oscillator of two and a half million volts. . . . The [impression] presents a sight marvelous and unforgettable. One sees the experimenter standing on a big sheet of fierce, blinding flame, his whole body enveloped in a mass of phosphorescent wriggling streamers like the tentacles of an octopus. Bundles of light stick out from his spine. As he stretches out his arms, thus forcing the electrical fluid outwardly, roaring tongues of flames leap from his fingertips, objects in his vicinity bristle with rays, emit musical notes, glow, grow hot. He is the lantern of still more glorious actions, which are invisible. At each throb . . . myriads of minute projectiles are shot off from him with such velocities as to pass through adjoining walls. He is in turn being violently bombarded by surrounding air and dust. He experiences sensations which are indescribable.
>
> —NIKOLA TESLA[4]

Tesla's life in the Gay Nineties is one of astounding accomplishment, a true star among stars living the high life at the Waldorf-Astoria, where he now resided, friend and business partner to the owner, John Jacob Astor, largest landowner in the city. There, Tesla dined with the leading women and men of the day: playwrights, writers, musicians, swamis, presidential candidates, architects, financiers, kings and queens, inviting the crème de la crème to his laboratory. A scholar and multilingual, according to his friend, *Century* editor Robert Underwood Johnson, Tesla "is one of the most cultivated of men . . . [with] extensive knowledge of the great classics of Greece, Italy, Germany, France and England" as well of course as the Slavic countries. Particularly fond of the poetry of Leopardi, Dante or Goethe, it was Johnson's belief that "he could take up any portion of . . . *Faust* and continue the quote [by memory] textually page by page."[5]

The inventor was hobnobbing with the elite. Here is a letter from one of Tesla's backers, Edward Dean Adams, the man most responsible for setting up the power station at Niagara Falls. Through the years, Adams invested at least $100,000 in various Tesla enterprises:

> June 28th, 1897
> 46 East Houston Street
> New York City
>
> My Dear Dr. Tesla,
>
> On Sunday next, I leave with my family for Niagara Falls en route via Buffalo, the Great Lakes and Duluth to Yellow Stone Park where we expect, with some friends, to pass about ten days. I most cordially invite you to take this trip with us, feeling quite sure the voyage on the lakes as well as the exhilarating mountain air of the Yellowstone Park will give you sunshine and nature's ozone with new scenes, which together will give you recreation. May I now count upon the pleasure of your company?
>
> Sincerely yours,
> Edward Dean Adams[6]

Not only was Tesla a highly creative inventor, he also impressed greatly the leaders of the day. The following letter is from Henry Fairfield Osborn, dean of the faculty of pure science at Columbia College, New York City, to Seth Low, president of the college (soon to be Columbia University). Osborn was not only appreciative of Tesla's scientific accomplishments, he was taken with the man as well.

I have especially upon my mind [a matter] which I think will appeal to you very strongly . . . [regarding] Nikola Tesla. . . . There seems little doubt that Mr. Tesla is the leading electrician. . . . I spent an afternoon recently with Tesla, and regard him as one of the most distinguished men I have ever met. I happened to meet Professor Crocker shortly afterwards, and learned from him that he had spoken to you in regard to giving Tesla an Honorary Degree. I would like to support this in the most earnest manner.[7]

A modern approach to understanding Tesla's personality was undertaken recently by Watson, the IBM computer that has been programmed to beat chess masters, play and win the TV game show *Jeopardy*, and simulate human consciousness. After analyzing the inventor's journals, patents, speeches, and articles for "underlying emotions and writing styles," Watson concluded that at his heart, Tesla saw himself as an artist.[8] IBM has billed this "discovery" as a new and rare insight, but Tesla called himself an artist at the speech he gave at Niagara Falls, and at times, compared himself to the great painter Raphael. Nevertheless, this tack of having a computer look for key words in the writings of a subject is a fascinating idea sure to uncover important clues about various individuals Watson sets "his" sights on.

Tesla researcher Martin Hill Ortiz studied *Trow's New York Business Directory* to get a better idea of his various businesses starting in 1889: the Tesla Electric Company with Alfred S. Brown, the electrical engineer who first "discovered" Tesla and brought him to banker Charles Peck with capital listed at $300,000; the 1901 Nikola Tesla Company with William Rankine, key developer of the Niagara Falls project, with capital listed at $500,000; the 1906 Tesla Machine Company, capital at $300,000 with directors including Langdon Greenwood and William Andres; Tesla Electro-Therapeutic Company, circa 1909 to 1912 with capital listed as $400,000 with directors Archibald Langford, Morton Bogue, and Ward Pickard; Tesla's International Propulsion Company capitalized at $1 million with Henry Quimby, John F. Valieant, and James Eagan; and the Tesla Ozone Company for 1911 at $400,000 with W. Kelley, A. Macmurdo, and C. M. McKeever as directors. In the 1920s, Tesla also started producing motion picture equipment for his Cinemachinery MFG Corporation, which he capitalized at $250,000.

Apparently, each of these companies petered out, with the figures for capital invested based on projections rather than actual amounts. In his 2016 book *Wireless: The Life, Work, and Doctrine of Nikola Tesla*, Branimir Jovanovic states that, in general, many of these companies "were merely attempts to create a legal framework for certain business plans which never materialized."

One of Tesla's most successful offshoot companies was his Tesla Propulsion Company, which was partially funded by Harris Hammond and his father, John Hays Hammond Sr., around 1911, and also J. P. Morgan Jr., around 1914. This company developed pumps and turbines that Tesla would later refine in the 1920s with Allis Chalmers and Pyle National Corporation and another turbine production company that ran as a partnership with John H. Hedley, who founded both an American and British turbine manufacturing company, with plants in Bridgeport, Connecticut, and Providence, Rhode Island. A wealthy businessman, Hedley owned two large ships, the *Alabama* and the *Den*, which both had won World Cup championships for motor-yacht racing.

Although Hedley wanted Tesla to install his revolutionary jet-propelled turbines aboard his ships, Tesla needed more time to perfect them and talked Hedley into trying to develop and manufacture hydraulic machines, pumps, compressors, and turbines at his plants in Bridgeport and Providence. According to Jovanovic, the Tesla archives in Belgrade have over two thousand documents related to these Tesla-Hedley endeavors. Unfortunately, each of these inventions took longer to perfect than expected, and over time, their partnership fell apart, with Tesla losing about $10,000 and Hedley about ten times that much.[9]

As with most of these endeavors, Tesla kept his sense of humor and remained "perfectly confident of success, but the expenses have been so great that I am reminded of the statement of Fabricius after obtaining a victory over the Carthaginians—One more victory like this and we are lost."[10] In general, his many inventions were indeed sound; however, perfecting them was another matter, and the many failures he had along the way often resulted in lawsuits, some brought by him, but most brought on by business associates who were very unhappy with the numerous instances of their complete failures in the marketplace.

THE WIZARD'S LAB

The following letter, written in 1894, captures the zeitgeist of Tesla's reputation and personality from another observer, Marion Crawford, one of the most prolific novelists of the day. A contemporary of Twain and Kipling, Crawford had been invited to be photographed at Tesla's laboratory by Robert Underwood Johnson, assistant editor of *The Century* magazine and Tesla's closest friend. Crawford was the nephew of Julia Ward Howe, the famous abolitionist, fighter for woman's suffrage, and author of "The Battle Hymn of the Republic." He penned this letter to his wife, Elizabeth Berdan, daughter of Hiram Berdan, a well-known Union Army general from the Civil War.

> Yesterday when I was at *The Century* about business, Johnson proposed to take me down in the afternoon to the laboratory of Tesla, the Serbian electrician who is revolutionising everything here by bringing two hundred thousand horse power from Niagara Falls into the City of New York. He has thrown Thomas Edison completely into the shade.
>
> [Tesla] has discovered a new way of handling the dynamo, so as to create what he calls electric fields. Within these fields, an electric lamp will burn without <u>any</u> wires to connect it with the dynamo, and many he carried about like a candle. When you are too far from the "field," the lamp goes out. It is like magic.
>
> The room is electrified by moving a switch in the corner, and things you are holding in your hands, *without wires* to them, begin to glow. As this is a history-making discovery, Tesla wished to take historical photographs of the experiments and Johnson at once collected the three highest celebrities he could lay hands on, to wit, Mark Twain, Joseph Jefferson the famous old actor and myself. We all went down together, Jefferson pretty interested in the whole thing, Clemens grimly amusing and I very curious about it.
>
> The great Tesla is a man of thirty seven, about my height, but even thinner, and painfully narrow shouldered, though very erect. He has a keen, tired face with a big nose, and much smooth black hair, with very thoughtful eyes.

The room was darkened and Jefferson was photographed first, holding a lamp over the field produced by the electric Tesla. The photograph [was taken with] a flashlight, so as to get all the surroundings, as much as the glow of the little lamp. All went on well, when on the third photograph, the [photographer's] magnesium flash light . . . blew up with a terrific explosion, smashing the ground glass of the door behind him to atoms, and twisting the metal reflector of the light out of shape. Both Jefferson and I, prepared for anything, supposed that the explosion was part of the "experiment" and did not move a muscle, but there was considerable confusion afterwards and Johnson said he had had a bad shock to his nerves.

After that, as nobody was scratched—by a miracle—we were photographed by [artificial] daylight, and were shown various marvelous experiments. One way of measuring electricity is by so many "volts." Two thousand are used at Sing Sing to execute criminals. But by increasing the [frequency] enormously the current becomes harmless to the human body, and I yesterday had for some time a current calculated at *seven hundred thousand* "volts" going through me, in the course of the experiment. Tesla believes that it is healthy, and I daresay it is. At all events the portraits of Twain, Jefferson and myself will go down to posterity in connection with Tesla's early experiments!

[This] pleased the man himself and Johnson, so I made no objection, [and thus] I fancy that the whole, with an article, will be in an early number of *The Century*.

—MARION CRAWFORD, April 27, 1894[11]

May 2, 1894

Dear Johnson

One of the photos is simply immense. I mean the one of [Joseph Jefferson] showing him alone in the darkness. I think it is a piece of art.

Yours sincerely,
N. Tesla[12]

As astounding and history making as these experiments with cold wireless light bulbs were, the wizard's work with X-rays were perhaps even more startling. Claiming he could make a "man's body luminous . . . [so that he could] distinctly see his skeleton," and proving it with startling X-ray photographs, this act led newspapers to proclaim that Tesla "CAN SEE THROUGH A MAN!"[13]

RISING STAR

It was these kinds of achievements that attracted Spanish-American War hero and Medal of Honor recipient Richmond P. Hobson into Tesla's orbit, writing to his fiancée that the world-famous inventor greeted him by giving him "a kiss on the cheek as once before."[14] Hobson gained worldwide recognition because of his quick thinking during the Spanish-American War. On June 4, 1898, the *New York Times* reported the sinking of the Merrimac, which had "made a dash into Santiago Harbor, Cuba, "under lively cannonade of fire" in attempts to attack the waiting Spanish armada. The ship was sunk and "an officer and engineer and six seamen" were taken as prisoners and held in the dungeons at Morro Castle. The *Times* article, which came out just a day after the event, captured the imagination of the reading public. When they found out, the following day, again in the *Times,* that the ship was sunk on purpose to lock the Spanish armada into the harbor and that the officer was Lieutenant Richmond P. Hobson, he gained overnight world acclaim.[15]

As editor of *The Century* magazine, Robert Underwood Johnson was able to publish Hobson's account when he was released from the Spanish prison and made his way to New York. The story was so popular that the issue recounting his ordeal became an instant best-seller!

Rather quickly, a deep friendship arose between Hobson and Tesla as they also dined with Robert and Katharine Johnson. Several years later, Hobson, soon to be a U.S. Congressman, helped Tesla line up a deal to place a wireless system on Navy ships, but Tesla's great undertaking at Wardenclyffe interfered. With the wizard now facing mounting financial debts, Hobson decided to get married.

May 1, 1905

My dear Tesla:

I know it will please you to hear of the great happiness that has come to me. Miss Grizelda Houston Hull . . . has consented to become my wife and the wedding has been [planned] for May 25th. . . . Do you know, my dear Tesla, you are the very first person, outside of my family that I thought of and which the ceremonies will be very simple. I wish to feel you present in standing close to me on this occasion so full of incoming in my life.

Indeed, I could not feel the occasion complete without you.

Sincerely yours,
Richmond Pearson Hobson[16]

Hobson referred to Tesla as having "a cosmic mind that sweeps the universe . . . with a kind of cosmic intuition," and their friendship continued until Hobson's death in the 1930s.

Tesla had close friendships with other luminaries as well, such as Sir William Crookes and Lord Kelvin, whom Tesla invited to his laboratory when he came to the States.

15 Eaton Pl.
Langdon
May 20, 1902

Dear Mr. Tesla,

I do not know how I can ever thank you enough for the most kind letter of May 10, which I found in my cabin in the Lucania, with the beautiful books which you most kindly sent me along with it: *The Buried Temple* [by Maurice Maeterlinck], *The Gospel of Buddha* . . . the exquisite edition of Rossetti's *House of Life*, and last but not least, *The Century* magazine for June, 1900 with the splendid and marvelous photographs on pages 176, 187, 190, 191, 192, [and the] photo[s] of electrical lessons.

We had a most beautiful passage across the Atlantic, match[ing] the finest I have ever had. I was trying hard nearly all the way, but quite unsuccessfully, to find something definite as to the functions of ether [with] respect to plain, old-fashioned magnetism. Apropos of this, I have instructed the publishers at [Macmillan], to send you at the Waldorf, a copy of my book . . . on *Electrostatics and Magnetism.* I shall be glad if you will accept it from me as a very small mark of my gratitude to you for your kindness. You may possibly find something interesting in the articles on atmospheric electricity which it contains.

Lady Kelvin joined me in kind regards, and I remain,
Yours always truly,
Kelvin

P.S. Thank you so warmly for the beautiful flowers.[17]

The duo discussed how to send a signal to the hypothesized Martians. They philosophized about the purpose of life. Tesla had given Kelvin a copy of Édouard Schuré's *The Great Initiates,* a masterwork on the lives of Rama, Krishna, Hermes, Moses, Orpheus, Pythagoras, Plato, and Jesus, a book that has been described as "a spiritual adventure of depth and intensity."

We see here great evidence of Tesla's spiritual awareness and interest in the highest state of consciousness. One of the most important teachings of these individuals is that the universe has purpose, which is a theme of Aristotle's that also runs through Maeterlinck's works.

When it comes to women, Tesla's letters between himself and Katharine Johnson often read like love letters. The following provocative letter from Katharine may refer to a photograph Tesla took of his hand, which he displayed in the technical journals in relation to his new method of fluorescent lighting. This was a time when palm reading was all the rage.

June 6, 1898

Dear Mr. Tesla,

I want very much to see you [tomorrow evening], and will be really disappointed if you do not think my request worthy of your consideration. . . .

You must save this evening for us. After this date I am going away to Washington for a visit, so if anybody cares to see Mrs. Filipov?

When you come tomorrow evening, we'll talk about the

hand which is before me now but which is doomed to seclusion. . . . I cannot stand it. It is too strong, too virile—when I enter the room without thinking it makes me start—*It is the only thing in it.* . . . You must try again and make your hand be as large and grand as it is.

Faithfully yours,
Katharine Johnson[18]

This is just one of many alluring letters from Katharine, and she flirts here with Tesla as she also tries to get him married. In one exchange, about a year after Tesla's arrival back in New York, Katharine gets interested in spiritualism and tries to contact the inventor through thought transference. She sets up a special time to think about Tesla and then a few days later writes to him to see if he received the telepathic message. Tesla replies, "My dear Mrs. Johnson . . . [on that day, at the very moment] I never thought of you [not] even for a moment," signing the letter, "Sincerely, Millionaire Kid."[19]

OF INTEREST TO WOMEN

Nikola Tesla, whose name has been associated widely with study and application of electricity, often surprises his friends by displaying more or less knowledge of a great variety of subjects. . . . "The other evening I took the prettiest kind of girl out to dinner."

Struck with the wide range of his knowledge . . . in sheer admiration the lady at last said, "How very much you know, Mr. Tesla! I had rather expected to find you a man of one idea, but instead . . . you are more conversant with more topics than any person I ever met. . . ."

"I simply have to keep myself posted upon everything going on," the inventor replied, "or I wouldn't stand a ghost of a chance with these bright college girls."[20]

Tesla claimed he was celibate, that "science was his only mistress." Was he homosexual or did he have repressed homosexual thoughts or impulses? Perhaps. Obviously, Tesla had a highly complex personality. As a biographer, how far down can one really dig? At his heart, Tesla was an inventor, moving as rapidly as he could to help design and, in that sense, choreograph the new age. He also enjoyed the high life. The precise details of his sexual life will probably never be known, but what is important to keep in mind was that Tesla maintained strong friendships with numerous leading men and women of the day. Clearly, there was something about Tesla that attracted them, more so than just the fact that he was the leading wizard of the day, performing take-your-breath-away experiments at his mysterious and fanciful lab.

Nevertheless, as popular as Tesla was, there were also numerous individuals, particularly electrical engineers, who not only disliked Tesla but also did their best to undermine him. And that included competitors as well as members of the Westinghouse Corporation whose entire livelihoods were dependent on Tesla's genius. One need only read some of the Elihu Thomson correspondence, Carl Hering's articles as editor of *Electrical World and Engineer* (who backed Dobrowolski over Tesla in the invention of the AC polyphase system), or Westinghouse man Lewis Stillwell, whose writings on his boss, George Westinghouse, placed his friend Oliver Shallenberger's name above Tesla's for the invention of both the induction motor and hydroelectric power system, when all Shallenberger did was notice that a spring tended to spin when placed in a magnetic field. Or one can read any number of other biographies on Elihu Thomson or Charles Steinmetz or the autobiography of Michael Pupin to see how Tesla's name was stripped time and again from his accomplishments. Longtime associate Fritz Lowenstein said it this way:

> April 18, 1912
>
> My dear Mr. Tesla:
>
> I attended the Marconi meeting last night, in company with illustrious society. . .
>
> Mr. Marconi gave the history, as he sees it, of wireless up to this date. [He] does not speak any more of Hertzian Wave Telegraphy, but accentuates that messages he sends out are conducted along the Earth. Pupin had the floor next, showing that wireless was due entirely to one single person. . . .
>
> In a brief historical sketch, [Steinmetz] maintained that while all elements necessary for the transmission of wireless energy were available, it was due to Marconi that intelligence was actually transmitted. . . . That evening was, without any

question, the highest tribute that I ever have heard paid to you in the language of absolute silence as to your name.

Sincerely yours,
Fritz Lowenstein[21]

What we do know is that over time, as Lowenstein tells us, Tesla became a nonperson, the quintessential elephant in the room, ostracized by many of his former competitors, not because of his personal life or mysterious sexual proclivities, but because his work was so central to their careers, and, for various reasons including jealousy and the downside of capitalism, it simply was not in their interests to credit Tesla's role in the advancement of their profession.

3

Gay Nineties

March 2, 1895

My dear Tesla

I cannot thank you too much for your kindness in showing us all your wonderful experiments the other day. They made a deep impression on me, as they did everyone, and I am going to see them again some day, if you will let me.

Sincerely yours,
Stanford White[1]

Bar none, Stanford White was the premier architect of the day and perhaps the premier architect of the modern age. One of the founders of the prestigious architectural firm McKim, Mead & White, White not only designed the original Madison Square Garden, considered by many to be the most beautiful building ever to grace New York City, he also designed the original Penn Station, numerous churches, both the Metropolitan and Players Clubs in New York as well as the mansions of the robber barons there and in Rhode Island, the wedding cake–like Rosecliff Mansion and the futuristic and highly original Tennis Hall of Fame in Newport, the castle-like Towers in Narragansett, just a few miles from my house, which happens to be my favorite building, and the regal white marble capitol in Providence, an edifice that not only greatly resembles its older brother in Washington, DC, but also boasts the third largest unsupported dome on the planet.

Less than two weeks after receiving White's letter, Tesla's South Fifth Avenue laboratory burned to the ground. Claiming losses of over $50,000, Tesla stated in the press, "I am in too much grief to talk. What can I say. . . .

Everything is gone. I must begin over again." [2] But to his friends, he presented a different feeling.

> Marcy 25, 1895
>
> My dear Mrs. Anthony,
>
> It was very kind of you to write me for I needed such proof of sympathy. Please accept my most sincere thanks.
>
> I have gone through an experience which can not fail to leave some trace, but I am not broken in spirit, quite the contrary, my energies have risen to the occasion.
>
> My new work is progressing fast and soon, I hope, you will hear from me more pleasant news.
>
> Sincerely yours,
> N Tesla. [3]

Over time, Stanford White, or Stanny as he was called by his friends, would change his salutation to Tesla from "sincerely yours" to "affectionately" to truly reflect the sense of camaraderie that this odd duo shared.

A charismatic sensualist, Stanny was not only an architect extraordinaire, he was also a trendsetter, purchasing from Europe artwork and tchotchkes for his clients, importing this new game for the masses called golf, and reshaping the urban landscape to reflect America's rise as the new superpower, spearheading humanity's breakthrough to the modern era with the simultaneous curtailing of the horse and buggy, giving way to the corresponding spread of automobiles, the railway system, and even the promise, through the experiments of Samuel Langley, of human flight.

Leading the charge on the electrical front were dueling wizards, Tom Edison from Menlo Park, New Jersey, who invented the best electric light, a machine that talked and reproduced the calls of birds and the sounds of music and another that produced pictures that moved; and Nikola Tesla, whose system of electrical power distribution lit the Chicago World's Fair, held the promise of harnessing Niagara Falls, and sent wireless signals to remote-controlled vessels. Tesla demonstrated his phantasmagoria to literally thousands of onlookers at lecture halls in Philadelphia, New York, Chicago, and St. Louis, and his laboratory became a magician's lair where so many of society's elite came to be mesmerized and entertained.

March 12, 1896
The Waldorf

Dear Mr. Tesla,

It would give me great pleasure to have you dine with us this evening at seven o'clock. I must return to Washington tomorrow, Wednesday, my son and I want to see you.

Hoping to have that pleasure,
I am yours very truly,
Phoebe A. Hearst[4]

Finding a new location on Houston Street in sight of the Brooklyn Bridge, paradoxically, Tesla received aid in the interim from Edison, who lent his former employee space in New Jersey while Tesla located and then constructed his new place.[5]

The list of people who dined at Delmonico's or Sherry's and then made their way back to Tesla's lab was a veritable Who's Who not only of the Four Hundred but also of historical giants in their own right, from multimillionaires and authors to poets, painters, robber barons, musicians, war heroes, and politicians, including John Jacob Astor and his alluring wife, Ava Willing; Robert Underwood Johnson and his wife, Katharine; Johnson's boss at *The Century*, Richard Watson Gilder, and his wife, the painter Kay Gilder, founder of the Arts Student League; newspaper men and journalists like Charles Dana of the *Sun*, Arthur Brisbane from the *World*, Joseph Collier, and William Randolph Hearst and his mother, Phoebe; novelists Marion Crawford, Rudyard Kipling, and Mark Twain; Tesla's *Electrical World* editor T. C. Martin; Spanish-American War heroes Richmond P. Hobson and Teddy Roosevelt; Roosevelt's sister Corinne Robinson, one of the founders of the Metropolitan Museum of Art, and her husband, Douglas Robinson; railroad magnates like Austin Corbin, James H. Hill, and Chauncey Depew; the naturalist John Muir; African explorer William Astor Chanler; the highest paid performer of the day, pianist Ignace Paderewski; conductor and composer Anton Dvorak, author of New World Symphony; playwright Marguerite Merington; electrical competitors and colleagues Charles Scott and Albert Schmid of the Westinghouse Corporation; Peter Cooper Hewitt, who, like Tesla, invented fluorescent lighting; the thespian Joseph Jefferson; the oil baron Colonel Oliver Payne; socialites like Mrs. Wolcott, Mrs. Griswald, and Mrs. Winslow; Daisy Maud Gordon and her husband, the very wealthy British army officer Major Walter De S. Maud; Mary Mapes

Dodge, author of *Hans Brinker*; August St. Gaudens, the country's leading sculptor; and Lord Kelvin.

One of the more illustrious individuals Tesla interacted with in the 1890s was the newspaperman and soon to be "foremost novelist" Theodore Dreiser, who started his career working as a journalist in Chicago and then transferred to St. Louis to write for the *Globe-Democrat* and the *St. Louis Star.* In 1893, the editor of the *Star,* Tobias Mitchell, assigned Dreiser to interview a potpourri of newsmakers ranging from "third rate spiritualists, . . . mountebanks and quacks" to VIPs who came to town, including pianist Ignace Paderewski; Theosophist Annie Besant; prize-fighter turned actor John L. Sullivan; and "a scientist of standing," Nikola Tesla. "Verily," Dreiser wrote, "I was sent to get their views on something—anything or nothing really, for Tobias seemed a bit cloudy as to their significance, and I certainly had no clear insight into the matter yet. One of my favorite thoughts . . . was to ask them what they thought of life, its meaning, since this was so much uppermost in my mind at the time."[6] In 1900, Dreiser wrote the acclaimed novel *Sister Carrie,* a Horatio Alger–type story about a country girl who gets somewhat corrupted. When she enters city life, she sheds a benefactor/lover who, in retaliation, commits suicide as she moves on to become a successful actress. It was the basis of the highly successful *A Star Is Born* movie franchise, (the most recent of which starring Bradley Cooper and Lady Gaga). It has been suggested by Dreiser experts such as Tom Riggio, editor of the *Dreiser-Mencken Letters,* that one of the minor characters, Bob Ames, might have been based in part on or inspired by Nikola Tesla. In a rather long soliloquy, Ames comments that "if you have powers," they should be "cultivated" through the exercise of the will. "If you want to do most, do good. Serve the many."[7] Certainly these traits pair well with Tesla's actions and philosophy.

Dreiser was an ardent sensualist who had numerous affairs. Also an author, he boldly described them in lurid, fictitious form. In 1909, one such affair cost him his marriage. That same year, Dreiser wrote Tesla requesting that he write an "educational" article for the "on-coming generation" for one of the magazines he was editor-in-chief of: the *Delineator,* the *Designer,* or the *New Idea.* Dreiser entices Tesla by informing him that their combined circulation was 1.8 million. "I should very much appreciate it if you would give this proposition your serious consideration and . . . personally let me know."[8] Tesla's response is missing, and it is quite possible that since they both were New Yorkers, Tesla's answer may have been given in person or by phone.

Throughout his career, Dreiser was continually influenced by aspects of Tesla's life. In 1925, after a ten-year hiatus, Dreiser published arguably his most important work, *An American Tragedy*, which was, in part, influenced by the story of William Kemmler, who in the early 1890s was the first man to die in the electric chair.

In this "tragedy of desire," the protagonist/antihero, Clyde Griffith, allows his pregnant lover to drown so he can move on to a more alluring gal. Like Kemmler, his actions result in his execution by the electric chair.[9]

A few years later, Dreiser had his secretary contact Tesla to inform him that his offices for the *American Spectator* had moved and that he could now be reached at the new phone number: Lexington 2-6130. "I shall of course," the secretary writes, "transmit to Mr. Dreiser immediately any arrangement you wish to make so that he can meet them."[10] The implication through these letters is that Tesla and Dreiser maintained cordial relations through the duration of their lives.

Another interesting lifelong relationship Tesla had was with the astronomer George E. Hale (1868–1938), the man who built the huge forty-inch telescope at the Yerkes Observatory in Williams Bay, Wisconsin. Tesla met Hale at the Chicago World's Fair in 1893, where Hale's gigantic telescope was put on display. A strong friendship between them arose, and it is most likely that Hale arranged for Tesla to view the planets and stars at this meeting.

In 1896, Hale invited Tesla to Wisconsin to speak at a conference and enjoy another viewing of the stars at the Yerkes Observatory. "Should you be able to attend and participate in such a conference," Hale wrote, "the meetings would be most interesting and profitable with ample time for discussion. . . . Are there any objects which you would especially like to experiment with the 40 inch telescope?" Hale ended the letter, "Hoping you can arrange to be here [to meet with other physicists and astronomers] in the various pleasant summer homes on the shores of the lake. . . . I am very truly yours, George E. Hale."

Due to pressing matters, including Tesla's extensive studies with X-rays and unfortunately also an attack of the grippe, Tesla graciously declined the generous offer.[11] But a decade later, on June 4, 1908, he wrote to Hale at his "Solar Observatory" at Mount Wilson, California. "Learn[ing] with pleasure of your forthcoming book *The Study of Stellar Evolution* from which I expect to derive much needed information. I have greatly regretted that since our meeting in Chicago years ago, we have never been able to get together. Your work interests me very much and I am heartily in sympathy with you." Tesla ended the letter, "Please do not fail, the next time you are in New York, to call on me and give me an opportunity to exchange a few ideas. . . . Sincerely yours, N. Tesla."[12]

Heavily influenced by Hale's work and finally having another face-to-face meeting circa 1912 in Manhattan, Tesla was particularly taken by Hale's study of the Sun as well as by his photographs of so-called "island universes," which by this time were renamed spiral galaxies. In the early 1890s, before Tesla's first major laboratory burned to the ground, Tesla hypothesized that "the production of light [was created] by the vibration of the atmosphere. According to the inventor, the light of the Sun is the result of vibrations in [the] ether. His idea is to produce on Earth vibrations . . . as intense of that to the Sun. The inventor had already done something towards accomplishing this end when the fire occurred." Tesla told *Current Literature*, "I have come to the conclusion that sunlight is produced by five hundred trillion vibrations of the atmosphere per second. In order to manufacture the same kind of light, it will be necessary to produce an equal number of vibrations by machinery. I have succeeded up to a certain point, but am still at work on the task."[13]

The frontispiece of Hale's book on the evolution of the cosmos was a spectacular reproduction of the Andromeda galaxy. This image made such an impact on Tesla's thinking that he came to fashion one of his letterheads as a spiral galaxy, which he probably linked to his discovery of the rotating magnetic field and thus to the issue of cosmology or birth of the cosmos.

Many years later, in the mid 1930s, concerning a "subject, I have devoted much of my time to," Tesla told the *New York Times* that he had invented "a new small and compact apparatus by which energy in considerable amounts can now be flashed through interstellar space to any distance without the slightest dispersion." Tesla revealed that he considered conferring "with my friend George E. Hale, the great astronomer and solar expert, regarding the possible use of this invention in his own researches." However, instead, Tesla sought to offer the invention to a French research committee attempting to study a way "to communicate with other worlds," thereby hoping to win a monetary reward should such a device prove successful.[14]

4

Electric Bath

Bathing by electricity is the latest development of that science, due to recent experiments of Nikola Tesla, the famous American electrician. "The busy man's bath" is the way Mr. Tesla himself describes this "bath," which is produced by passing a current of millions of volts through the body. He has himself passed a current of 2,000,000 volts, alternating at a rate of 300,000 or 100,000 times a second, through his own body, and the effect is, he declares, to cause all impurities to be thrown off the skin. The mechanism used is exceedingly simple, consisting merely of an insulated metal platform, on which the individual stands and holds in his hand an electrode, which is connected by a wire with an oscillator. The invention is also described as a powerful tonic.

—The Daily News,
Perth, Washington, Friday, February 3, 1899

Tesla's electric bath caught the fancy of the *World*, and they sent a journalist in the fall of 1897 to meet with the wizard. Created as a feature story that would run in the Sunday edition for October 31, the article revealed that the inventor had "devised a reclining chair in which the subject lies, perfectly nude. Two wet sponges are applied to [a lady's] feet, two to her hands. The current is turned on; thrills, tingles awaken the skin and then . . . from her, all the dirt which covers the skin . . . flutters away, also, the hideous microbes, [are] fairly shocked into quitting their hold."

Tesla told the reporter that about seven years earlier, he had purchased the most powerful microscope on the market. He would then study the surface of a person's skin before and after showering. "People who bathe themselves thoroughly in soap and water every day have an idea that they are utterly free from microbes and that their skin cannot be made any cleaner. If these people had only the opportunity to gaze for a moment through

56

a powerful microscope such as mine they would be utterly astonished to see the millions of germs swarming over every inch of their bodies. These germs besides making such a hideous sight are eating up the vitality and freshness and destroying the healthy particles of the skin at a rapid rate." This prompted the inventor to conceive of a way to rid oneself of such destructive microbes.

Tesla thereupon performed an experiment for the journalist. In the center of the room, the inventor had placed a large copper ball that he had painted black a week or two in advance of the meeting. "Through this big globe he ran a powerful electric current which caused the paint to scale off, leaving the globe clean and fresh. That is precisely what the wizard of the wire proposes to do with these women to make them more fair."

What Tesla was suggesting was the fantastic idea that a person's body would be disinfected in a much more thorough way through his electric bath as compared with taking a shower with soap and water! "The presence of these microbes being the cause of so much injury," Tesla told the *World*, "there is every reason to believe that the skin of a perfectly clean woman can be made to retain through old age to her death the bloom and freshness of youth, the color and softness of girlhood and the vigor and pliableness of the skin which every woman possesses by ridding her of these hideous microbes."[1]

Tesla's electric bath was one of a number of medical devices he invented. As early as 1891, Tesla designed electrotherapy machines for sending mild electric currents through the body for various healing purposes, which were advanced by the French physician Jacques d'Arsonval, whom Tesla met in Paris shortly thereafter. In 1896, Tesla patented an ozone generator (#568,177) for the purpose of manufacturing ozone. Noticing its curative affects, he created, around 1906, the Tesla Ozone Company, which helped spawn ozone therapy clinics as a way to disinfect wounds, cure the body of various ailments, and kill viruses. One such method he used was an ultraviolet (UV) light that "worked to ozonate the air such that it acquired antiseptic properties. . . . When filtered through olive oil, the ozonated air could be inhaled to treat lung diseases" and other maladies of the skin.[2] Even today, ozone therapy, which involves the injection into the bloodstream of pure oxygen mixed with about 3 percent ozone, has been used to kill viruses such as Ebola and COVID-19.[3] In the early 1930s, Tesla's work with electrotherapeutic high-frequency equipment was found to have "highly beneficial results" in cancer treatment, according to Dr. Gustav Kolischer of the Mount Sinai Hospital.[4]

Perhaps it was Tesla's electric bath that drew Daisy Gordon into the wizard's orb. An equestrian, golfer, and stunning beauty, Elizabeth Daisy Gordon was a wealthy socialite from the prominent Gordon family of Cincinnati. The granddaughter of the industrialist W. J. Gordon, who gave Gordon Park to the city, Daisy was closely associated with another Cincinnati beauty, Carrie May Harrington, the daughter of a wealthy owner of coal mines. These two prettiest girls in town vied for the attention of Daniel Rhodes Hanna, the son of Great Lakes shipping magnate and coal and iron ore entrepreneur Mark Hanna, who would go on as a senator in 1896 to be the driving force behind William McKinley's rise to the White House. Although Daisy was taken with Dan, Carrie May beat her to the punch and eloped with the young Hanna in 1888.

Often traveling to Europe and stopping off at the Waldorf-Astoria en route, Daisy entered into an ardent friendship with Tesla, which picked up steam in November of 1897. Daisy thanked Tesla for "your very kind invitation to dine . . . [at Sherry's and] see your wonderful and interesting laboratory which I am looking forward to with no end of pleasure." [5] And their friendship continued for the next several years, with Daisy ordering a photograph of herself to be sent to Tesla and even possibly visiting him when he set up his laboratory in Colorado Springs, as she enjoyed traveling there. [6]

One of the biggest scandals of the day was when Dan Hanna got fed up with his wife Carrie May, who was self-schooling their three children and charging her husband tens of thousands of dollars to cover her lavish trips to London, Rome, Venice, and Paris, where she bought the latest fashions. "Dan complained that he had few opportunities to visit his children," and thereby attempted to wrestle them from their mother by finagling with a judge to get his father, Mark Hanna, to take over as legal guardian. Pinkerton cops went to snatch her, but Carrie May had taken off for New York and was supposedly staying at the Hotel Savoy. Eluding the cops, she booked herself and her children into another hotel under a pseudonym and then snuck them on board an ocean liner bound for Europe.

Carrie May let the Hannas know that she would rather go to the North Pole than return to Cleveland and give up her children, so the Hannas relented, removed the warrant for her arrest, and allowed her back to the States. After divorcing Dan and backing a political party that opposed the now Senator Hanna, she remarried and moved back to New York and took up residence at the Waldorf-Astoria. [7]

The scandal was all the rage, and there is no doubt that Tesla, like everyone of the day, talked about it, particularly because Tesla was so closely associated with Carrie May's high school friend, Daisy, who had now married Major Walter De S. Maud, "a dashing British Army Officer and Englishman of considerable wealth."[8] When Daisy left for England on her honeymoon, Tesla sent to the ship a bouquet of flowers and one or more books to read by the latter-day Nobel laureate and author of *Quo Vadis*, Henryk Sienkiewicz.

"We had a bad voyage, and I should have read more than I did, but what I read I thoroughly enjoyed and thank you a thousand times for your thoughtfulness and kindness to me," Daisy wrote.[9] "I am delighted with the book you sent me and the charming sentiment expressed on the front page."[10]

Often writing with "the hope to have the pleasure of seeing you again," in one letter, Daisy noted "the dinner was delightful, and the time spent in your laboratory was interesting and charming. I have so many things to thank you for," Daisy wrote, ending another letter with her gratification that the individuals she introduced Tesla to had "the pleasure of shaking the hand of such a genius as you are!"[11]

On December 8, 1898, Daisy invited Mrs. Wolcott, who was, most likely, the wife of the former acting governor of Massachusetts, to join her to dine at the Waldorf with Tesla. Unfortunately, Mrs. Wolcott caught a cold and the dinner and the ensuing soiree at the wizard's lab was deferred, and then attended by Daisy and Mrs. Wolcott and also Countess Hélène de Pourtalès, the American wife of a count who was a boat captain for the King of Prussia. Two years later, the countess would go on to win a gold medal at the 1900 Olympics as a crew member sailing for the Swiss boat *Lérina*, and become the first female to do so.[12]

> Plaza Hotel
> December 15, 1898
>
> Dear Mr. Tesla,
>
> . . . I did not have time to thank you for your beautiful dinner Thursday night nor <u>half</u> tell you how much I enjoyed it. I shall be here until Saturday night when I leave for Cleveland. Should you have a moment to spare, I will be delighted to see you again.
>
> Sincerely yours,
> Daisy Maud[13]

Daisy followed up this letter with a note wishing Tesla a happy Christmas, writing, "It was a pleasure seeing you again and I know I have in you such a true friend." Ending the letter with "Kindest remembrances, believe me . . . Daisy Maud."[14] But curiously, her stream of correspondence ended abruptly just six months later, approximately a year before Daisy divorced her husband and quickly married, of all people, Dan Hanna, the very man who had married and divorced her high school friend Carrie May! Moving back to Cincinnati, Daisy had two children with the gadabout and future owner of the *Cleveland Leader* in a marriage that lasted seven years and netted the vivacious socialite $100,000 in their divorce settlement.[15] While Daisy moved for the next ten years to Scotland, where she lived in a castle, and also to Paris, and was married for a third time, to Franklin Dwight Pelton, Dan would go on to marry two more times, spend another fortune in alimony payments, and die of kidney failure at the age of fifty-five in 1921.

Unfortunately, Daisy's fate was actually worse. An ex-military man, Pelton died in 1913, barely over the age of fifty, leaving her a widow. Moving back to a Fifth Avenue apartment overlooking Central Park, Daisy became active in the war effort, but passed away just a year later, in 1919, possibly a victim of consumption or the flu, having suffered an entire year before her passing.[16]

Tesla's correspondence includes many other luminaries who also witnessed the wizard at his laboratories, starting with his first major one on South Fifth Avenue circa 1890 until it burned down in 1895, or its replacement, which was just a few blocks from the Brooklyn Bridge at 46 and 48 Houston Street, circa 1896 to 1901. This list includes industrialists Clarence McKay and Simon Guggenheim, Swami Vivekananda, actress Sarah Bernhardt, land baron James Warden, dime store magnate F. W. Woolworth, Emily Vanderbilt, artist extraordinaire Edwin Austin (Ned) Abbey, Louis Comfort Tiffany, President Grover Cleveland, and literally hundreds of others.

In the case of Tiffany, he invited the inventor to numerous luncheons throughout the 1890s and early 1900s, and through the years, Tiffany also sent the inventor notices of the weddings of his daughters, and Tesla may have attended one or more such happy events. The following letter refers to an invitation to visit Tiffany's estate in Cold Spring Harbor, located at the eastern tip of Long Island, several train stops past Wardenclyffe. By this time, they had already known and dined with each other for nearly twenty years.

> 1136 Woolworth Building
> Laurelton
> Cold Spring Harbor
> Long Island, New York
> May 16, 1914
>
> My dear Mr. Tiffany,
>
> I cannot resist writing and thanking you for the exquisite entertainment of yesterday. It was a treat, splendid and unique such as only an artist gifted with most delicate perception and master of form and color display could offer. I have carried away a vivid impression.
>
> With kindest regards,
> Yours very sincerely,
> N. Tesla[17]

Another close friend of Tesla's was Phoebe Hearst, the well-known social-ite and mother of William Randolph Hearst, who at the time was an arch-competitor via his *New York Journal* with Joseph Pulitzer and his *World*. Tesla, of course, was featured in both newspapers. In March of 1896, in anticipation of building a laboratory there three years later, Tesla traveled out to Colorado Springs to conduct wireless experiments using electronic equipment to implement the principle of resonance through ground trans-mission via tuning forks or autoharps and microphones out at Pikes Peak, where Tesla transmitted the song "Ben Bolt" for a distance of four miles. This experiment was perhaps the first documented instance of transmit-ting music by means of wireless and established for Tesla the proof he needed to realize that the same essential mechanism could be used in the evolving wireless telephone system he was working on.[18] Three months later, Phoebe invited Tesla to return to the West and join her and her son on a trip to Yosemite. Unfortunately, the wizard was too busy to get away, so she enabled him to enjoy the trip vicariously through several letters she wrote while on the journey.

WESTERN UNION TELEGRAM:

July 11, 1896, Yosemite California

To Nikola Tesla
Gerlach Hotel, NY

Music of Falls better than Bleecker Street trucks. Lightning on mountain peaks for you to play with giant tree at Mariposa christened "Tesla."

C. R. Anthony and Phoebe

P.S. Regret you are not here.[19]

Tesla's friendship with Phoebe and William Randolph Hearst continued throughout their lives, and Phoebe was also a good friend of Katharine Johnson's. In 1902, running as a progressive Democrat in support of unions, Hearst was elected to Congress. Even though he missed 168 of 170 roll calls, Hearst was still able to finagle a second term. A force to be reckoned with, at the close of his second term, the congressman and newspaper mogul ran for governor of New York but lost to Charles Evan Hughes in 1906.

In 1919, Phoebe passed away at the age of seventy-seven, and Tesla wrote to William, saying, "Please accept my heartfelt condolences. Mrs. Hearst will lie in our memories as one of the noblest and most loveable women."[20]

Using his inheritance to begin construction of the Hearst Castle, Hearst began making movies and sought Tesla's advice on creating a lightning effect on camera. Tesla replied, "Since my talk with you yesterday, I have devoted thought to the matter and found a way of producing a very striking effect which would not endanger the heroine in the least. . . . Moreover, it will be a genuine display of force, and not a mere pictorial effect such as can be readily recognized by an intelligent audience. I will endeavor to perfect the device in my mind to the point of application and will then be pleased to give you a full explanation."[21] Hearst "worked out some electrical effects in [his own] laboratory" and thus did not take Tesla up on the offer, but their friendship remained intact.[22]

PART II

Wardenclyffe

February 17, 1905

Dear Mr. Morgan,

Let me tell you once more. I have perfected the greatest invention of all time—the transmission of electrical energy without wires to any distance, a work which has consumed ten years of my life. It is the long sought stone of the philosophers. I need but to complete the plant I have constructed and in one bound, humanity will advance centuries. . . . Help me to complete this work or remove the obstacles in my path.

Faithfully yours,
Nikola Tesla (LOC)

5

Wardenclyffe

LONDON, FEB. 21, 1901: *[In a cable dispatch] it is stated that Mr. James Galbraith, an agent of Nikola Tesla, left London today for Lisbon to establish a receiving station on the Portuguese coast at the 40th parallel of latitude, which will be in communication with a Tesla transmitter located on the New Jersey coast. This will be the first practical application of Tesla's long-distance wireless system.*[1]

In May of 1899, Tesla moved out to Colorado Springs to construct a wireless laboratory. Taking the Westinghouse people up on their offer to provide free electricity, he constructed a barn-like structure about seven miles from town, at the beginning of a long plain, at the edge of the foothills of the mighty Rocky Mountain range with Pikes Peak in the background. Unfettered by the confines of New York City, Tesla constructed numerous Tesla coils and receiving equipment and erected a two-hundred-foot-tall radio tower to test his theories. There, he created ball lightning and thunderbolts in excess of sixty feet in length!

In his 1919 article "Can Radio Ignite Balloons?" Tesla describes the great danger associated with his experiments in Colorado Springs. Admitting that "fires of all kinds and explosions can be produced by wireless transmitters," he himself created such a fire and was forced to crawl to safety, lucky to not have burned down his entire lab.

When he fired up his transmitter to undertake experiments, Tesla claimed that electrical equipment twelve miles away registered the effects! One of the greatest mysteries is exactly how far away from the lab he was when he illuminated light bulbs which were stuck into the ground. Using "less than five or ten percent" of the capacity of his equipment, Tesla stated that he "excited . . . lighted incandescent lamps at a considerable distance from the laboratory," which he estimated to be about one hundred feet away. By watering the ground around the lab to increase conductivity in one experiment when exciting his "large transmitter coil, 51 feet in diameter . . . butterflies were carried around in a circle as in a hurricane and could not get out, no matter how they tried," sparks were created "in the sand when one walked at some distance from the building . . . and a horse at a distance of perhaps *one-half a mile*, would become scared and gallop away. . . . When using damped waves the roar was so strong that it could be plainly heard ten miles away." And of course, Tesla also blew out the power station which he had to fix, which was seven miles away.[2]

By monitoring lightning storms at distances of six hundred miles from his lab, Tesla concluded that his receiving equipment would be adequate for wireless communication over long distances. He launched high-altitude balloons to try unsuccessfully to transmit wireless impulses through what today is called the ionosphere, and then turned his attention to using the ground as the medium for transmitting messages.

By monitoring a monster lightning storm in early July, Tesla claimed to have measured the resonant frequency of the planet. Thus, he constructed equipment that took into account the size of the Earth, the speed of light, and the planet's resonant frequency to send what he claimed were impulses to the antipode, or opposite side of the planet, where they rebounded back, and he also claimed to have transmitted electrical power by wireless means to illuminate light bulbs hundreds of feet and perhaps several miles from his laboratory.

Having concluded that he had also received pulsed frequencies from outer space, perhaps from intelligences from a nearby planet such as Venus or Mars, Tesla returned to New York in early 1900, convinced that he could now construct a wireless plant near New York City and use the facility to send wireless messages to Europe and even illuminate the upcoming Paris exposition.

During the Christmas season of 1900, the inventor was invited to the home of J. Pierpont Morgan, and after some discussion and a challenge to Morgan to not be "close-fisted," the inventor was able to wrangle a deal whereby Morgan provided $150,000 for the construction of Tesla's wireless operation. Having now hooked the biggest fish on Wall Street, the wizard proceeded to purchase land out on Long Island sixty-five miles from New York City from "James Warden, director of the North Shore Industrial Company. Warden, who was in control of an 1,800-acre potato farm along Long Island Sound, provided Tesla with 200 acres adjacent to Route 25A. The inventor was also given the option to purchase the remaining parcel.

Perhaps to sweeten the deal, or in lieu of certain other arrangements, the site was named Wardenclyffe, after the owner and a housing development Warden was constructing, and a post office under that appellation was established in 1901 on April 2nd. Five years later, in 1906, the name was officially changed to the Village of Shoreham." [3]

Tesla hired the architect Stanford White of McKim, Mead & White to design the laboratory and transmission tower, and White, who happened to have a home just several miles from Wardenclyffe in the village of St. James, in turn hired W. D. Crow to do the construction. With White's help, Tesla set out to design a "Radio City," with plans to eventually hire as many as 2,500 workers and technical people to run the worldwide wireless enterprise, build a "model city" to house the workers, and construct for himself a "palatial abode" which would overlook the entire property. "The laboratory will draw men from the highest scientific circles and their presence will benefit all of Long Island," Tesla told the press.[4]

> "It's a mighty fine tower," said one good farmer . . . last week. "The breeze up there is something grand of a Summer evening, and you can see the Sound and all the steamers that go by. We are tired, though, trying to figure out why he put it here instead of Coney Island." [5]

Tesla moved out to Wardenclyffe and rented a cottage from Warden himself set on a high bluff that overlooked Long Island Sound. As one of those incredible coincidences, the Tesla cottage sat right next to another cottage owned by the woman suffragist iconoclast Elizabeth Cady Stanton. Born in 1815, with a granddaughter later married to radio pioneer Lee DeForest, Stanton, who was eighty-five years old when she met Tesla, recorded their meeting in her diary.

> **WARDENCLYFFE, L.I., JULY 12, 1901**. *We are down here on the Sound for the Summer. Nikola Tesla has his laboratory near us. He said to me the other day: "it is possible to telegraph to all parts of the Earth without wires." Think of it! Where will the wonders of science end?*[6]

Two other individuals who knew Tesla at that time were R. Hartley Sherwood, who became president of the Central Indiana Coal Company, and Theresa V. P. Krull, a journalist who reminisced about this period in an article published four decades later.

In 1902, Tesla was completing the timber and steel "tower" or conical scaffolding, hidden in the Long Island woods near a summer shore-hamlet and its small railroad station. Paths [from the] tower and big one-story brick laboratory [actually it had two stories] led . . . inquiring reporters . . . out [to] the open shore, across a ravine from quiet Wardenclyffe Inn [next to where] Tesla had a lone cottage. He dined at the Inn as did Mr. Warden's dozen cottagers.

Some of us stayed at the Inn. This writer first saw Tesla there dining alone at his own small table, a tall, singular man with unforgettable eyes. Daily he astounded a beach "gallery" with his swimming feats. He had "floated all day in the Danube." Sometimes Tesla would linger after dinner at the Inn. Everyone, children included, watched for such evenings. He held us spellbound over what "wireless" would yet do. Thus the writer first heard that some day Tesla and others would control from shore a boat without a human aboard, direct[ing] its behavior at will.[7]

According to some anecdotal evidence, Tesla not only frequently swam in the Sound but did indeed also bring his telautomaton down and controlled it from the bluff that was situated in his backyard, high above the water. Krull reminisces about "some favored mortals [who] have seen Tesla wreathed in light-streams of unbelievable voltage emanating 'halos' for sometime after." And then she describes an adventure she undertook with another girlfriend and James Warden's son when Tesla was away from the property, whereby they entered Tesla's lab where "neither then nor now could a visitor describe the many appliances," and then they ascended to the top of the tower, took some photos, and dropped down into its depths.

With lantern [in hand] . . . we descended [where] we uncovered a hole where a narrower wooden stair led down a "well" with sides of corrugated metal, pitch-dark save in lantern radius, and eerie from faint dripping far below. Without lingering, we arose and emerged, only to find Mr. Tesla conversing with workmen! Mutual surprise! Tesla did not scold, but gently admonished Warden that the well was no place for "za ladies." Admonished as gently as he is said to have insisted on harboring pigeons storm-blown to his hotel window.

With World War I, this strange "tower" yielded to the ban on private stations. But Tesla never yields working. He still lives in a New York hotel, is still a bachelor, still keeps abreast through the press, has no secretary, so is hard to catch. With shining Faraday, practical Kelvin, stolid Steinmetz, canny Edison, glamorous Marconi, and other immortals, American-citizen Tesla still towers over the radio world.[8]

Another story concerned Tesla's relationship with Dewey Lewin, a local farmer from nearby Wading River. Sometime around 1903, Tesla took a carriage out to Lewin's farm and gave him several glass bulbs that had wires protruding from them. Most likely, the bulbs also had stakes of some sort that enabled them to be connected to the ground. This all suggests that Tesla was testing his wireless system with the locals to see if he could illuminate light bulbs at distances of several miles from the plant. According to legend, two of these bulbs remained in Lewin's possession until the 1960s, when they were broken and thrown away.[9]

Initially, Tesla's plan with Morgan was to construct a tower 90 feet high. However, when he realized that Marconi was pirating his apparatus, Tesla decided to double the size of the tower and increase its total length to 600 feet by running metal piping down a total of 420 feet, because Tesla thought that size would best fit various resonant properties of the planet. However, finances were limited, and so the wizard constructed a tower 187 feet high with a well 120 feet deep, for a total length of 307 feet, and then, at about six stories below, Tesla sent out an additional 300 feet of metal tubing to grip the Earth with about sixteen lateral pipes of varying lengths, some 15 to 20 feet each.[10]

Concerning its technical details, Tesla's own writings on the magnifying transmitter from his autobiography *My Inventions* are an excellent source. Written for Hugo Gernsback, who serialized it in 1919 in *Electrical Experimenter,* the memoir appeared two years after the precious tower was demolished. Rather than stay in town to face this awful tragedy, the inventor went on the road, developing his bladeless turbines at Pyle National in Chicago and at Allis Chalmers in Milwaukee. He also went to Boston to sell speedometers and odometers to the Waltham Watch Company, and he traveled to Philadelphia in the mid-1920s in attempts to market a number of flying machines to the U.S. Navy. Tesla, of course, would also return to New York frequently, where he kept an apartment at the Hotel Pennsylvania, at which he would dine with the Johnsons or Richmond P. Hobson and also meet with Gernsback and his artist Frank R. Paul, so that Wardenclyffe could at least be completed on paper in an artistic way in an exciting magazine. The following section, taken from pages 176 to 178 of the June 1919 issue of *Electrical Experimenter*, explains in vivid fashion the precise design and goal of his transmitter. The opening paragraph, often overlooked or misunderstood, describes the look and function of the unusual mushroom-shaped cupola that dominated the top of the tower:

The Magnifying Transmitter
NIKOLA TESLA

In the first place, it is a . . . wireless transmitter . . . in which the Hertz-wave radiation is an entirely negligible quantity as compared with the whole energy, under which condition the damping factor is extremely small and an enormous charge is stored in the [cupola, the] elevated capacity. Such a circuit may then be excited with impulses of any kind, [magnified when the charge is released at a more rapid rate than when accumulated. In this way impulses are produced] even of low frequency [employing] continuous oscillations like those of an alternator. . . .

It is a *resonant transformer* . . . accurately proportioned to fit the globe [and constructed to produce] the wireless transmission of energy. Distance is then absolutely eliminated, there being no diminution in the intensity of the transmitted impulses. It is even possible to make the actions increase with the distance from the plant according to an exact mathematical law. . . .

By its means, for instance, a telephone subscriber here may call up and talk to any other subscriber on the Globe. An inexpensive receiver, not bigger than a watch, will enable him to listen anywhere, on land or sea, to a speech delivered or music played in some other place, however distant. These examples are cited merely to give an idea of the possibilities of this great scientific advance, which annihilates distance and makes that perfect natural conductor, the Earth, available for all the innumerable purposes which human ingenuity has found.[11]

Concerning the importance of the ground connection, Tesla told his patent attorneys in 1916 that in a wireless system most of the energy radiating from the antenna is "absolutely wasted . . . because electromagnetic wave energy [in the air] is not recoverable [in any appreciable quantities], while [the Earth] current is entirely recoverable. . . . [For successful transmission, by] maintaining the Earth in electrical vibration . . . [I] suppress [the electromagnetic waves] as much as possible, and intensify the [Earth] current to perform any work at any point of the globe."[12] From his 1905 patent, Tesla explains the multidimensional aspects of his Wardenclyffe tower:

> It is practicable to shift the nodal and ventral regions of the waves at will from the sending-station. . . . In this manner the regions of maximum and minimum affect may be made to coincide with any receiving station or stations. . . . *The projections of all the nodes and loops on the Earth's diameter . . . are all equal* . . . [so] the entire globe would be subdivided into definite zones of electric activity.[13]

The local papers reported that Tesla's compound was fenced off in an area comprising about three acres. *The Babylon Signal* erroneously reported that Tesla had constructed a receiving station in Scotland and that commercial operation of a transoceanic system was imminent. The *Brooklyn Times* reported that "Mr. Tesla has maintained rigid privacy about his method and plans. . . . The visitor who walks into the grounds and approaches either the machine shop or the tower . . . is met by an employee who explains in polite but forceful language that this is private property and that Mr. Tesla does not care to have visitors at the place."[14]

The tower, made all of wood, with a wooden staircase, rose to a height of 187 feet, weighed 55 tons, and was constructed with fifty thousand bolts. "The July 1902 issue of the *Echo* [stated that] Tesla had his hands full with managing of crew, machinery and the need for more capital from his investor, J. P. Morgan." There were two different crews working on the tower, which at this point was 150 feet tall, sans cupola. The careless action by one of the crew in releasing a support that came crashing down caused the second crew to walk off the site and return to their base in Port Chester, and this delay "proved to be very costly to Nikola Tesla."[15]

THE TUNNELS

You see the underground work is one of the most expensive parts of the tower. In the system that I have invented, it is necessary for the machine . . . to have a grip on the Earth so that the whole of this globe can quiver, and to do that it is necessary to carry out a very expensive construction. I had in fact invented special machines, but I want to say this underground work belongs to the tower.[16]

—NIKOLA TESLA

One of the best descriptions of the tunnels was written by Natalie Steifel, editor of *Distant Sparks*, a Long Island newsletter on wireless communication. She describes the "the tunnels, which were built under the great tower [as] a great mystery." Steifel correctly describes "a well [which] was dug below the tower 120' deep and 12' square, lined with 8' timbers." She then speculates that a spiral stairway that went down the full twelve stories "encircled a telescopic steel shaft" that supposedly had the ability to be either raised to the full height of the tower or lowered with air pressure. So far, no photographic evidence can be found to support this ingenious feature. She continues the description:

The *Port Jefferson Echo* reported in February of 1902: "The staircase leading down into this subterranean chamber is partially completed, and next week a force of workmen will begin the driving of a series of four small tunnels, each 100 feet long transversely across the bottom of the well. As these tunnels will be below the water level, some skillful engineering will be required to carry the work through." The March 1902 *Patchogue Advance* [noted that these] "four tunnels will be driven out a distance of 100 feet each to the north, south, east and west. The particular use to which all this is to be put is one of the mysteries of the wireless system."

The *Echo* continued, "Mr. Tesla's energy is pushing the work of construction forward and the fact that the boilers, engines and heavy machinery need only the finishing touches to make the power available, is an assurance that within a very brief period, he will be transmitting messages across the ocean through his wonderful wireless system. . . ."

Leland I. Anderson mentioned that four stone-lined tunnels going out in various directions . . . were used to establish ground connection (transmission) for the tower and not to be walked in.[17]

Steifel goes on to write that "the tunnels gradually rose to the surface into brick, igloo-shaped mounds. Some people remember seeing these mounds at the edge of the Tesla property, near the present Fire Department." This term *igloo* is an unusual word, which, as far as I know, first appeared in this context in the literature in 1996 when I published *Wizard*. I chose that word because I went on-site to Wardenclyffe, circa 1984, walked the property, photographed the cement foundation of the tower, and saw the façade of the laboratory before a guard rushed out of the building and literally kicked me off the property.

On the way to my car, across the street, I saw a beehive-type brick mound at the edge of the Fire Department property nestled in an untouched area of brush. It was about three feet high and seemed most likely to be an air shaft. Since the red brick structure resembled an igloo, that's how I described it in *Wizard,* which was nine years before the Steifel article was published.[18]

While filming *The Tesla Files,* Prometheus Films hired Hager Geoscience, a ground-penetrating radar crew, to scan the property. From the image that resulted, two lines leading towards the base of the tower appeared, one for electrical power and one for water. Ten to fifteen feet down was a concrete slab that cordoned off a dumpsite used by Agfa Corporation, who owned the property. Beneath that could be seen Tesla's central shaft. Then about fifty feet down, Tesla's "Earth grippers" fan out as lateral spokes, forming an elaborate ground connection. Tesla claimed a total of three hundred feet of tubing. We saw about a dozen, from ten to fifty feet in length each. Beneath the spokes are four tunnels, three about one hundred feet in length. The tunnel in the front runs east to west, roughly parallel to Route 25A, which is right in front of the property. The other tunnels crisscross in the back, with a final smaller tunnel in the rear that seems to be a separate room. The reason for the tunnels is unknown. Perhaps they were testing rooms. The original radar image stops at the tunnels, but according to Tesla's notes, assuming the tunnels to be seventy feet down, the central shaft would run another fifty feet. According to Travis Taylor, Tesla would have also had to install a pump at the base 120 feet down, and maybe additional pumps in the tunnels, as water eventually would leak into these areas.

Jane Alcorn, the head of the Wardenclyffe project, said she was in touch with an individual who remembers as a boy walking into one of these tunnels with Tesla sometime in the early 1900s. Alcorn also interviewed another gentleman who remembered seeing Tesla bring his remote-controlled boat down to Long Island Sound, where he set it in action. The image of the wizard using his remote-controlled equipment to propel his submarine-like boat offshore excites the imagination. The Wardenclyffe tower and laboratory are about a mile and a half from the Sound, and it was a straight run, exiting north, that went directly to a bluff that overlooked the sound where Tesla's cottage was situated.

Tesla loved to swim, and in the summer, he could be seen climbing down from the bluff about forty steps to the beach, where he would either take a dip or experiment with his submarine telautomaton. *The Buffalo Evening News* reported that in May of 1899, Tesla took his unique vessel to Chicago to demonstrate its abilities before the Commercial Club. The article reads somewhat like a scene from the movie *Frankenstein*, for there was "witnessed the strange sight of a miniature torpedo boat 20 feet from the seat of power [where it showed] every sign of life. The propeller whizzed around, the rudder moved and the boat was lighted throughout by electricity. To climax all, a shot was heard, the bow of the little wooden model poured forth flame and smoke, and all the effect and dramatic action of a torpedo leaving its place was observed by the onlookers. All this without any connecting wires!"[19] According to Randy Hagerman, his grandfather worked for Tesla on the construction of the tower, and clandestinely observed Tesla demonstrating his telautomaton to military personnel, most likely in the summer of 1902 or '03. For a quarter, one of the neighborhood boys would swim out and retrieve the boat so that Tesla could further explain the details of how it worked.[20] In the case of his robotic boat, batteries on board were the source of power. However, Tesla's ultimate plan was to transmit power wirelessly to boats, automobiles, and airplanes from his great radio tower.

Except for several articles dating from July of 1903, there is no direct evidence that Tesla ever "fired up" the tower to the point of sending impulses to the other side of the globe, let alone a few miles, or that Tesla overtly measured the effects. One would think his Wardenclyffe notes would have data on such tests, and yet this does not seem to be the case. Tesla's 1916 deposition states he did not undertake any long-distance experiments with detecting equipment from Wardenclyffe. He seems to hedge, but suggests he did measure effects up to ten miles away.

STRANGE LIGHT AT TESLA'S TOWER

> From the top of Mr. Tesla's lattice work tower on the north shore of Long Island, there was a vivid display of light several nights last week. This phenomena provoked the curiosity of

the few people who live near by, but the proprietor of the Wardenclyffe plant declined to explain the spectacle when inquiries were addressed to him.

—*New York Herald Tribune,*
July 19, 1903

Referring to an early photograph of the Wardenclyffe tower that was shown to me by Tesla expert Robert Golka, he pointed out that there is, in fact, a stepped lead coming from the ground that does rise up 150 feet to the top of the tower, which in this instance did not yet have the cupola. This lead appears just to the left of the wooden ladder, rising up alongside it. One would assume that this lead was present in July of 1903 when Tesla fired up his behemoth with the cupola in place, but this step-like lead is not present in other photographs of the tower taken just a year or so later.

Tesla was known to have stayed at Stanford White's Box Hill home in St. James, which was sixteen miles from the site (and is still owned by the White family). So it would have been relatively easy to undertake a sixteen-mile experiment from Wardenclyffe to Box Hill or a sixty-five-mile experiment to New York City. Any experiments he did do involved primary and secondary coils located *in his laboratory* sent by underground cables to the tower, as opposed to nonexistent oscillators on the main shaft of the tower. There is no photographic or any other evidence to suggest that Tesla ever placed oscillators on the actual tower, and no evidence that he ever installed a permanent central metal shaft running from the cupola to the ground. Since the tower was never completed, perhaps that is the reason why no long-distance experiments were ever conducted.

> The body of the Earth is a good electrical conductor insulated in space. It has an electrostatic charge relative to the upper atmosphere beginning at an elevation of about 50 kilometers. When a second conducting body, directly adjacent to the Earth, is charged and discharged in rapid succession, this causes an equivalent variation of Earth's electrostatic charge resulting in the passage of electric current through the ground.[21]

The role of the ground in electrical experiments dates back to at least 1838, when Munich University Professor Carl August Steinheil "demonstrated that the Earth could be used as the return conductor." This concept of using the Earth as a return was also employed by Samuel Morse in the

1840s when he set up a forty-mile telegraph line from Washington, DC, to Baltimore. It became apparent, however, that the return circuit was more efficient when the ground was wet, for instance, after a heavy rain. Nevertheless, Steinheil's discovery that a grounded return circuit worked, and this helped reduce the cost of construction of worldwide systems.[22]

The idea of conveying telegraphic messages by means of wireless began with Mahlon Loomis, who in 1864 successfully sent signals between hilltops in the Blue Ridge Mountains of Virginia by using kites, a ground connection, and primitive galvanometers as the transmitter and receiver. This instrument, which detected small changes in magnetic fields, registered a signal every time the transmitter was "plugged" and "unplugged" into the ground. Loomis performed this in the presence of two congressmen and patented this watershed experiment.

The Loomis scheme was so well received that he obtained a $50,000 grant from Congress to exploit his "aerial telegraphy." Tesla, of course, knew of Loomis' work. The Serbian inventor created a similar setup with the use of kites and a ground connection and even used some of the same language as Loomis, who talked of "harnessing the wheelwork of nature." Concerning the question of priority in the invention of the wireless, Loomis is the first comprehensive inventor of the technology.[23]

In the early 1880s, much like his predecessors, William Preece, an electrical engineer for the British Post Office, realized that the *Earth itself was an integral component in the successful implementation of any wireless system*. After isolating the role of the Earth as either a primary or secondary circuit, Preece used telephone receivers as detecting devices and concluded that "on ordinary working telegraph lines, the disturbance reached a distance of 3,000 feet, while effects were detected on parallel lines of a telegraph 10 to 40 miles apart in some sections of the country." Preece's work of detecting Earth currents, which was duplicated by Western Union engineers in the United States, significantly influenced the theories expounded by Tesla.[24]

Other key features of any wireless system besides the ground connection would include the principle of resonance or tuned circuits, an antenna, the ability to produce continuous electromagnetic waves, and also a means for transmitting and receiving wireless messages. Preece had visited Edison in New York in the mid-1880s to see firsthand Edison's "grasshopper telegraph," which was "a device for jumping messages from dispatch stations to moving trains by means of induction or resonance." Edison also worked with precursor radio tubes, which were actually dual-filament light bulbs that displayed a flow of current between the prongs, Preece having named this the "Edison effect." Other precursors to Tesla included Heinrich Hertz, Oliver Lodge, and Édouard Branly. A French professor of physics, Branly studied Peter Munk of Sweden's 1835 experiments using Leyden jars to send electricity through carbon and metal chips. This work was expanded by Samuel Varley from England, who in 1856 noticed that electrical charges affected the conductivity of metal powder inside glass tubes.

Combining this information with the Edison effect, Branly noticed that the gap of Hertz's tuned circuits could be replaced by a glass-enclosed tube that contained finely scattered metallic particles. When a wireless current passed through the tube, the particles aligned themselves along the path of the gap and closed the circuit. A light tapping reopened the circuit until transmission occurred once again. Lodge perfected Branly's 1890 discovery of particle cohesion and labeled it the *coherer*. Tesla further perfected the coherer by using special metal powders developed by Alexander Popov, a Russian inventor, and Dr. H. Rupp of Stuttgart, who realized that a constant rotation of the coherer would increase its sensitivity. Tesla employed these coherers at Colorado Springs when he tracked thunderstorms at distances exceeding six hundred miles and picked up signals that he theorized came from outer space.

> While the tower itself is very picturesque, it is the wonders hidden underneath it that excite the curiosity of the little [hamlet]. In the centre of [the] base, there is a wooden affair very much like the companionway on an ocean steamer. Carefully guarded, no one except Mr. Tesla's own men have been allowed as much as the briefest peep. . . .
>
> Mr. Scherff . . . told an inquirer that the [shaft entrance] led to a small drainage passage built for the purpose of keeping the ground about the tower dry; [but] the villagers tell a different story. They declare that it leads to a well-like excavation as deep as the tower is high, with walls of masonwork and a circular stairway leading to the bottom. From there, they say, the entire ground below has been honeycombed with subterranean [tunnels that extend in all directions].
>
> They tell with awe how Mr. Tesla, on his weekly visits . . . spends as much time in the underground passages as he does on the tower or in the handsome laboratory where the power plant for the world telegraph has been installed.[25]

In the Tesla scheme, the inventor points out that because signals are amplified at the accepting end by highly sensitive receivers, electrical engineers have mistakenly presumed that their present transmission system was the most practicable. However, by suppressing the electromagnetic radiations, boosting the ground connection, and controlling the span of the wavelengths, energy could be pinpointed with almost no loss to any location, and thus amplification of the signal on the receiving end is no longer such a critical issue.

"It is true," Tesla points out, "that electricity is displaced by the transmitter in all directions, equally through the Earth and air. . . . But energy is expended only at the place where it is collected and used to perform some work." [26]

With regards to the ground connection and taking into account the radar images obtained during the filming of *The Tesla Files* that depicted Tesla's earth grippers and network of tunnels, we can ascertain that Tesla did indeed place a complex staging area seven stories below the surface of the planet. What is so interesting about this new and rather incredible finding is that except for Tesla's testimony during litigation with the Waldorf-Astoria hotel authorities, who destroyed the property, no information whatever has ever been uncovered as to the precise reasons for what was done there.

Since so much work was put into this area of the tower, one can't help but wonder why no treatise of Tesla's has surfaced that explains the need for nearly four hundred feet of tunneling. What seems likely to me is that climbing seven stories down to monitor the grounding rods can be an exhausting trip, so it would seem that Tesla was planning on bringing down testing equipment, a generator, and so on to have a place to work underground when he launched his wireless operation. Some of Tesla's goals including using liquid nitrogen or some comparable compound to augment the ground connection and also create a means for studying various aspects of the planet. For instance, Tesla talked about monitoring the movement of ships, which would be some form of global radar, and he also talked about finding precious metals through a process known as telegeodynamics. Perhaps Tesla planned to map the mineral content of as much of the Earth as he could once his wireless apparatus was operational.

According to Professor Jovan Cvetic at the University of Belgrade, "The Wardenclyffe tower was partially a commercial and partially an experimental facility. Tesla was not certain about some solutions since that tower was something much bigger and different from the Colorado Station. He tried to find the best way to excite the ground oscillations. He had to make experiments and this can be seen from his notes after the Colorado campaign." [27]

For our show *The Tesla Files*, Travis Taylor, our rocket scientist/physicist, studied Tesla's notes and was thereby able to take a large Tesla oscillator, attach it to the ground, and illuminate light bulbs at modest distances from the transmitter. To complete the experiment, when the ground connection was disconnected, the bulbs went out, but when the ground was "plugged back in," the bulbs again were illuminated. Taylor's experiment clearly confirmed Tesla's contention that the Earth was indeed a viable way to transmit electric power.

Taylor's second experiment involved the transmission of electric power from one Tesla coil to another identical receiving coil. The receiving coil was in turn connected by wire to a small boat that was successfully powered in this manner. Due to limited time and resources, the experiment was somewhat crude, but it did successfully demonstrate precisely Tesla's contention that by using a ground connection, a sending tower can transmit power to a receiving tower, which can then use that electrical energy to power a vehicle.

One major question regarding Tesla's plan has to do with the structure of the transmitting wave. Apparently, Tesla was transmitting a guided surface wave that worked by conduction. With regards to how this setup could power airplanes, Tesla stated, "The current passes through the Earth, but an equivalent electrical displacement occurs in the air." This contention certainly adds another level of complexity to its overall operation. Tesla added that his "perfected system of positive selectivity [will act like a] time lock on a bank safe, so complete will be the individualization of the currents." [28]

Power, according to Tesla, would not be freely available everywhere, as is commonly believed, unless one had a receiving device tuned precisely to the particular idiosyncratic wavelengths generated. Information and small amounts of power, however, probably would be available freely from the mass media broadcasting aspect of the device in much the same way a car radio picks up local stations.

Once energy was "jumped" to a receiving tower situated by a suburban center, it could be stored in the bulbous crest of the tower or transmitted to mechanical devices in a variety of ways, including the propagation of energy in straight lines through space, by means of wires, by setting up alternations between the ground and the elevated terminal, or by transforming the energies to higher frequencies and distributing them through the natural medium. The key would be to determine ahead of time the exact span of the wavelength so that electrical power would be accurately transferred to any key location or major city on the planet by altering the wave to any specific length: for example, 790 miles to Chicago, 2,900 miles to San Francisco, or 3,600 miles to Paris.

> The fundamental difference between the broadcasting system as now practiced and the one I expect to inaugurate is that at present the transmitter emits energy in all directions, while in the system I have devised, only force is converted to all points of the Earth, *the energy traveling in definite paths determined before-hand.* Perhaps the most wonderful feature is that the energy travels chiefly along an orthodromic line, that is, the shortest distance between two points at the surface of the globe, and reaches the receiver without the slightest dispersion, so that an incomparably greater amount is collected than is possible by radiations. I have thus provided a perfect means of transmitting power in any desired direction more economically and without any such qualitative and quantitative limitations as the use of reflectors would necessarily involve.[29] (emphasis added)

Tesla's idea was to construct a network of Wardenclyffe-like towers round the planet, each in a resonant relationship to one another, much like how tuning forks resonate with each other. Thus, each tower, or resonant transformer, could act like both a sender and receiver of wireless information. The bulbous top of the tower was not for transmission, but rather for collecting the energy so that it could be driven down into the Earth and sent out by means of conduction. By changing the length of the electromagnetic waves and taking into account the size of the planet and the speed of light, Tesla planned to pinpoint the collected energy to any specific location. Each tower was comprised of three coils, a primary, a secondary and extra coil as explained below:

1.   **Resonant Transformer** comprising:

     a.   **Thick Coil** of shorter length and fewer turns that acts as the primary in the transmitter and the secondary in the receiver.

     b.   **Thin Coil** of longer length and many turns that acts as the secondary in the transmitter and as the primary in the receiver. The length of the thin coil "should be one quarter of the wavelength of the electric disturbance in the circuit" based on size of the planet and the speed of light.[*]

c. **Extra Coil** for modulating, stabilizing, and individualizing the frequencies.

d. **Magnetic Core** attached to the Earth and the elevated terminal.

2. **Power Source** deriving energy from coal or a waterfall.

3. **Ground Connection** consisting of a deep metal shaft going down 50 or more feet into the Earth and "earth-grippers" or grounding rods radiating out, like spokes of a wheel into the Earth for better transmission.

4. **Container of Liquid Air** (197°F), which causes "an extraordinary magnification of oscillation in the resonating-circuit[s]." (Patent #685,956)

5. **Elevated Terminal, a** cupola or bulbous top for accumulating stored charge/condenser, 187 feet off the ground.

6. **Total Length** of the circuit approximately three hundred feet (plus an additional three hundred feet of grounding rods).

---

[*] By way of illustration," Tesla calculated the frequency at 1/925 of a second "in a circuit 185,000 miles long and each wave would be 200 miles in length." Therefore, his secondary would be one-quarter of that figure or "fifty miles in length." (Patent #649,621, filed 9/2/1897, NT 1956, 293.)

6

House of Morgan

Top-hatted . . . and plump, brandishing a vast cigar and warding off photographers with a walking stick, J. Pierpont Morgan brusquely ran his empire from his rolltop desk in the small, elegant Morgan headquarters at 23 Wall Street. Inspiring awe, swelling crowds parted for him as he made his way down the street to save the stock exchange in the Panic of 1907. Tycoons turned to jelly when caught in his fierce glare. Kings envied his collection of Old Masters and young mistresses. . . . In its prime the House of Morgan behaved like a sovereign state, commanding more wealth than most countries hoarded in their treasuries, engaging in shadowy diplomatic missions, writing the script for war or peace, propping up governments or toppling them.[1]

Ultimately, in order to understand Nikola Tesla, one must come to grips with the viability of his fantastic scheme at Wardenclyffe and, thus, his relationship with J. Pierpont Morgan. Tesla suggests in his autobiography that the venture was largely "philanthropic." A significant percentage of Morgan's great wealth was derived from his stake in General Electric, a company whose existence was based on an *entente cordiale* with Westinghouse to obtain Tesla's electrical power distribution system in exchange for electric trolley patents Westinghouse needed. Thus, for Tesla to portray this new deal as philanthropic was certainly in accord with Tesla's worldview, that the highly rich should give back to the community. However, realistically, this view had no basis in reality.

There is no comparable figure in the world today like J. Pierpont Morgan. If we were to combine Bill Gates, Elon Musk, Steve Jobs, and Ted Turner, we would have just part of who Morgan was. Here was a magnate who controlled banking, insurance, shipping, railroads, the telephone, electric power distribution, lumber, mining, rubber production, and the steel indus-try. Morgan was often depicted as a force more powerful than the king of England, the kaiser of Germany, and the president of the United States combined.

In Tesla's speech at Niagara Falls, he depicted himself as "an artist," and he did indeed discuss the idea of philanthropy. When he entered into an agreement with Morgan, no doubt he was looking at the world through rose-colored glasses, but careful study of their arrangement lays to rest any serious conclusion that philanthropy had anything to do with Morgan's involvement. They had a contract, wireless was the future, and Morgan had decided to back Tesla over Marconi. In fact, Morgan was so savvy that he not only attached Tesla's wireless patents, he also added the lighting patents to the deal. The upside was enormous when one considers that the market in fluorescent lights was gargantuan, potentially worth hundreds of millions if Morgan sought to exploit it.

The problem with Tesla was that he overreached. Once he found out that Marconi was pirating his apparatus, he changed plans and constructed a tower twice the size that he had agreed upon with Morgan, the goal being to increase the range of the tower so that he could capture a larger share of the market. This major change took place while the financier was on his yearly jaunt to Europe, and thus without Morgan's consent. While all this was happening, the stock market crashed, mostly because Morgan was in a bidding war against Ned Harriman for control of the Northern Pacific Railroad, and Harriman almost bought the railroad out from underneath the Wall Street behemoth. Then, at the very end of the year, Marconi suc-ceeded in transmitting a message across the Atlantic, etching his name into the world of the immortals.

Having run out of the $150,000 that Morgan had provided, and having raised an additional $50,000 through relatives and other investments, Tesla was still short a good $100,000. His only solution was to either get Morgan to give him the funds to complete the tower or allow the inventor to raise the additional monies by bringing in additional partners. But because Tesla had breached their initial contract and because of Marconi's success on a much more meager budget, Morgan decided to withdraw his support.

In this cutthroat world of international multimillion-dollar business deals, had Morgan been a mere philanthropist as Tesla suggests, he would have allowed other investors to provide the additional funding the inventor required to complete Wardenclyffe, because the goal of philanthropy is

precisely that, an altruistic venture. The simple fact of the matter is that a
very large percentage of Morgan's wealth was directly related to Tesla's previous inventions in electrical power distribution and in the sale of motors,
generators, and every electrical appliance you can think of. It was not that
Morgan thought that Tesla's plan was folly. Quite the opposite: He feared
that Tesla might succeed. And if Tesla did succeed, what would happen
to Morgan's lumber businesses, copper mines, and rubber plantations, and
most importantly, to his precious General Electric, let alone AT&T? If Tesla
could transmit electrical power without using wires, how could one put a
meter on that? Was it worth sacrificing all those industries to gamble on
Tesla's wireless scheme? Morgan decided it was not.

Tesla received $10,000 from Thomas Fortune Ryan, with a presumed
promise for the balance of $90,000. But after meeting with Morgan, Ryan
went into the insurance business instead.[2] And so, Tesla turned to Henry
Clay Frick.

In a detailed letter to Frick at his offices in the Trinity Building, downtown, Tesla outlined the full scope of his wireless enterprise, including a
stunning photo of his laboratory and tower with a handwritten note in
the space where the sky would be. "This World Telegraphy Plant," Tesla
wrote, "will deliver electrical energy at the rate of ten million horsepower
and transmit telegraphic and telephonic messages to *any* terrestrial distance with *absolutely* the same precision and facility," ending the typed
portion as follows:

"Although the commercial possibilities of this advance are infinite, I am
appealing to you in the same spirit I have arranged to Mr. Morgan and in
which he has responded." And then Tesla moved in for the close. "Nature
has given you the power for managing business and making money. . . . You
and Mr. Morgan would take for yourself and accord to me whatever share
you like. . . . Just a little of that you have in abundance," he wrote to the
multibillionaire (in today's dollars), "combined with the exquisite and rare I
have will enable us to advance the world for centuries."[3]

Frick, whom Tesla ran into frequently as they both had suites at the Waldorf, kept delaying a response, posing to Tesla that "there must be a reason"
why Morgan had not contributed the balance necessary to complete the
project. Tesla recapped his past achievements in electrical power transmission, handed him copies of his wireless patents and several articles outlining
the full scope of the plan, and then figured out a way to frame the answer
which would avoid the key issue, namely that Tesla had constructed a larger
tower than Morgan had agreed to funding.

"Perhaps you are wondering why it is that Mr. Morgan does not advance all the capital necessary. The explanation is very simple Mr. Frick. I interested Mr. Morgan with the distinct understanding that he would not be called upon for more. He has treated me as only a great man can but he should not be expected to break his word. Will you please see him and come to some decision whatever it be."[4]

Frick did, and just as Morgan had dissuaded Ryan from funding the balance of the project, he dissuaded Frick and so Tesla was forced to write the Wall Street monster once again.

> February 17, 1905
>
> Dear Mr. Morgan,
>
>> Let me tell you once more. I have perfected the greatest invention of all time—the transmission of electrical energy without wires to any distance, a work which has consumed 10 years of my life. It is the long sought stone of the philosophers. I need but to complete the plant I have constructed and in one bound, humanity will advance centuries.
>>
>> I am the *only man* on this earth *to-day* who has the peculiar knowledge and ability to achieve this wonder and another one may not come in a hundred years. . . . Help me to complete this work or else remove the obstacles in my path.
>>
>> I was heartily glad to see you in such splendid health yesterday. You are good for another 20 years of active life.
>
> Faithfully yours,
> N. Tesla[5]

We should keep in mind that since Tesla lived in the Waldorf-Astoria, as a matter of routine, he often hobnobbed with numerous megawealthy individuals. John Warne "Bet-a-Million" Gates, like Henry Clay Frick, lived at the Waldorf, and another multimillionaire, John Jacob Astor, who owned the hotel, had funded Tesla's previous foray at Colorado Springs. With the Wall Street monarch's backing, August Belmont, who had funded Westinghouse, had just been reimbursed $35 million to construct a new interborough subway, and it ran on nearly a dozen gargantuan Tesla turbines. In the case of Frick, when Morgan bought out Andrew Carnegie to create U.S. Steel, the first billion-dollar company, Frick, as Carnegie's manager, walked away with about $60 million. To be surrounded by such wealth, much of it overtly generated by the wizard's previous creations, and to be shut out by the very man who benefited the most, J. Pierpont Morgan, was just too painful for Tesla to bear.

On the surface, Morgan "wished Tesla success," but actually, he ensured Tesla's failure by not only not supplying additional funds to complete the tower but by also blocking other multimillionaires from investing. Once the Tesla deal was dead, there was a run on the stock market. Copper stocks were manipulated, and that sparked the Panic of 1907. I have suggested that since Tesla's wireless plans were now defunct, the market understood that copper would again become a valuable commodity, and that lay at the heart of the panic.

A number of banks tumbled. Teddy Roosevelt, now president, placed Morgan in charge of raising and disseminating funds to bail out the stronger ones, but when Morgan refused to help his friend Charles Barney, who ran the Knickerbocker Trust Company, Barney grabbed a gun and committed suicide.

Dear Tesla,

Mrs. Filipov is well and both [she and I] are happier for having seen you Xmas night. When the millionaires desert you the humble hearth of poverty will always have a welcome [home] for you.

Faithfully yours,
RUJ [Robert Underwood Johnson] [6]

NERVOUS BREAKDOWN

For the last seven years I have had wireless telegraphy for breakfast, dinner and supper and have so many ideas on the subject myself, that I am beginning to suffer from indigestion. [7]

Unbeknownst to the public, Tesla went through a year of private hell, topped off by the assassination of his friend and partner to Morgan in the construction of Madison Square Garden, as well as his architect at Wardenclyffe, the inimitable lothario Stanford White. Having courted the premier Gibson Girl, Evelyn Nesbit, when she was just 16, White, who was 47, had met her at the Garden Theatre when she was featured there as a Florodora Girl. In what was considered by many the "crime of the century," White was killed

six years later by Evelyn Nesbit's deranged husband, Harry Thaw, a bil-
lionaire in today's dollars, who got off three shots at a party at the rooftop
restaurant at the Garden, killing White instantly. Waving his gun in the air
while proclaiming that he had "shot the beast," Thaw was arrested, and thus
began his lengthy trial, his demure wife writing Thaw love letters, sitting
dutifully near his side. He tried using the ruse that young Evelyn had been
violated, but in actuality, Thaw, who had a penchant for whipping prosti-
tutes, was simply jealous of the debonair man about town, and he paid the
price by spending a good part of the rest of his life incarcerated in a mental
institution.

> October 13, 1904
>
> Since a year, Mr. Morgan, there had been hardly a night
> when my pillow is not bathed in tears, but you must not think
> me a weak man for that. I am perfectly sure to finish my
> task, come what may. I am only sorry that after mastering
> all the difficulties which seemed insuperable, and acquiring
> a special knowledge and ability which I now alone possess,
> and which, if applied effectively, would advance the world a
> century, I must see my work delayed.
>
> In the hope of hearing from you favorably, I remain,
> N. Tesla[8]

Had it not been for the evidence I found through the disintegration of
Tesla's handwriting trail for that period around 1906, and corrobora-
tion from contemporary letters from Tesla's secretary George Scherff
and from his editor T. C. Martin attesting to the inventor's "frightful"
physical state, the myth that the Waldorf dandy came through somewhat
unscathed would have persisted.[9]

> February 2, 1906
>
> Dear Mr. Morgan,
>
> Please see Mr. Frick. . . . He is going to call on you. Time is
> flying.
>
> Yours sincerely,
> N. Tesla[10]

The magnitude of Tesla's suffering during the collapse of his Wardenclyffe dream was not only kept from public view, it was also kept from Tesla's biographer, John O'Neill, who knew Tesla for forty years. Coincidentally, if not synchronistically, another intimate friend, Katharine Johnson, also suffered greatly that same year, 1906, and the collapse of the dream of Tesla's Wardenclyffe global enterprise may have played a role.

KATHARINE JOHNSON

In September of 1906, while still in Europe, Katharine wrote, "Dear Robert, I am not well. This you must be prepared for. . . . Something vital has come out of me in that long illness and convalescence in Florence. . . . You must not look for a realization of what you are now dreaming of—that is long past. I know I am changed. I am sadder—I have a feeling of the descending order, a twilight one must be resigned. . . . One must look deeper, further to go to want to live. Perhaps I shall find a little bit of my whole self at home." [11]

In fact, upon her return to the States, Katharine actually garnered the nerve to visit the Wall Street mogul at his office at 23 Wall Street, but Morgan refused to see her.

Tesla, however, continued his lifelong friendship with Morgan's children. The fact that J. P. Morgan Jr. would later invest in Tesla's turbines suggests that essentially no one other than Tesla and Morgan knew the true level of the animosity that existed between them. To his good friend Anne Morgan, Tesla wrote:

> March 31, 1913
>
> My dear Miss Morgan,
>
> Words cannot express the sorrow I have felt at the news of your father's death. Not until I realized that he was no more did I appreciate how deeply his forceful image was impressed on my memory.
>
> All the world knew him as a genius of rare powers, but to me he appears as one of those colossal figures of the ages past which marks an epoch in the evolution of human thought.
>
> Sincerely yours,
> Nikola Tesla [12]

Had it not been for the fact that Tesla kept his correspondence with the Wall Street magnate, there would be no way to understand the true suffering

Tesla endured because of Morgan's obstinate decision to not only cut off all funding for his precious tower but also to *block* all other investors from helping Tesla succeed. The fact that Tesla kept up appearances speaks to the complexity of his nature and his need to maintain a front so that he could continue to seek other investors after Morgan's passing. And it was due to this public persona, when it came to Morgan, that the truth about their relationship was kept from public view.

Hiding the private hell he had really entered, Tesla emerged in 1907 with an anger he did not quite know how to contain. In rapid succession, the ailing wizard emerged with a plethora of mean-spirited proclamations that displaced his aggression, shifting it from the mighty financier to the public at large. Not only would he claim that he could take down any building in a city by placing an oscillator on a major support beam, but, as the *New York Herald* proclaimed, "Tesla Will [Also] Destroy an Army a Minute" by generating "twenty million horse power to fight the battles of the future."[13] And if that wasn't horrific enough, the inventor further claimed that, with the right equipment, he had the ability to create tidal waves that would sink any ship or destroy an armada and that, given "a few weeks" time, using the same principle of resonance, he could also split the very planet in two![14] This was a new Tesla, angry, petulant, spiteful, and vindictive. But with Morgan, as ever, he held his tongue.

An interesting insight into the mindset of the mighty financier can be found in Jean Strouse's 1999 biography *Morgan: American Financier*, which suggested that he had misgivings about having destroyed Tesla's grand vision. In a personal conversation with his secretary Belle Greene one night, while a bit drunk and somewhat morose because time was running short, the aging financier wrestled with some of his more difficult decisions. Morgan could see the world of wireless developing by this time, and it is indeed possible that he had second thoughts.

> The night before Morgan sailed for Europe in 1908, "he told me bits of his life," Greene writes in her diary, "of his unful-filled hopes . . . failures and disappointments . . . of how he had always sought to be a builder—not a wrecker in the world of things—of how he had tried to put ambition, as such, behind him and accomplishment in the betterment of mankind . . . before him. . . . So completely [was I] attuned to him—that I

> did not realize until long after . . . that my hand was aching
> and sore from his grasp"[15]

Compare this passage to Tesla's final letter, out of about sixty, that the
inventor sent to Morgan before he completely fell apart. Could this missive
have haunted Morgan?

> Every opportunity is there, [Mr. Morgan]. I have high
> regards for you as a big and honorable man . . . [but] there
> is greater power in the leaf of a flower than in the paw of a
> bear. That is as much as I'll ever say. . . . You are reputed as a
> builder of properties, but if you prefer in this instance to chop
> down poles . . . go ahead.[16]

Tesla's insights in this brief passage are breathtaking.

"May gravity repel instead of attract, may right become wrong. Every
consideration no matter what . . . must founder on the rock of your brutal
resolve. . . . Will you not listen to anything at all? Are you to let me perhaps
succumb, lose an immortal crown . . . because you once said no? I tell you I
will return your investment a hundredfold," Tesla writes to Morgan in this
poignant plea.[17]

No one knows for sure why Morgan insured that Wardenclyffe would fail.
But that is certainly what he did. It was one thing for Morgan to pull his funding; it was another to use all his power to dissuade others from bailing Tesla
out, and that is precisely what occurred. I hypothesize that the main reason
was Morgan's inability to envision how he could earn a profit, because in
those days there would be no way to put a meter on a wireless scheme. Understanding that the future would mean a worldwide interconnection of wireless
communication, Tesla was trying to tell Morgan that revenues would derive
in a different way. For instance, for many years, radio and TV were free,
yet both venues generated billions of dollars, mostly through advertising. But
perhaps Morgan also questioned the feasibility of the enterprise. More than
a hundred years later, the debate on this point still continues. The question
is, Was Tesla's design at Wardenclyffe viable? Part of the answer is contained
in the previous chapter on Wardenclyffe, but another key can be found in the
following chapters on Tesla's invention of remote control, on how he helped
Telefunken expand its wireless station out at Sayville, New York, on Long
Island, in 1915, and in a full understanding of the litigation between Tesla, the
U.S. Navy, Telefunken, and Marconi.

PART III

Wireless

I am glad to say that ... I have devised means which will allow us ... the transmission of power from station to station without the employment of any connecting wire. .... Some of the ideas I have expressed may appear to be hardly realizable; nevertheless, they are the result of long continued thought and work. With ideas it is as with dizzy heights you climb. At first they cause you discomfort and you are anxious to get down, distrustful of your own powers, but soon the remoteness of the turmoil; of life and the inspiring influence of the altitude calm your blood; your step gets firm and sure and you begin to look—for dizzier heights.

—NIKOLA TESLA,
speech at the Ellicott Club Banquet,
Niagara Falls,
January 12, 1897

7

Remote-Control Robotics

As early as 1898 I proposed to representatives of a large manufactur-
ing concern the construction and public exhibition of an automobile
carriage which, left to itself, would perform a great variety of oper-
ations involving something akin to judgment. But my proposal was
deemed chimerical at that time and nothing came from it.

—NIKOLA TESLA[1]

With regards to Tesla's invention of telautomatics or remote control and even remote-controlled robotics, Tesla discusses in a
patent he never submitted, titled "Methods of Controlling Automata at a
Distance," the particulars of how he planned to activate distant instruments
and vehicles (e.g., submersibles, torpedoes, aircraft, and automobiles):

> In certain uses of such automata, perfect safety against willful
> or accidental disturbance is desirable, if not essential. In such
> cases, I take advantage of the principles of the art of individ-
> ualization which I have originated and set forth in my patents
> Nos. 723,188 and 725,605. . . . If required, I vary arbitrarily
> . . . groupement [*sic*] or order of succession [of frequencies] . . .
> securing the desired isolation of the controlling vibrating energy
> in both aspects: non-interference, and non-interferability.[2]

This device, which is an early radio-guidance system, if used in war, was actually given two opposite monikers, "devil automata" for torpedoes or aerial
bombardment, and yet also "a machine which would abolish all war!"[3]

At the time of writing, this technology involved line-of-sight control by
the operator for an instrument of war. However, this scheme could also be
used for any remote-controlled device, such as drones which are now used
for photographing homes, for overhead action shots in making motion pictures, remote-controlled toy cars and airplanes and so on.

While working on his telautomaton, "government officials . . . from Japan [and . . .] a number of [other] scientific men visited me [at my Houston Street lab]. . . . To these I gave the most complete information on all the patents [I had on remote control]." Reflecting back six years later in 1904 during the war in Manchuria between Russia and Japan, Tesla further stated, "It would not surprise me if by this time [the Japanese] had succeeded in producing models of my automaton." Tesla was referring specifically to his remote-controlled dirigible torpedo, which journalists were speculating the Japanese were using to destroy the armies and ships of their enemy, the Russians. This device, which Tesla saw as a means to make war obsolete, would "behave like a blindfolded person obeying directions received through the ear." And yet, as megalomaniacal as this plan was, it was only part of his vision![4] As early as 1898, Tesla said it this way:

> Long ago I conceived the idea of constructing an automaton which would mechanically represent me, and which would respond, as I do myself, but of course, in a much more primitive manner to external influences. Such an automaton evidently had to have motive power, organs for locomotion, directive organs and one or more sensitive organs so adapted as to be excited by external stimuli. . . .
>
> Whether the automaton be of flesh and bone or of wood and steel, it mattered little, provided it could provide all the duties required of it like an intelligent being.[5]

As a prototype, Tesla's brand new life-form was "embodied," in his words, with a "borrowed mind," his own. But also, it would be able to learn![6]

> [It will be able] to follow a course laid out or . . . obey commands given far in advance, it will be capable of distinguishing between what it ought and what it ought not to do . . . and of recording impressions which will definitely affect its subsequent actions.[7]

BINARY CODE AND THE THINKING MACHINE

On December 29, 1897, Ernest Wilson and Charles John Evans created a British patent for remote control that predated Tesla's patent #613,809, "Method of and Apparatus for Controlling Mechanisms of Moving Vessels or Vehicles," applied for on July 1, 1898, and granted four months later.

Although Wilson and Evans (U.S. patent #663,400) predated Tesla by several months, his invention was more complex. For instance, the oscillating frequencies used in radio-dynamic control are all based on the Tesla coil and Tesla frequencies, whereas Wilson and Evans stipulate the use of Hertzian waves, a mechanism later proven untenable. The steering component for the Wilson and Evans design involved the implementation of two receivers at right angles to each other. Presumably, if transmitter A were activated, the vehicle would move to the left, and if transmitter B were activated, it would move to the right.

Tesla's scheme, on the other hand, was much more elegant. It involved the use of a combination of frequencies to activate a series of relays, one of which was specifically for steering the rudder, set up ahead of time to be pointed in one direction. "For example, the rudder could be steered to the left when the electromagnetic wave was received and to the right otherwise." By using a *counter-spring* and switching the mechanism on or off, the boat could be steered in any direction.[8] Jack Hammond noted that this fundamental "on-off" scheme became the "forerunner of the 'AND' principle" inherent in the computer chip of modern computers. In a move of pure genius, Tesla had reduced the complex mechanism of steering a boat to a binary system. The simple act of having the current either on or off, corresponding to 1 and 0, became the basis for the complex steering mechanism and for what Tesla envisioned as a prototype of the first *thinking* machine.[9]

Aside from working the rudder and starting and stopping the propeller "by electronic impulses transmitted across the room without wires," Tesla added additional relays when he brought his telautomaton to the Commercial Club in Chicago. This complex invention actually employed an early prototype of a logic board.

While in Chicago en route to Colorado Springs Tesla lectured before a crowd that included Cyrus McCormick Jr., president of McCormick Harvesting Machine Company; German manufacturers Baron Cornelius Heil and Von-Bismarck-Platke; publisher Charles Scribner; and department store mogul A. G. Selfridge. There, he astounded the audience when he got his remote-controlled torpedo boat to motor in various directions, turn its lights on and off, perhaps in response to different questions, such as one blink for yes and two blinks for no, and "fire five guns on board."[10] Tesla then really stretched the imagination of his audience when he claimed that he could perform these same operations from implementing a wireless transmitter in America to control a boat in Paris for the upcoming 1902 Paris Exposition!

"I can apply this system of control to any type of vessel and of any size," Tesla pontificated, but added that if used as a remote-operated torpedo "in the operations of war," he conceded that "the radius of control would usually be limited by the range of vision of the operator."

Tesla noted that his telautomata responded as humans and animals do, to the reception and activation of an outside influence, suggesting "the practicability of providing a large number of receiving currents . . . fifty or a hundred or more, each of which might be called up, or brought into action whenever desired without the others being interfered with. This made it possible for one operator to direct simultaneously the movements of many bodies, each of which might have a distinct duty to perform, as well as to control the action of various devices located on the same body."[11]

This was to be the first of a new race of mechanical thinking beings, responding as humans and animals do, to the reception and activation of an outside influence.

This idea of a binary system of switches could also be expanded. In Tesla's case, he could not only steer his telautomaton but he could also turn its lights on and off and shoot out a simulated torpedo. Eventually, Tesla planned for the remote-activated boat to also become a submersible, and it is also known that the wizard continued to develop his robot, although the extent of the development of his new "species" remains unknown.

In terms of the etiology of the invention, the author Edward Bulwer-Lytton (the novelist who first coined the phrase "It was a dark and stormy night") had introduced the concept of "Vril power" which would be used by self-activated automata in his book *The Coming Race,* first published in 1871. However, the word *robot* did not become part of the lexicon until 1920, more than two decades after Tesla introduced his telautomaton, when the word was coined by Czech playwright Karel Čapek in his 1920 play "R.U.R.," standing for "Rossum's Universal Robots." The word *robot* was derived from the old Slavonic word *robota,* which means "servitude."[12] In Spain, this invention was applied as a means of steering an airship by remote control in 1901 by Leonardo Torres-Quevedo, whose switches not only steered the ship but also controlled its speed.[13]

Tesla unveiled his telautomaton during the height of the Spanish-American War at Madison Square Garden in 1898 to a group of private investors. The main arena, filled with water, was used as a giant swimming pool. For the electrical show, prototype models of the Spanish Armada and American battleships were placed in the simulated sea, and Tom Edison's son, Tom Jr., working for Marconi, who was not present, used remote control to detonate bombs planted on board the mock Spanish ships.

However, Marconi had yet to perfect selective tuning. Thus, when Tom Jr. pressed the button, not only did he destroy one of the Spanish ships, he also blew apart a desk in the back room where additional bombs were stored. Fortunately, no one got hurt.[14] Tesla on the other hand, had a private showing of his telautomaton before such investors as John Oliver Ashton and his father, who was head of Okonite Wire and Cable, the largest cable company in America; Cornelius Vanderbilt; J. Pierpont Morgan; J. D. "Jack" Cheever, a Wall Street broker and principal from Okonite Wire and Cable; his associate Willard Lyman Candee; and also John Hays Hammond Sr., who, according to John O'Neill, invested $40,000 in the venture.[15] Another major investor in Tesla's work was John Jacob Astor. But "Colonel" Astor had armed his gigantic yacht, the *Nourmahal*, with a naval brigade and four machine guns. Having donated it to the war effort, he was down in Cuba at the time, protecting his property, overseeing aspects of the war.[16]

Along with Hilbourne Roosevelt, Candee and Cheever had organized the first telephone company in New York in 1877, installing five lines throughout the metropolitan area. A captain in the army and a crack marksman, Candee was also the vice president of the Electric Club of New York and the first to install a telephone line across the Brooklyn Bridge when it was erected in 1884. While running a lighting company in New Jersey and serving as director of the Franklin Avenue Railroad, along with Cheever and Frank Jones, Candee started Okonite Cable and Wire, making his real fortune when Okonite secured a mammoth contract with the Westinghouse Corporation to provide wire and cable for the Niagara Falls operation.[17] One can only imagine the mixed emotions the wire and cable bigwigs must have had as they watched firsthand as Tesla inaugurated an entirely new technology that did away with their product.

Caught up in Spanish-American War fever, Tesla did not just promote his telautomaton as a remote-controlled boat or even, astonishingly, as a new race of thinking beings. No, Tesla's boat was also conceived as a mechanism to end all war.

With plans to go down to Cuba to launch a telautomaton loaded with explosives to "annihilate" the Spanish Armada, Tesla proclaimed, "War will cease to be possible when all the world knows tomorrow that the most feeble of all nations can supply itself immediately with a weapon which will render its coast secure of the united armadas of the world. Battleships will cease to be built, and the mightiest ironclads and the most tremendous artillery afloat will be of no more use than so much scrap iron. And [thus these] irresistible . . . instrument of destruction . . . can be exerted at any distance by an agency of so delicate, so impalpable a quality that I feel that I am justified in predicting that the time will come, incredible as it may seem, when it can be called into action by the mere exercise of the human will."

Lamenting that he would prefer to be remembered as the inventor who would abolish bellicose confrontations, the wizard of war nevertheless continued to expand on the possibilities. Applying "this system of control to any type of vessel" of any size, Tesla went on to describe the truly diabolical idea of loading a boat with hundreds of tons of explosives which could be detonated "even a mile or so away," so that a monster wave would ensue that "would overwhelm even the biggest ship ever built."[18]

Billed in the December 7, 1898, issue of *Electrical Review* as the "Genius of Destruction," Tesla had claimed that his submarine boat, "loaded with torpedoes," could set out without a crew, "make its devious way along a surface through dangerous channels and mine beds, watching for its prey . . . discharge its deadly weapon and return to the hand that sent it."[19] Commenting on this article, *Electrical Review* stated that "the Genius of Destruction" would seem to have, then, two aims: "It creates evil, but mostly good [because] it could hasten the utopian dream of the abolition of war."[20]

As the forerunner to the now seemingly ubiquitous drone strikes occurring in this era in the Middle East, the promise of Tesla's "devil automata" spread all the way to Europe, where articles on the invention were read by Mark Twain, who was staying with friends in Vienna.

November 17, 1898

Dear Mr. Tesla,

Have you Austrian and English patents on that destructive terror which you have been inventing?—And if so, won't you set a price upon them and commission me to sell them?

I know cabinet ministers of both countries—and of Germany, too; likewise William II. I shall be in Europe a year, yet.

Here in the hotel, the other night when some interested men were discussing means to persuade the nations to join the Czar & disarm, I advised them to seek something more sure than disarmament . . . to contrive something against which fleets and armies would be helpless and thus make war impossible.

I did not suspect that you were already attending to that,

and setting ready to introduce into the Earth permanent peace and disarmament in a practical and mandatory way.

I know you are a very busy man, but will you steal time and drop me a line?

Sincerely yours,
Mark Twain[21]

Tesla's telautomaton was still in a very primitive state, and so the deal with Twain was never consummated. Nevertheless, Tesla's view for ending human conflict was played out with his conclusion that machines would fight instead of men and, further, that if all countries had such advanced weaponry systems, war would become a suicidal venture and so no country would ever go to war. This was a controversial stance embraced by some and completely rejected as a cynical rationalization by others.

One way or another, this incredible paradigm-shifting invention contained within it not only the concepts of remote control, robotics, and artificial intelligence, but also selective tuning; the seeds of encryption and protected privacy; cell phone technology (that is, the ability to create an unlimited number of wireless channels); and such everyday devices as the garage door opener, wi-fi, and the TV remote.

THINK TANK

September 29, 1916

Dear Mr. Miessner,

In an article in *The Century Magazine*, a copy of which I am forwarding . . . I have related the circumstances which led me to develop the idea of a self-propelled automaton. My experiments were begun some time in 1892 and from that period on until 1895 in my laboratory at 35 South 5th Avenue, I exhibited a number of contrivances and perfected plans for several complete telautomata.

After the destruction of my lab by fire in '95, there was an interruption in these labors which, however, were resumed in '96 in my new lab at 46 East Houston Street where I made more striking demonstrations. . . . [Beginning the following year I constructed] complete automata in the form of a boat . . . described in my original patent specification #613,809 . . .

[circa 1897] often exhibited to visitors who never ceased to wonder at the performances. . . .

In that year, I also constructed a larger boat which I exhibited . . . in Chicago during a lecture before the Commercial Club [in 1898] . . . limiting myself to mechanisms controlled from distance but to machines possessed of their own intelligence. Since that time I have advanced greatly in the evolution of the invention and think that the time is not distant which I shall show an automaton which, left to itself, will act as though possessed of reason and without any willful control from outside. Whatever be the practical possibilities of such an achievement, it will mark the beginning of a new epoch in mechanics.

Tesla went on to describe the use of "individualized control; that is, one based on the co-operation of several circuits of different periods of vibration, a principle" that was developed in his patents #723,188 and 723,189 of March 1903. This device, Tesla stated, was "demonstrate[d] before the Chief Examiner, Seeley" in Washington, DC, in 1898.

In my . . . investigations in Colorado from 1899–1900, I developed, among other things, two important discoveries . . . described in my patents #685,953 and 119,732 . . . taken out at a later date. These two advances make it possible to supply an automaton great amounts of energy and also to control it with utmost accuracy when it is entirely out of sight at any distance.

During the past few years, I have devoted much of my time to the perfection of a small, high speed vessel [embodied] with certain new means of propulsion in an endeavor to produce a most effective weapon of defense, such as would seem to be at this time of paramount importance to the United States.

I may be able to respond to your request to furnish you one or two illustrations . . . for publication in your book which I hope will prove a complete success.

Yours very sincerely,
N. Tesla[22]

In 1911, Tesla formed a partnership with John Hays Hammond Jr., son of industrialist John Hays Hammond Sr. Jack, as he was called, had known Tesla since he was a boy, as his father had helped fund Tesla's remote-controlled boat. Having worked in the patent office, Jack became a student of invention and would eventually compile hundreds of patents on everything from a microwave oven to musical instruments to radio-guidance systems.

A graduate of Yale University and a protégé of Alexander Graham Bell, Jack became enamored of Tesla's 1903 patent on selective tuning. In 1909, during his senior year, he asked his father to arrange a meeting with the "Serbian High Priest of Telautomatics," and in that way Jack, just twenty-one years old, took a train down from Yale to meet the venerated inventor at his offices at the Metropolitan Towers, at the time the tallest building in the city. Jack praised Tesla for his "prophetic genius patent" because it provided the means for creating an unlimited number of private wireless channels, and it is quite possible that Tesla reciprocated by traveling up to Gloucester, Massachusetts, where the Hammonds resided, to see Jack's crudely built remote-controlled robotic dog on wheels.

Jack began working at the patent office and also talked Tesla into forming a company together called the Tesla-Hammond Wireless Company. "In thinking of this name," Jack wrote, "I have followed Emersonian advice, and as you can see have hitched my chariot to a star." [23]

Coming from a wealthy family, literally living like a prince, Jack would go on to commission a naked statue of himself with a fig leaf strategically placed, and, about a decade later, in the 1920s, erect a medieval castle as his home, complete with parapets, towers, and a drawbridge, a bulky entrance door dating from feudal times, narrow winding staircases made of stone, a central atrium shining down onto a swimming pool, secret passageways, and a basement auditorium that housed a gigantic organ—all of which overlooked the bay in Gloucester, Massachusetts. And it was there, at the castle, during the Roaring Twenties, that Jack invited the Hollywood glitterati for nights and weekends.

Initially, Tesla told Hammond that "the Tesla-Hammond combination looks good to me, but we should go at it with some circumspection." Thus the partnership, sadly, ended up not in "telautomatics" but rather in the design and perfection of turbines. Tesla's wireless patents were still controlled by Morgan, and that was a key problem, but Tesla was probably also reluctant to share all he knew in that capacity. [24]

In retrospect, this restricted arrangement by Tesla appears shortsighted as it ultimately monkey-wrenched the partnership. Nevertheless, torpedo engines were a sizeable market, and the duo had pending deals with the United States armed forces and foreign governments, particularly with the Germans and Japanese. Unfortunately, the impending war destroyed this foreign market, but also, Tesla was having difficulty perfecting his specialized engine.

Hammond, who over the course of a lifetime would amass over five hundred patents, proceeded simultaneously to set up a military think tank in Gloucester in radio-dynamics, and he hired Fritz Lowenstein, Tesla's closest engineering acolyte, and also Benjamin Franklin Miessner (1890–1976), a fellow Yale graduate from the School of Engineering. Although Hammond continued to help Tesla develop the turbines, his real passion was in remote control, so it seems that his relationship with Tesla was hampered, to say the least.

Operating in secret for the military, the think tank worked on perfecting its equipment. In Lowenstein's case, this meant further developing the radio tube known as the audion. And while this was happening, Hammond installed a more complex remote system in a sizeable boat that he named the *Natalia*, after his daughter.

Certainly, the development of an efficient motor to be placed in torpedoes had lucrative potential. In fact, Tesla had a pending contract to sell thousands of them to the Germans. Also working to reconfigure the motor to fit into airplanes and, on a much larger scale, to power ocean liners, Tesla kept attempting to expand the business. Since all of the correspondence between Tesla and Hammond concerns the motors, there is substantial reason to believe that their relationship unfortunately stayed clear of radio-dynamics.

Fearful of stepping on the toes of his partnership with Morgan, Tesla may have also understood that the perfection of radio-dynamic control was many years away, so it made more sense to deal with the nuts and bolts of developing his turbines.

One way or another, even that avenue was fraught with unforeseen difficulties. Tesla required enormous sums to perfect his turbines, and this caused a tremendous strain on their relationship. Jack, who had apparently kept his job at the patent office, was working on writing a thesis on the history of the radio and remote control. Yet, at the same time, his associate, Miessner, was putting together his own dossier on the same subject, so this would result in two separate treatises on essentially the same topic.

While still helping to fund Tesla's turbines, Hammond also began working with Elmer Sperry to develop a reliable way to connect a Sperry gyroscope to a steering gear, a mechanism that came to be known as the autopilot. Sperry, who had attended Tesla's famous 1891 lecture at Columbia College, and whose famous invention was, in fact, based completely on Tesla's rotating magnetic field, provided the stabilizing compass while Hammond provided the radio-dynamic technology.

Since Hammond was working in secret for the military, this allowed "Fritz Lowenstein's contribution in making the DeForest grid audion [radio tube] into an amplifier and oscillator" to stay out of the press and out of the eyes of competitors.[25] Lowenstein's receiving device or advanced audion, also known as a "controller," was a real improvement on Hammond's radio-guidance system, which, at the time, was being developed for the torpedo for the Navy. Although the work was top secret because of its military implications, Lowenstein was able to get this aspect released, and he later sold the patent to AT&T for $150,000. Lowenstein was so thrilled with the deal that he made a copy of the check and often showed it to disbelieving colleagues.[26]

After leaving Tesla's employ, Lowenstein began in 1910 working for the Radio Telephone Company of Newark, on developing a wireless telephone. "When the Radio Telephone Company failed, Lowenstein started his own laboratory on Nassau Street in New York City," but that enterprise failed and so he began working for Hammond.[27] In 1911, Hammond traveled to Germany while Lowenstein held the fort in Gloucester. Two years later, Lowenstein returned to his home in the Czech Republic to confer with his family and probably check out the competition. Meanwhile, Hammond tried to tie up loose ends, so he invited Lee DeForest to Gloucester to discuss the invention. The meeting was short, and DeForest soon returned to the West Coast to perfect his version of the audion while seeking legal help to protect it from such interlopers as Lowenstein and the new kid on the block, Columbia University graduate Edwin Armstrong. DeForest approached AT&T with the help of John Stone Stone, another well known wireless inventor and president of the Institute of Radio Engineers. (Stone's mother's maiden name, by coincidence, was also Stone.)

Other big competitors included Reginald Fessenden (1866–1932), who attempted to set up a transatlantic wireless company only to have it collapse in a storm, and the growing German wireless behemoth Telefunken, which was operating in the United States under the moniker of Atlantic Communications Company. With offices located in the small tower of the Trinity Building at 111 Broadway, down by Wall Street, Telefunken hired, in 1909, Lloyd Espenschied (1889–1986), who stayed there for a year. Espenschied, who would amass over one hundred patents and go on to work for AT&T and Bell Telephone Laboratories, would, forty years later, pen a comprehensive article on the history of wireless communication and radio-dynamic control, "Discussion of History of Modern Radio-Electronic Technology."

Competing against Telefunken, Hammond, now in a partnership with Tesla, provided wireless sets for the Navy in 1912 as he cultivated a way to create a plurality of carrier channels of intermediate frequencies. However, the Navy was still reluctant to get financially involved, particularly when war broke out, "as long as [it] raged only in Europe." Meanwhile, Hammond continued to partner with Sperry to further develop the autopilot and at the same time continue to perfect means for directing said torpedoes from stationary sites such as crow's nests or even blimps.[28]

Still collaborating with Lowenstein, Hammond continued to develop high-frequency tubes, which came to play a significant role in the perfection of the radio. In Hammond's words, "This early work for radio-dynamics was conducted in secret in the same sense that the Manhattan Project in atomic energy was conducted in secret. . . . Parts of the work were kept from Congress by the military . . . [and] certain patent applications were placed in the secret archives of the Patent Office." Military applications in the defense of the nation were the issue.[29]

Miessner, another key member of the "Lowenstein-Hammond" think tank who would later go out on his own and invent electrical musical instruments and, for the military, directional microphones for aircraft and submarines, had apprenticed at the Naval Radio Station in 1911 before working at Gloucester. At this time, Miessner wrote to both Hammond and Dr. L. W. Austin of the U.S. Navy the following letter about the problem of Tesla's patent priority, penned at the very same moment that Hammond was partnered with Tesla!

> February 6, 1912
>
> Dear Mr. Hammond,
>
> Mr. Lowenstein has kindly loaned me your letters to him concerning the new selective system. . . . I believe the invention a good one. It, at least, is a very good way of dodging the Tesla and Fessenden patents.[30]

It is possible that neither Miessner nor Hammond knew that at the turn of the century Tesla had successfully sued Fessenden for patent infringement in somewhat secret litigation. However, Lowenstein testified in the case on Tesla's behalf, admitting on the stand that he did not know anything about tuned circuits until Tesla "showed me [their] great value . . . and then I understood what tuning was." [31] A dozen years later, Fessenden was still considered a pioneer in the field.

Either way, believing that he would be able to help improve the radio-guidance system, Miessner followed up with a second letter, which routinely made use of Tesla's lingo for his remote-controlled vessel:

> I am glad to learn what Dr. Austin thinks about your new selective system and the probability of its being applicable for commercial as well as tel-automatic work. [32]

Hammond's invention, which was patented over a decade later, explains that "if a circuit is tuned to a periodicity of 100,000 oscillations per second [and another] current is tuned to 105,000 oscillations per second, a third circuit can be made to respond to the difference, namely 5,000 cycles per second [cps]." This "beat frequency" can be changed, say to 4,000 cps or 4,500 cps and so on. [33] This invention, or one very much like it, greatly resembles Tesla's invention, whose selective tuning system involved the use of an extra coil. As explained in patents #723,188 and #725,605, Tesla states that separate channels and protected privacy can indeed be created by altering the combination of frequencies in a variety of ways such as by manipulating "their order of succession." Interestingly enough, this fundamental idea was expanded a generation later by Hedy Lamarr, the Austrian-born, multitalented Hollywood actress, who in 1940 sought a solution to the problem of an enemy such as the Nazis jamming signals of a remote-activated device such as a drone or torpedo. Working with the avant-garde composer George Antheil, Lamarr patented the idea of spread-spectrum frequency hopping, which protected wireless signals through multiple channels that constantly changed or "hopped" from one to the next. Based on the work of Tesla and Hammond, the idea simply was that if one channel was jammed, a variety of other channels would continue communication. [34]

This device was installed onto Hammond's remote-operated boat, the *Natalia*, in 1914, a year after Hammond's breakup with Tesla. It was at that time that Jack invited General Erasmus M. Weaver Jr. of the U.S. Coast Artillery and other key Navy personnel to witness a test run. Lieutenant Decker remembers that "the boat was allowed to steer itself." After entering the harbor at Gloucester after a run up from Boston, "the operator on shore made us perform various fancy curves . . . and steer around a few spar buoys. During the whole of the test, while I was on the boat, I did not see a single thing go wrong."

However, "problems in the creation of secret channels were made apparent in 1915 and 1916 when the USS *Dolphin* successfully interfered with a torpedo launched by Hammond over distances of several hundred feet, but the Hammond system was successful when the torpedo was launched farther away." The War Department wanted to therefore maintain visual contact with the weapon, so Hammond began working on radio-dynamic control systems that could be directed from "airplanes and drones."[35]

Clearly, Hammond did everything he could to work with Tesla, crediting him in the literature and surrounding himself with Tesla acolytes. Unfortunately, in 1913, Tesla had a falling out with the budding wunderkind. The problem was money. Tesla kept requiring more funds to perfect his turbine, and most likely through the recommendation of Hammond's father, who was funding this research, Jack turned off the spigot. A few years later, after Morgan's passing, Tesla would obtain additional development funds for the turbine from J. P. Morgan Jr. and then later from Pyle National and also Allis Chalmers, and it seems that over time cordial relations were resumed between Hammond and Tesla. In 1915, at a New York meeting of the American Institute of Electrical Engineers, although Hammond wasn't present, Tesla stood next to Lowenstein and DeForest in the well-known group photo that included John Stone Stone, Jonathan Zenneck, Karl Ferdinand Braun, David Sarnoff, and others, and the following year, Tesla wrote a rather detailed letter to Miessner to help him with his book on radio-dynamics (see pages 99–100). Tesla also continued a business association with Lowenstein, receiving a small percentage of the proceeds of certain Lowenstein patents.[36]

Continuing to develop his radio-guidance system, Hammond would go on to install the Sperry gyroscope in torpedoes, ships, airplanes, and aerial drones as the duo continued to perfect the autopilot. However, as Espenschied points out in his detailed 1959 article, even though a tremendous amount of work went into perfecting remote-controlled weaponry, astoundingly, radio-guided torpedoes and missiles were never used in either world war.

Although Tesla and Hammond played such key roles in the development of remote control and radio-guidance systems, when it came to the topic of ESP, or "wireless" communication between brains, they differed greatly! Initially, Tesla was somewhat open to the topic after conferring with Sir William Crookes during his trip to London in 1892, when Crookes told Tesla of his telepathy experiments that he had conducted with Sir Oliver Lodge and Lord Rayleigh, all of them members of the British Society of

Psychical Research; however, shortly thereafter, the Serbian wizard came to completely reject the idea.

Hammond, on the other hand, fully embraced the topic and spent a considerable amount of time, particularly in the late 1940s and early 1950s, studying the work of Eileen Garrett, who was one of the greatest psychics of the twentieth century and the founder of the Parapsychology Foundation in New York. A medium as well as a telepath, Garrett traveled up to the Hammond castle in 1948 with psychic researcher Andrija Puharich, and there they conducted telepathy experiments while she was housed in a Faraday cage, which screened out electromagnetic waves. Since she successfully picked up various targets, the implication was clear: ESP operated in a realm separate from the normal electromagnetic spectrum.[37]

Obviously, Hammond was serious about his study of psychic phenomena, while Tesla, the true father of wireless communication, completely eschewed the idea.

A key similarity, however, aside from their shared passion for invention, was the wish and perhaps the need to invite the glitterati of the day to their respective enclaves. While Tesla invited so many stars of the Gay Nineties to his lab, Hammond invited comparable VIPs from his era, the guest list to his castle including such notables as Ethel and John Barrymore, Mary Astor, Burgess Meredith, Helen Hays, George Gershwin, the Ringling Brothers, Guglielmo Marconi, Tom Edison Jr., Monty Wooley, Leopold Stokowski, John D. Rockefeller, Clarence Birdseye, Cole Porter, and Noel Coward.[38]

Whether Tesla ever visited Hammond's castle is not known. Having toured the palace on several occasions, I can only imagine and hope that Tesla did indeed make the delightful journey. In 1933, Tesla referred specifically to John Hays Hammond Jr. on several occasions, particularly in regards to suggesting that the government hire Hammond to use his remote-controlled drones to be armed with explosive devices so as to fly into the center of tornadoes to disarm them and "break up the whirl." Under Hammond's guidance, Tesla wrote, "The fear of danger from tornadoes and the great losses of life and property can be effectively combated by these means."[39]

8

Marconi and the Germans

*It is my intention to describe fully the efforts I have made in order to
tune or syntonize the wireless system . . . [and correct] a large amount
of inaccurate and misleading information being published.*

*I first constructed an arrangement which consists of a Leyden jar
or condenser circuit in which included the primary of what may be
called a Tesla coil [and a] secondary [which] is connected to Earth or
aerial conductor. The idea of using a Tesla coil to produce oscillations
is not new. It was tried by the [British] Post Office [with Sir William
Preece] . . . and also suggested in a patent specification by Dr. Lodge
dated May 10, 1897 and by Professor Ferdinand Braun . . . in 1899.*

—GUGLIELMO MARCONI,
June 15, 1901[1]

Guglielmo Marconi's greatest rival in the legal arena may have been
Nikola Tesla, but in the battleground of the marketplace, it was
Telefunken, the German wireless concern. Although Marconi had patents
in Germany, the Telefunken syndicate had too many important connections on its home front and easily maintained a monopoly there. Formed in
1903 through a forced merger to quell a vicious patent priority battle under
orders of the kaiser, the Braun-Siemens-Halske Company, headed by Karl
Ferdinand Braun, combined with Arco-Slaby, headed by longtime Tesla
associate Adolf Slaby, to create Telefunken. This new larger German concern was, without doubt, the number two competitor in the world. However,
it didn't start out that way.[2]

A decade earlier, in January of 1892, Tesla had sailed for Europe to
present his lecture on high-frequency phenomena to the Royal Society in
London and the Société Française des Electriciens in Paris. Arriving in
London on January 26, Tesla was met at the harbor by Sir William Preece
and stayed at Preece's home during his time in England. Educated at Kings

College in London, Preece was chief engineer of the British Post Office, which meant that he was in charge of the telegraph system.

Already by this time, Preece had been aware of the possibility of what would become wireless telegraphy. As far back as 1879, Preece, along with William Crookes and James Dewar, was witness to the experiments in Morse code transmission of "aerial telegraphy" by David Edward Hughes. This insightful Welsh scientist had noticed that electrical sparks could be picked up by a telephone receiver at distances of five hundred yards. Predating Heinrich Hertz by a decade, but not publishing his results, Hughes also invented a spark-gap transmitter and developed a coherer before Édouard Branly and a crystal radio detector before Braun.

"Elihu Thomson recognized the Hughes claim to be the first to transmit radio." However, Hughes himself said, "with characteristic modesty," that Hertz's experiments were "far more conclusive than mine" and that "Marconi's efforts at demonstration merit the success he has received. . . . The world will be right in placing his name on the highest pinnacle, in relation to aerial electric telegraphy." In honor of these achievements, the Royal Society created the Hughes Medal, first given to J. J. Thomson in 1902 for his advancements in the electrical sciences.[3]

Recognizing the role that the ground played in transmitting signals between different telegraph lines, in 1889, Preece successfully sent wireless Morse-coded messages to a parallel line receiving station a mile away, and these kinds of experiments were also conducted by Sir William Crookes before Tesla's arrival in 1892.

During Tesla's watershed lecture, wherein he demonstrated selective tuning through the ability to illuminate different fluorescent bulbs when different resonant frequencies were transmitted, Tesla took the time to "praise a name associated with the most beautiful invention made: it is Crookes! . . . I believe that the origin [of my advance] . . . was that fascinating little book [on radiant energy] which I read many years ago."[4]

Basing his work on the Crookes tubes, Tesla displayed before the Royal Society what could be considered the first radio tube. In an evacuated bulb, Tesla had placed an inner vacuum tube, and within this chamber, the inventor generated a beam of light that was so sensitive that it could respond even to the twitch of a muscle, let alone to the geomagnetic torque of the Earth. Tesla speculated that this beam would rotate counterclockwise in the Southern Hemisphere because it rotated clockwise in the room in England.

This cathode ray tube would not only become the basis for a variety of radio tubes and wireless lighting, but also, when transmitted onto a screen, such as an oscilloscope, the basis for X-ray equipment, radar, and the TV tube, and also an invention that reacted to geomagnetic forces.

Tesla's lecture was so outstanding that James Dewar cornered and then cajoled the reluctant inventor into giving a second demonstration so that Lord Rayleigh could witness it as well, and the following day, Tesla complied.

When Marconi, at the age of twenty-two, came to town four years later, Preece tried to interest Marconi into adapting, with Tesla's permission, some of his equipment, but Marconi declined.

"Some time after the experiments with the classical Hertz devices, conducted under the auspices of the English Imperial Post Office, Sir William Preece, then head of the department, wrote me a letter conveying the information that the tests had been abandoned as of no value, but that he believed good results possible by my system," Tesla wrote. "In reply, I offered to prepare two sets for trial and asked him to give me the technical particulars. . . . Just then Marconi came out with the emphatic assertion that he had tried out my apparatus and that it did not work. Evidently he succeeded in his purpose, for nothing was done in regard to my proposal.

"He furthermore declared . . . that wireless communication across the Atlantic was impossible because there was a wall of water several miles high between the two continents which the rays could not traverse. But subsequent developments showed that he had used my system in secret all the time, received the plaudits of the world and accepted stolidly even my own congratulations, and it was only a long time after that he admitted it."[5]

After a few setbacks, one way or another, Marconi's system began to become more efficient. Along the way, he hired Preece and also Preece's close associate Ambrose Fleming (who had also attended Tesla's lectures) to help him develop his wireless apparatus. Since Lodge's work predated Marconi's by about a year, after a long legal dispute, the best course of action was to purchase Lodge's patents, which Marconi did in 1911.

In 1897, Preece invited Adolf Slaby and Alexander von Arco to London to partake in Marconi's experiments to send wireless messages across the English Channel, and shortly after that, in the same year, Slaby and von Arco transmitted a wireless message from Schöneberg, a suburb of Berlin, to Rangsdorf, which was a distance of 13.7 miles.

Over the next several years, Marconi continued to hone his equipment and increase the distance between sender and receiver. Certainly one of Marconi's greatest coups, while Tesla was holed up in Colorado Springs, was to report the America's Cup race of 1899 from ship to shore for the New

York Yacht Club, not far from New York City off of Sandy Hook, New Jersey. On October 16, Marconi was able to transmit details of the race from fifteen miles offshore at a rate of about fifteen words a minute. By day's end, he had transmitted over four thousand words to the offices of the *New York Herald*, who were thus able to report the race almost in real time. Shortly after, the Marconi Company began installing wireless equipment aboard ships, not only those from the United Kingdom but also on Belgian, Italian, and even German carriers.

At the same time, Marconi also set his sights on attempting to transmit an impulse across the Atlantic. Yet, in the very days that Tesla was writing to tell his benefactor, J. Pierpont Morgan, that he was capable of creating an unlimited number of wireless channels, Marconi didn't know that multiple channels were even possible! Referring specifically to 1901 and Marconi's transatlantic experiment, his head engineer, Richard N. Vyvyan, said it this way:

> Some combinations gave good results, some poor, but an arrangement that gave good results one night, the next gave very different results, and thus it became difficult to say for certain whether any improvement or progress was being made. . . . We knew nothing then about the effect of the length of the waves transmitted governing the distance over which communication could be affected. We did not even have the means or instruments for measuring wavelengths, in fact we did not know accurately what wavelength we were using.[6]

Compare this astounding admission with a typical passage from Tesla's Wardenclyffe notebook for the same time period, for instance, on April 28, 1901:

> [The] very slow frequency . . . charge passes chiefly on the surface as far as I have found and therefore, tak[ing] the diam[eter] of Earth . . . , the half wavelength should under above assumption not exceed 12,700,000 π meters. Taking speed of electricity 320,000,000 meters per sec. we would have about only $32 \times 10^7 : 254 \times \pi \times 10^5 = 4$ cycles per second approx. I shall first consider the case as assumed although it has many practical disadvantages. Higher frequencies will be subsequently treated.[7]

Tesla had created a prototype planet, which he referred to as a "sphere," to perform experiments and calculations with various wavelengths to find nodal points and pinpoint them, which he then applied to the Earth itself. In this instance, Tesla had decided that the half wavelength should not exceed the diameter of the Earth, which he calculated at 12,700,000 meters (or 7,891 miles).[8]

His notebooks are simply filled with advanced calculations, many much more complicated than this. He also was using undamped or continuous wave frequencies to enable complex forms of electrical energy to be transmitted along multiple channels through surface waves (e.g., text, voice, pictures, and even power) whereas Marconi was sending Morse code, using a primitive Hertzian spark-gap transmission system predominantly through the air. Because these were pulsed frequencies, Marconi was incapable of creating multiple channels and his system could never form the basis for the transmission of voice or pictures.

Unfortunately for Tesla, he could not convince Morgan of the viability of his plan. The reasons are many. First off, Tesla constructed a larger tower than they had agreed on, changing the deal from a ninety-foot tower to one twice that height. Second, there was the issue of the need to construct a receiving tower somewhere in Europe, most likely in England or Portugal, so costs right there could be doubled. It is also possible that either Morgan did not believe that Tesla was capable of doing what he said he could do, or the reverse, that he feared Tesla could have been able to transmit electric power without wires, as this was his ultimate goal. Had Tesla succeeded on that front, in the short run, this would have created a huge financial burden for Morgan, as he had rubber plantations in Africa and copper mines out West, he controlled the shipping and railroad lines to transport rubber and copper to industrial sites, and he also owned great forests used to make telephone poles. Simply put, he wanted wires.

Then, again, Marconi was succeeding with much less expensive equipment, symbolized in monumental fashion not only in December of 1901 when Marconi rocked the world by sending the first transatlantic message, but again on January 18, 1903, when President Teddy Roosevelt transmitted a wireless greeting from Marconi's Wellfleet Station out on Cape Cod to the king of England. This event so irked Tesla that he was moved to write directly to "His Excellency" at the White House to complain to Roosevelt that he was lending his name to a patent pirate![9]

Since Tesla's system was overtly being studied by electricians at Telefunken, they were in a better position to advance the art than Marconi, even though Marconi had the name and a wider distribution.

Problems began early in 1903 between Telefunken and the Marconi concern when Marconi refused to accept a wireless message sent from the kaiser's brother because it did not come from a Marconi station. On the face of it, this might seem petty of Marconi. However, his refusal may have been in retaliation to the German policy inaugurated by the kaiser to refuse to recognize wireless advances submitted by individuals from other countries.

This was also the case with Tesla, and it created a moral conflict for Telefunken's Adolf Slaby, who considered Tesla "the father of wireless telegraphy." By not recognizing patents from foreigners, this allowed Telefunken to adapt all of Tesla's wireless technology without compensating the inventor, and because it was condoned by the Fatherland, it did not have to be done in secret.

Like every other non-German inventor, Tesla was painfully aware of the kaiser's policy of condoning and even promoting patent piracy. And so, he continually sought legal redress in Germany and Austria because the upside would have been enormous. It was no surprise that he continued to hit a brick wall, and the onset of the Great War certainly did not help the matter.[10]

In the case of Marconi, he simply pirated what he needed and pretended it was his. Since Preece tried to talk Marconi into purchasing Tesla's equipment legally, no doubt Marconi's tactics must have been of great concern to someone of Preece's stature, particularly because Preece had tried to get Marconi to do the right thing. Ambrose Fleming, who had been friends with Tesla and also corresponded with him through the mail, was now also working for Marconi. Later, Fleming said it this way: "Marconi had genius of a certain kind, but he over-reached himself in thinking that he could appropriate the whole credit for wireless."[11]

While Tesla worked in the early 1900s to complete a single wireless station at Wardenclyffe, both Marconi Wireless and Telefunken expanded their enterprises, establishing hundreds of wireless stations on as many civilian and military ships as possible and also on every continent in as many countries as they could.

It would seem that the Swedish Academy of Sciences, aware of this great competition, decided to honor both entities by awarding the Nobel Prize in physics in 1909 to both Guglielmo Marconi, a half-Italian/half-British subject, for sending the first transatlantic wireless message in 1901, and to his German counterpart inventor in wireless, Karl Ferdinand Braun, for his marked advances in the development of the cathode ray tube, which led to advances in radar and television, and his work on the crystal detector or crystal diode rectifier, which picked up the ambient wireless transmissions and converted them for use in radio.

When one studies the history of these inventions, in the case of the crystal detector, it is clear that Braun was anticipated by David Edward Hughes, the Welsh scientist discussed earlier, who first experimented with crystal radio detectors. Hughes had discovered that a "tungsten needle" set very close to a carbon block would vibrate in the presence of electromagnetic waves. Braun found similar properties in "the mineral galena, which was a natural rectifier."

In the case of AM radio, which was invented a few years later by Edwin Armstrong, "carrier waves have frequencies of about a million cycles a second. But the modulations that carry the information—the voices and the music—are of relatively low frequency, ranging from about twenty to twenty thousand cycles. The high-frequency oscillations can be rectified [read out of the carrier wave], but many rectifiers are inefficient for the low-frequency oscillations. The rectifier acts to separate information. Hence building a radio receiver boils down, in a certain sense to making a rectifier." [12]

With regards to the crystal radio set I constructed with my dad, circa 1960, like all crystal radio sets of this design, there were no batteries and no plug to an electrical outlet. All energy to drive the radio stemmed from the energy collected from the broadcast stations themselves. Since my radio only began to work when my father set up the ground connection (through the radiator), the implication was clear to me as a twelve-year-old, namely that the real power did not come from the broadcasting tower alone, but rather, the transmission through the ground was an essential component.

There is little doubt that Tesla was emotionally affected by the decision by the Nobel committee to award secondary inventors for work he saw clearly as his, yet, from Marconi's point of view, he had no understanding as to why he should share the award at all. [13]

As wireless apparatuses became more efficient, competition between Marconi Wireless and Telefunken reached a new height and lawsuits began to fly. Although Marconi Wireless had achieved a recent coup in Spain, in America, Telefunken gained an edge when it constructed two enormous transatlantic systems in Tuckerton, New Jersey, and Sayville, New York.

The situation at Tuckerton, however, was complex. Located along the New Jersey coast, one hundred miles south of New York City, the Tuckerton plant was designed by Rudolf Goldschmidt, a German electrical engineer who first built a sister one-hundred kilowatt station at Eilvese, Germany, and then sent the prefabricated parts to his German workers at Tuckerton, where the wireless tower rose to a staggering height of 820 feet and thus was the second tallest construction in the world next to the Eiffel Tower. Like the structure at Sayville, the Tuckerton facility used the patents and designs of Tesla. In July of 1915, the Serbian wizard iced a deal, with Telefunken paying him royalties that amounted to $15,000 for the year for the use of his equipment and expertise.[14] Ironically, although completely a German construction, the plant was actually built for the Lafayette Radiotelegraphic Station located at Croix-d'Hins, near Bordeaux, in France. In other words, the Tuckerton site was slated to be a transatlantic French operation.

Once World War I began, the Germans were unwilling to cede the station to the French, and once America entered the war, they also stopped paying royalties to Tesla. He tried to sue, but naturally, he was unable to continue to receive compensation. Added to that was the simple fact that the designer, Goldschmidt, a former London employee for the Westinghouse Corporation, had sold his patents in 1912 to Marconi! Thus, Marconi thought that he had a right to the Tuckerton facility, and one *New York Times* article even suggested as much.[15]

Tesla was in a competitive field. Although fairly recognized by the German companies operating in the United States, he faced a very different situation from Marconi and various American concerns. Once his lectures and patents had been published, numerous electrical engineers emerged with their own wireless schemes. Some inventors simply pirated Tesla's work, while others copied the pirates or developed their own schemes, but Tesla did his best to combat them all. One of the more successful early wireless pioneers was Reginald Fessenden, who as early as 1900 had set up a series of radio towers to report the weather, and he also was one of the first to transmit the sound of voice. It soon became apparent that Fessenden's use of tuned circuits was based entirely on Tesla's work, so the latter sued Fessenden in a lengthy, but somewhat secret litigation, and won.[16]

Dealing with Goldschmidt was a lot more difficult because of the unfair patent laws set up in Germany.[17] For nationalistic reasons, Tesla had been prohibited from obtaining his rightful royalties there, but Slaby never hid the fact that he considered Tesla the patriarch of the field. Thus, when Telefunken came to America, Slaby sought out his mentor, not only on moral grounds but also for gaining a legal foothold against Marconi and to obtain the inventor's technical expertise.

A meeting was held between Tesla and the principles of Telefunken's American holding company, the innocuous sounding Atlantic Communications Company, located at 111 Broadway. Present was the director, Dr. Karl George Frank, "one of the best known German [American] electrical experts," and his two managers, Richard Pfund, a frequent visitor to Tesla's lab, who was head of the Sayville plant; and "the Monocle," Lieutenant Emil Mayer, head of operations at Tuckerton.[18]

Although Slaby died in 1913, his admiration for Tesla carried a lot of weight. The powerful German wireless station at Sayville opened on January 17, 1914, when the first wireless message between Germany and the United States brought heartiest greetings from the kaiser to President Wilson,[19] and it was at this time that Tesla was hired by Frank to help in the further development of their wireless plant.

A graduate of the University of Munich, with a doctorate in philosophy and chemistry, Frank claimed U.S. citizenship, but most likely had dual citizenship with Germany. With offices in the Hudson Terminal Building as well as on Broadway, Frank not only ran the Sayville wireless station, he also was in charge of commercial sales for Siemens and Halske.

Frank's head of operations, Richard Pfund, was also of German heritage. An electrical engineer who spent a good amount of time with Tesla, Pfund, who had apparently attended Tesla's watershed talk on high frequency phenomena and wireless communication at Columbia College in 1891, had begun his apprenticeship in the field of wireless working for the Marconi Company. In 1903, he helped install a Marconi Wireless plant in Alaska.

Forming a liaison with Tesla was a coup for the Germans, not only because it solved part of the dilemma of having used Tesla's technology without compensating him; it also gave them the edge in upcoming litigation in priority battles against Marconi, and most importantly, more than anyone, the Germans appreciated Tesla's expertise.

Realizing the situation, Tesla sought assurances from Frank that he could obtain German patents, and Frank replied that he "did not see any reason why" this could not be achieved.[20] Tesla thereupon asked for an advance of $25,000 and royalties of $2,500 per month, but settled for $1,500 per month with a one month advance. The inventor met with Pfund to discuss the turbine deal with the kaiser and also to fix a wireless transmitter that Atlantic was using at the Manhattan office, and then he had plans to go out to the Sayville station, where he could institute his latest refinements in order to boost its capabilities.[21]

Shortly thereafter, Tesla traveled out to the other German wireless station, located in Tuckerton, New Jersey. There he negotiated an even more lucrative deal. Even though the Tuckerton site was constructed by the German company Hochfrequenz-Maschinen-Aktiengesellschaft für Drahtlose Telegraphie, known more frequently as Homag, this wireless station was slated to be sold to the French concern Compagnie Universelle de Telegraphie et de Telephonie Sans Fils.

Since "the French Supreme Court had upheld Mr. Tesla's claim against the French Marconi Company . . . the Tesla Company filed suit against the Marconi interests in [America]." Tesla's argument was that Marconi had "infring[ed] on two of his patents . . . [and he was] demanding a settlement on the royalty basis." Fearful of a costly legal suit, Homag agreed to pay Tesla "5% on the gross income" from revenues of Tuckerton, but they would not stipulate which specific Tesla patents they were using.[22] Legally, this was not the best outcome for Tesla, but financially, the deal was highly lucrative, as one way or another, Homag was recognizing Tesla for his wireless invention, and through the French courts the inventor was also receiving concrete vindication over Marconi as well.

With regards to Sayville, once it was up and running, "it became a beacon for amateur wireless enthusiasts . . . who could tune their home-made sets to the station's nightly press dispatches," said Hugo Gernsback, editor-in-chief of *Electrical Experimenter*. Gernsback would follow this up with an interview with Tesla, and he would also have the inventor write a series of articles on this and related topics.

Tesla was finally back in the news and back in the money. Now, he could see a route towards resurrecting Wardenclyffe through proof that his theories were correct. With increased capital he could begin paying off his back rent to the Waldorf-Astoria, which was now approaching $20,000. To give himself some time, Tesla transferred the mortgage of Wardenclyffe to George C. Boldt, the hotel's manager, with the oblique understanding that he would get the property back as soon as he covered his debt.

Possibly when John Jacob Astor was alive, Tesla was allowed to live at the hotel rent-free. Astor may not have been overjoyed with this possible development, as he was miffed a decade earlier when the wizard took his investment, which Astor thought was to be used for the development of fluorescent lighting but was used instead to fund the Colorado Springs laboratory. When Tesla formed his partnership with Morgan in 1901, he further ruptured this relationship when he assigned to Morgan the lighting patents that Astor thought he had a claim on, and Astor refused to help bail Tesla out when Tesla had exhausted Morgan's funds. However, in March of 1909, apparently back in Astor's good graces, Tesla wrote him a detailed letter telling the hotel magnate confidential information about his work on new turbines that he hoped would be used in locomotives, ocean liners, automobiles, and a variety of flying machines—including a jet-propelled dirigible and a combination helicopter-aeroplane. This "single, double or triple passenger aerocar of Professor Tesla . . . [could replace the normal automobile and be used] "for business or pleasure." [23] As a related prototype to the dirigible, there is tangential evidence that Tesla showed Astor a hydrofoil that he may have demonstrated on the Hudson River.

In April of 1912, Astor made the unfortunate decision to sail to Europe on the *Titanic*, and he went down with the ship. Tesla sent Astor's son Vincent a condolence letter and received the following reply:

> May 15th 1912
>
> My dear Mr. Tesla
>
> I wish to thank you for your letter of sympathy, . . . for your kind expressions regarding my father . . . and your thought of me at this time. I have often heard my father speak of you and I know what high regard he held [you in].
>
> Again, thanking you,
> Yours very sincerely,
> Vincent Astor [24]

Now, with Astor's tragic death, Tesla had no leverage to offset his sizeable hotel bill. Further, Tesla's fortunes were also affected by the onset of the Great War, which began two years later on June 28, 1914, with the assassination of Archduke Franz Ferdinand and his wife, Sofie, while this heir to Austro-Hungary's throne was visiting Sarajevo.

Since the archduke was assassinated by a Serb, this gave Austria a reason to invade the assassin's home country, but they were surprised when Russia came to Serbia's defense. Because France had an alliance with Russia, Germany feared an attack from the West since they were now aligned with Austro-Hungary. Thus, they jumped first by sweeping through Belgium to attack France, and this brought Great Britain into the malaise, because they had a treaty with France.

Still in its early stages, the war was not expected to last more than a few months. This meant that America remained an ally to both sides. Eight months later, in February of 1915, Karl Ferdinand Braun, a physics professor from Strasbourg, and Jonathan Zenneck, a physics professor from Munich Technical University, arrived from Germany to help out at the Sayville station and also to testify in the upcoming trial between the Atlantic Communications Company, which was installing wireless setups on U.S. Navy ships, and the Marconi Wireless Company, which was suing both.

> February 16, 1915
>
> My dear Dr. Frank,
>
> I am glad to learn that Braun and Zenneck are in this country and look to the pleasure of meeting these remarkable men soon.
>
> Yours sincerely,
> N Tesla[25]

To the *Brooklyn Eagle*, Zenneck was described as a code expert, which he denied, but he also was circumspect about his role at Sayville. "I came to America solely to testify in the Marconi suit," said Zenneck. "I am not a code expert, I stayed at Sayville merely to continue my studies."[26] The public, as well as the government, was not convinced. Assuming that the quote from the *Eagle* was accurate, it suggests that Zenneck was cagey because he was, indeed, brought to the States not only to testify but also to help in the reconstruction of the Sayville plant.

Zenneck, to the Americans, was an enigma. On the one hand, as professor of experimental physics from Munich Technical University, he was respected for writing *Electromagnetic Oscillations and Wireless Telegraphy*, which had become "the standard textbook" in the field. Yet, on the other hand, as a captain in the German army, Zenneck had been part of the brigade that stormed into Belgium on a killing spree en route to invading France at the outset of the war.

A student of wireless telegraphy under Braun as far back as the late 1890s, Zenneck had experimented with directional antennae, and he also had studied the work of Tesla, and thus was able to corroborate Tesla's proclamations on the importance of ground waves in efficient radio transmissions. "While working to explain Marconi's trans-oceanic results, [and applying Maxwell's equations, Zenneck] showed that a unique type of surface wave could travel along the interface between the ground and the air."[27] According to Tesla expert Dr. James Corum, "the distinguishing feature of the Zenneck wave was that the propagating energy didn't spread like radiation, but was concentrated near the guiding surface." Corum points out that Zenneck's associate Arnold Sommerfeld had shown that "an electromagnetic wave could be guided along a wire of finite conductivity, and Zenneck conceived that the Earth's surface would perform in a manner similar to a single conducting wire."

Working along similar lines today at their own trailblazing wireless power transmission plant, Viziv Technologies in Milford, Texas, James Corum and Ken Corum point out that Zenneck and Sommerfeld shared adjacent alpine ski lodges. Further, Sommerfeld, who invited so many of the giants of quantum physics to his lodge, (e.g., Bethe, Pauli, Heisenberg, etc.), thanked Zenneck in his watershed 1909 article, which distinguished between a Hertzian electromagnetic wave that propagated through radiation through the air and a Tesla or Zenneck ground wave. "Zenneck, himself, pointed out that it was actually Andre Blondel [who witnessed Tesla's lectures in Paris in 1892] and Ernst Lecher at Vienna who first intuitively speculated that radio waves were probably *surface guided waves* (like ocean waves) and not Hertzian waves (*radiation*)." [28]

Just a year after his meeting with Zenneck, Tesla reiterated his finding that the radiation of radio waves was almost completely lost when transmitted through the air but was "conserved through conductive action" when transmitted through the ground. Tesla had calculated at Sayville that they were wasting energy in electromagnetic radiations. "Those waves will dissipate only a few miles off shore," he told Zenneck, "whereas the energy that will reach Germany will be from your ground connection.

"The effect at a distance is due to the current energy that flows through the surface layers of the Earth," Tesla concluded, elated to add that this "has already been mathematically shown" by the respected physics professor Arnold Sommerfeld. "He agrees on this theory; but as far as I am concerned, that is positively demonstrated.

"When Professor Zenneck took me out to the Sayville antenna and gave me the particulars, I went over the calculations and found that at 36 kilowatts they were radiating 9 kilowatts in electromagnetic wave energy. They had, therefore, only 25 percent of the whole energy in these waves, and I told Professor Zenneck that this energy is of no effect—that they produce, by the current, differences of potential in the Earth, and [it is] these differences of potential [that] are felt in Germany and affect the receiver; but the [Hertzian] electromagnetic waves get a little beyond Long Island and are lost." [29]

What Tesla had done was explain to Zenneck that he had to increase the ground connections because that was the real reason why the impulses were traversing the planet. Shortly after this visit, the *New York Times* reported that "before the end of next week the new and more powerful apparatus will be working . . . [at the] Sayville [plant]."[30]

J. P. MORGAN JR.

Tesla now had the physical proof he needed to establish that his tower at Wardenclyffe was, indeed, on the right path. To augment this new triumph and reassert his dominance as the preeminent inventor in wireless, Tesla forwarded the entire French litigation proceedings whereby Marconi's work was overturned in favor of his to Jack Morgan, who was now in charge. At the same time, he negotiated with Frank in an attempt to help him obtain wireless patents in Germany and Austria. Simultaneously, he also charged Atlantic Communications a service fee for testifying for them and for using his existing patents that usurped Marconi's patents in the upcoming litigation that was about to take place in the U.S. courts. If Jack could help in the fight, the wireless enterprise, which they contractually shared, could be revived.

Jack, however, was not smitten by the same vision of destiny, and he graciously declined to become involved in the litigation or in Wardenclyffe in any way. However, he did continue to invest in Tesla's turbines, and he wrote Tesla several checks for $5,000 each and told the inventor to keep him informed of any progress.

Tesla returned to his turbines at Edison's Waterside Station with a new transfusion of 23 Wall Street blood. To reflect the reactivation of the resurrected alliance, the inventor set out to search for more fashionable chambers for his office. Within a few months, he took up residence on the fifty-third floor of the brand new Woolworth Building, telling Woolworth he would place a wireless transmitter atop the skyscraper at his own expense "to signal ships and railroad trains throughout the known world."[31] Decorated with a gold-leaf emerald-colored mosaic ceiling in the lobby and located by City Hall, near Wall Street on Park Row, the gothic-styled Woolworth Building soared above the city to the dizzying height of eight hundred feet, knocking the Metropolitan Towers out as the loftiest skyscraper in the world.

Although Woolworth apparently did not accept Tesla's offer, the inventor was invited to the gala opening and took the Johnsons along. The banquet began with the illumination of the building's eighty thousand light bulbs by President Woodrow Wilson, who pressed a button in Washington, DC, to set the building aglow. Tesla met with the mayor, Mr. "Dime-Store"

Woolworth, and other dignitaries, and then Katharine lured her two escorts into one of the twenty-four high-speed elevators to the roof, where they could gaze out over the sprawling megalopolis.

"Do not worry about finances, Luka," Tesla said confidently, referring to his friend Robert. "Remember, while you sleep, I am at work solving your problems." Johnson brought up the old AC polyphase debacle, and Tesla replied that there were "billions invested [in it] now. I won every suit without exception and had it not been for a 'scrap of paper,' I would have received in royalties Rockefeller's fortune, but just the same, I feel I am safe to invite you to dinner."

Tesla's wit and latest maneuver once again brought a needed smile to the oft-brooding Mrs. Filipov. As usual, when the wizard returned to her orbit, she seemed to step back from the veil. Robert, however, reiterated his concern that because he did not have a job, his and Katharine's home at 327 Lexington Avenue could go onto the market.

"Please take my words seriously," Tesla insisted. "Do not worry, and write your splendid poetry in perfect serenity. I will do away with all difficulties which confront you. Your talent cannot be turned into money, but mine is one which can be transformed into car-loads of gold. This is what I am doing now."[32]

During this period, Tesla continued to pay back monies he owed the Johnsons, as he labored with his new engines, his best leads coming from the Ford Motor Company in the United States, the Bergmann Works in Germany, a licensing agreement with the king of Belgium, and the development of a water fountain for Tiffany.

In a letter sent to Morgan, Tesla outlined his numerous strategies for achieving financial success and then made a plea for additional backing.[33] Morgan agreed to defer interest payments on the loans for developing the turbine, which were now up to $15,000, but decided against further investment. Tesla, however, simply required the funds and followed up the letter with a testimonial from Excellenz Alfred von Tirpitz, Secretary of State of the German Imperial Navy, "who has been requested to the German Emperor relative to the Tesla Turbine who is greatly interested in this invention." Von Tirpitz had "promised his Excellency that the machine will certainly be here on exhibition about the middle of January, so you know what that means." Tesla also informed Morgan that if the deal was consummated, Bergmann would come through with royalties on the turbine of $100,000 per year.

Considering Morgan's keen antipathy for the Germans, their connection to the Jewish banking houses, (he was notoriously anti-Semitic), and the longstanding policy of the House of Morgan to shun financial arrangements with Germany after they had double-crossed J. Pierpont many years earlier, it would seem unlikely that Morgan would reverse his decision. However, unlike his father, the son was able to compromise. He graciously changed his mind and forwarded another check for $5,000.

As Tesla awaited news from the Bergmann Works, he labored to perfect a new speedometer and new tachometer he had invented. With a market in the hundreds of thousands and a selling price around twenty-five dollars each, the potential for large profits was great, and Tesla offered the deal to Morgan. He declined and asked again for the interest payment due on the loan. However, Tesla had suffered a setback with the turbines, which now leaked due to pitted bearings, so he requested that Morgan either be patient or help with the bills to complete the turbine project or protect their other common interest at Wardenclyffe.

A HISTORY OF NAVY INVOLVEMENT

May 6, 1902

Dear Tesla,

I have brought the matter of your inventions in wireless transmission of electrical energy in connection with the naval exhibit at Buffalo to the attention of the Navy Department and you will get an official letter on the subject. . . . Why could you not send the model [telautomaton] I saw in your laboratory in '98 and your patent drawings and descriptions, even if you could send nothing else? . . .

I think this is a good opportunity for bringing your patents to the attention of the Navy without the usual formalities . . . and therefore my dear Tesla, do not fail in this matter of the first step towards their introduction.

R. P. Hobson[34]

As the Spanish-American War was winding down, Tesla made overtures to the U.S. Navy with the hope of installing wireless apparatuses on their ships. One of the reasons the United States won was because of the quick thinking of the naval officer Richmond Pearson Hobson, who had purposely sunk his ship in the harbor at Santiago Bay, blocking in the Spanish

Armada, thereby crippling their offense. Arrested for the bold move, Hobson became an international star during his incarceration by the Spanish, and afterwards when he returned to America, he was hailed a bona fide war hero. *Century* editor Richard Watson Gilder and his associate Robert Johnson secured the rights to relate this amazing story and thereby sold an additional hundred thousand copies of the magazine.

Hobson became part of the Tesla/Johnson inner circle, and a friendship was born that lasted their entire lives. From the letter above, it is clear that Hobson, who at the time was even considered a potential presidential candidate, was doing his best to resurrect the rocky relationship Tesla had with the Navy. The story actually began several years earlier. In 1896, A. R. Sulfidge, lieutenant commander of the U.S. Navy, discussed with Tesla the idea of having submarines communicate with each other "using sea water as the medium. The solution of this problem," Sulfidge went on, "would be of unspeakable benefit to our Navy, not only in war, but also in peace." [35]

Tesla had not yet developed his wireless system to the point where it was ready. However, three years later, the Navy persisted, via Rear Admiral Francis J. Higginson, who requested that Tesla place "a system of wireless telegraphy upon Light-Vessel No. 66 [on] Nantucket Shoals, Mass, which lies 60 miles south of Nantucket Island." Tesla was on his way to Colorado and was unable to comply. Further, the Navy did not want to pay for the equipment, but rather wanted Tesla to outlay the funds himself.

Higginson, who had visited Tesla in his lab in the late 1890s, wanted to help, but was placed in the embarrassing position of withdrawing his offer of financial remuneration because of various levels of bureaucratic inanity. Instead, R. B. Bradford, chief of the Bureau of Engineering and Development, said that the Navy would "be willing to test any finished device [but that] such tests must always be made without any expense to the Department." [36]

Insulted, Tesla nevertheless suggested that he would come to Washington to possibly demonstrate his remote-controlled boat and display his wireless system of sending "absolutely private messages". [37] The Navy reconsidered and suggested that Tesla do just that, suggesting Tesla send a wireless message from the Navy Yard at Tompkinsville to the Torpedo Station in Newport, Rhode Island, a distance of 130 nautical miles, and/or to the Navy Yard in Norfolk, Virginia, a distance of 370 nautical miles, and also demonstrate his wireless system on board a naval vessel. [38]

Tesla took the train to Washington to demonstrate the principles of his remote-controlled boat to the patent office, and he conferred face to face with the high command of the Navy, and then the two sides dickered for the next few months. Tesla promised to install six wireless transmitters at various Navy Yards in New York, Newport, and Virginia, and also on board several ships, but then he received a communiqué from Bradford that discussed various newspaper articles that called into question Tesla's abilities, suggesting that his focus was too scattered and that he planned to signal the planet Mars. Miscommunication continued, and by this time Tesla had formed his contract with J. Pierpont Morgan, and so in October of 1901, he wrote, "I am putting up at Wardenclyffe facilities for the manufacture of my apparatus and expect its introduction very soon. Besides, I am erecting a plant for the transmission of messages across the Atlantic. I am anxious to complete it in time to prevent the laying of a Pacific cable which with my present knowledge, and experience in electrical vibrations, I consider an unrational enterprise." Ending the letter, "Yours, very respectfully," Tesla abandoned the plan to put in smaller wireless stations at the various Navy Yards and concentrated all his efforts on Wardenclyffe.

Arguably, this decision (circa 1900–1901) was one of the biggest blunders of Tesla's career. Tesla thought that the plant at Wardenclyffe would usurp the need for smaller demonstrations. He simply did not have the capability to build his great wireless tower on Long Island and also comply with the construction of smaller transmitters for naval use, but ultimately, it was a serious error in judgment not to demonstrate his abilities on this smaller scale before the U.S. military when he had the chance, even if they were unwilling at this point to pay for it.

In June of 1900, before any deal was signed with Morgan, Tesla received a check for $5,000 from William Crawford, owner of the highly prized department store Simpson-Crawford.[39] It is uncertain what Tesla did with these funds, but he certainly had the means to finance a demonstration to satisfy the Navy, and it remains somewhat of a puzzlement as to why he never completed the challenge.

In 1902, the Office of Naval Intelligence called Commodore F. M. Barber out of retirement to be put in charge of the acquisition of wireless apparatuses for testing. Although still taking a frugal position, the Navy came up with approximately $12,000 for the purchase of wireless sets from different European and American companies. Based on their results, it was determined that the Slaby-Arco system outperformed the others, and the Navy ordered twenty sets, which as we now know, were based on the pirating of Tesla's design. Simultaneously, they purchased an eleven-year lease on the Marconi patents.[40]

9

The Great War

We sat in the inventor's offices near Fifth Avenue, looking at a picture of his new wireless telephone plant, designed to develop unlimited, undreamed of electrical energy for wireless transmission, a Merlin's tower, overshadowing the imagination. From his desk, Lord Kelvin watched us from his portrait....

"The next war will be fought with electricity," [Tesla exclaimed]. "Cannons will be impotent compared with what is to come. ... But [this] change will not ... decide this war, because it takes time to develop great, revolutionary things."

—EDWARD H. SMITH,
the *World*, January 30, 1916[1]

With the onset of World War I, the use of wireless became a necessity for organizing troop movements, surveillance, and intercontinental communication. While the country was still neutral, the Navy was able to continue its use of the German equipment up until sentiments began to shift irreversibly to the British side. Via the British Navy, Marconi had his transmitters positioned in Canada, Bermuda, Jamaica, Columbia, the Falkland Islands, North and South Africa, Ceylon, Australia, Singapore, and Hong Kong. His was a mighty operation. In the United States, the American Marconi division, under the directorship of the politically powerful John Griggs, former governor of New Jersey and attorney general under President McKinley, had transmitters located in New York, Massachusetts, and Illinois. One key problem, however, was that the Marconi equipment was still using the outmoded Hertzian spark-gap method.

Within two weeks after World War I began, Germany's transatlantic cable was severed by the British. The only reasonable alternative for communicating with the outside world and surmounting the dominance of the British Navy was through Telefunken's wireless system. Suddenly, the Tuckerton and Sayville plants became of paramount concern. The Germans obviously wanted to maintain their stations to keep the kaiser abreast of President Woodrow Wilson's intentions, but the British wanted them shut down.

In March of 1914, Marconi was made a "Senatore" in Italy, a distinguished man of science, and he spoke before the royal couple. In Great Britain, the land of his mother, in July, he was decorated by the king at Buckingham Palace. Now, the fight against Telefunken would be fought on military as well as commercial grounds, as it became clear that the Germans were using their plants to help coordinate submarine and battleship movements. The wireless lines also marked the burgeoning alliance forming between Italy and the British Empire.

As a pacifist, Wilson maintained a strict policy of neutrality, a position bolstered by Bull Moose Party war hero and former president Teddy Roosevelt, himself a contender for the upcoming 1916 election. Although the country was officially neutral, the sentiments of the majority of the American population were with England, particularly after Germany stormed through the peaceful kingdom of Belgium. Nevertheless, fully one-tenth of the population was of German stock, and their sentiments were mostly with the other side.

Even before the war broke out, there were numerous cries from Congress for the military to take over the wireless field. Coupled with an increase in acts of skullduggery by German agents, and Dr. Karl George Frank's unconvincing declaration that Atlantic Communications Company was not funded by Telefunken or the German government, there was a strong argument to commandeer the Tuckerton and Sayville plants.[2]

With a budget which would mushroom to over $500,000, Frank had begun construction of the Sayville wireless station in 1913, with its first communiqués back and forth to Cuba. As the secretary and director of Atlantic, Frank, surreptitiously, was also a representative of Telefunken, which was technically illegal, because the Sayville plant, like all wireless plants in the country, was supposed to be an American-owned operation. In late summer of 1914, "in Berlin [Frank] discussed with the representatives of Telefunken the business relations between that company and Atlantic. At that time Telefunken either owned all or nearly all of the stock of Atlantic. When asked what proportion of the stock of Atlantic was owned or held on behalf of Telefunken in 1914, Frank was unable to answer the question accurately."[3]

In a letter to Tesla dated September 17, 1915, Frank writes that his company is unable to manufacture Tesla's tachometers "as long as the war lasts." Frank goes on to inform the inventor that "there was apparently a misunderstanding, as of course, all understandings are subject to the approval of my home office."[4]

Having just returned from Berlin in the midst of war, Tesla's friend, the poet George Sylvester Viereck, courted Teddy Roosevelt and emissaries of President Wilson as a liaison to Germany. Simultaneously, Viereck, rumored to be Kaiser Wilhelm's illegitimate son, began a new publication with other leading German-Americans. Initially welcomed by the press, *The Fatherland*, secretly funded by the German high command, soon achieved a subscription base of one hundred thousand.

Ignoring Viereck's plea for neutrality, Wilson prepared a presidential decree "declaring that all radio stations within the jurisdiction of the United States of America were [to be] prohibited from transmitting or receiving . . . messages of an unneutral nature. . . . By virtue of authority vested in me by the Radio Act, one or more of the high powered radio stations within the jurisdiction of the United States . . . shall be taken over by the Government." Tuckerton was the first major takeover.

Throughout the beginning of the war, Tesla stepped up his legal campaign against Marconi as he continued to advise and receive compensation from Telefunken via their subsidiary, Atlantic Communications. As the country was officially neutral (America would not enter the war for another three years), the arrangement was entirely aboveboard. Nevertheless, few people knew about the German/Tesla link, although the inventor made no secret of it to Jack Morgan.

> February 19, 1915
>
> Dear Mr. Morgan,
>
> I am expecting to embody in their plant at Sayville some features of my own which will make it practicable to communicate with Berlin by wireless telephone and then royalties will be very considerable. We have already drawn papers.[5]

Camouflaged by the smoke screen of the American-sounding Atlantic Communications Company, Telefunken moved swiftly to increase the power of its station at Sayville. Located near the town of Patchogue out on the flats of Long Island, just a few miles from Wardenclyffe, the Sayville complex encompassed one hundred acres and employed many German workers. With its main offices in Manhattan and its German director, Dr. Karl George Frank, an American citizen, Telefunken was legally covered, as no foreigner could own a license to operate a wireless station in the country. (Thus, Marconi also had an American affiliate.) It was an easy matter for Tesla to confer with Atlantic in the city and also go out to Long Island to the site of the plant.

The headline in the New York Times on August 4, 1915, read "Tesla Sues Marconi on Wireless Patent, Alleges That Important Apparatus Infringes Prior Rights"[6]

As noted in the previous chapter, in the early spring of 1915, Tesla met with Jonathan Zenneck out at Sayville, walked the property, studied the situation, and explained in detail precisely what was wrong with the plant. They were wasting nearly three-quarters of their energy by transmitting their messages through the air. Within two months of Tesla's letter to Morgan, the plant at Sayville *tripled* its power. Using Tesla's theories on the importance of ground transmission and a double tuned circuit, resonating accoutrements were spread out over the land for thousands more feet. Thus, by shifting emphasis away from aerial transmission, as Tesla explained, Telefunken's output was boosted from thirty-five kilowatts to over one hundred, catapulting Germany into the number one spot in the wireless race. Now, complex messages could easily be transmitted at the speed of light across four thousand miles of ocean and terrain to Berlin and back again. The *New York Times* reported on their front page, "Sayville was becoming one of the most powerful transatlantic communicating stations in this part of the world." This event remains the single most important factor for supporting the premise that Tesla's overall scheme in wireless was viable.[7]

Calling wireless "the greatest of all inventions," Tesla made an additional appeal for legal assistance to Morgan. "Can you put yourself for a moment in my place?" he wrote the financier. "Surely, you are too big a man to permit such an outrage and historical crime to be perpetuated as is now being done by cunning promoters." Expecting to "receive satisfaction from the Government," as they had installed "$10,000,000 of [his] apparatus," the inventor also revealed that "the Marconi people approached me to join forces, but only in stock and this is not acceptable."

Again, Morgan declined assistance in protecting their commonly held patents. The Wall Street mogul, however, had not at all abandoned the field, as he was funding a college radio station near Boston at Tufts University.

The years preceding America's entrance into World War I contained an overwhelming quagmire of litigation involving most countries and virtually every major inventor in the wireless field. Aside from Tesla's priority battle, Telefunken was also suing Marconi, who in turn was suing Lee DeForest, the U.S. Navy, and also Fritz Lowenstein for patent infringement. It was at this time that Franklin D. Roosevelt, then assistant secretary of the Navy, began to do his own digging to find out what kind of case Marconi had against the U.S. Navy. "In the files of the Bureau of Steam Engineering," Roosevelt wrote:

> [There] is found a copy of a letter by Nikola Tesla to the Light House Board, under date of September 27, 1899, from the Experimental Station, Colorado Springs. This letter is evidently an answer to a communication from the Light House Board, requesting information as to Tesla's ability to supply wireless telegraph apparatus. . . . [It] may be made use of in forthcoming litigation in which the Government is involved . . . as the discovery of suitable proof of priority of certain wireless usages by other than Marconi [which] might prove of great aid to the Government.
>
> —FRANKLIN D. ROOSEVELT,
> acting secretary of the Navy[8]

MARCONI

In the spring of 1915, Marconi was subpoenaed by Telefunken. Due to the importance of the case, the Italian sailed off for America on the fastest ship on the high seas, the *Lusitania*, arriving in April to testify. "We sighted a German submarine periscope," Marconi told astonished reporters. As three merchant ships had already been torpedoed without warning by German U-boats the month before, Marconi's inflammatory assertion was not taken lightly.

The *Brooklyn Eagle* reported that this suit brought "some of the world's greatest inventors on hand to testify." Declared a victor in the Lowenstein proceedings by a Brooklyn district court judge, Marconi clearly had the aura of the press behind him. Nevertheless, he was beaten by the Navy in the first go-around, so this case against Telefunken, with all the heavyweights in town, promised to be portentous. Once and for all, it appeared, the true legal rights on the invention of the wireless would be established in America.

Aside from Marconi, there was, for his defense, Columbia Professor Michael Pupin, whose testimony was even quoted in papers in California. With braggadocio, Pupin declared, "I invented wireless before Marconi or Tesla, and it was I who gave it unreservedly to those who followed!" He continued, "Nevertheless, it was Marconi's genius who gave the idea to the world, and he taught the world how to build a telegraphic practice upon the basis of this idea. [As I did not take out patents on my experiments], in my opinion, the first claim for wireless telegraphy belongs to Mr. Marconi absolutely, and to nobody else." [9] One can only imagine Tesla's keen disappointment at seeing a fellow Serb so wantonly dismiss his true achievement.

Marconi's lawyers had a lot of arrows in their quiver. They had prevailed in a key lawsuit against Lee DeForest. In lieu of continuing to battle Oliver Lodge, they had purchased his patents, and they also owned those of Rudolf Goldschmidt, both giving Marconi a foothold in equipment concerning continuous wave transmission, which was precisely Tesla's invention.

The cornerstone of Marconi's defense was his original 1896 patent. First registered in England, this early date gave Marconi a true legal hold on priority. However, since it was based on a Hertzian spark-gap transmitter, it was useless for creating separate channels or for sending complex wireless messages such as those containing voice or pictures over long distances. All it could do was send Morse code along a single channel. So Marconi may have had a key early patent, but its details, when studied, revealed a primitive design completely unrelated to the more sophisticated present-day models of 1916, which were actually based on Tesla's patents, such as his patent #649,621 from 1897. This evidence would eventually cause the U.S. Supreme Court to rule, shortly after Tesla's death in 1943, Tesla's priority in what became the invention of the radio over Marconi.

Marconi's lawyers argued that even if aspects of Tesla's invention predated Marconi, Tesla never used his system for transmitting messages and thus he should not stand in the way of development of this art. Further, where Tesla was stressing the idea of transmitting electrical *power* by means of *conduction*, either through the high strata (e.g., the ionosphere) or through the Earth, Marconi was transmitting *information* over the airwaves through the mechanism of *radiation*. Also, although not necessarily stated overtly, there was the simple fact that Tesla's single transmitting tower was never used and had lain dormant for fifteen years, whereas Marconi had literally hundreds, if not thousands, of working wireless transmission systems at every corner of the globe. If Tesla's system was so good, how come it never worked?

Although Marconi was in court and slated to testify, he was called back to Europe before taking the stand because of the war. However, before he returned to Italy, Marconi did accompany Judge Veeder and other dignitaries to the Sayville plant where he tested the equipment and again concluded that Atlantic Communications had infringed on his work. Further, he did leave an extensive deposition, and it was reasonable to assume that this testimony could have been entered into the record. In this deposition, Marconi's attorney, Mr. Betts, stated, "The courts all over the world have held that Marconi was the first to devise a complete telegraph transmitting system utilizing effectively these Hertzian or electric waves."

Betts noted that Judge Veeder in America and Judge Parker in England had recognized Marconi's 1896 priority. And when Betts asked Marconi when he first got interested in the field, the Italian replied, "I began my studies in the field in 1894, aware of the works of Lord Kelvin, Clerk Maxwell and Heinrich Hertz. Early in 1895, I set up my wireless apparatus whereby I could transmit signals at distances of two miles. During the course of these various arrangements, I also tried what would be the effect of connecting one end of the transmitter to the ground and the other to an insulated conductor or a conductor connected to a pole raised in space."

Betts pointed out that Marconi had traveled to Great Britain to display his wireless system to Sir William Preece and other dignitaries on a roof in London and later at Salisbury in Wiltshire in July, August, and September of 1896. Articles in the *Daily Chronicle* and *Electrical Engineer* at that time confirmed these accomplishments. And it was at this time, in 1896, that Marconi registered his first patent.

Marconi testified that in 1901, he sent the first transatlantic wireless message, and in 1902, he was transmitting dispatches between Poldhu, England, and Kronstadt, in Russia, a distance of 1,500 miles. Shortly thereafter, Marconi began setting up companies in "France, Germany, Spain, Italy, Australia, South America, Russia and Canada, all of which carry on a successful system of wireless telegraphy using my invention." He added that his system now routinely sent messages 6,500 miles.

When asked about his background, qualifications, and accolades, Marconi said, "In recognition of my work in connection with wireless telegraphy, I have received nine Orders of Knighthood from various governments and European sovereigns. Besides other decorations, I am a member of the Institution of Electrical Engineers of London and a fellow of the Royal

Society of Arts of London. In Italy, I am an honorary member of the Italian Society of Engineers and Architects, and a fellow of the Royal Society of Rome and also a member of the Royal Academy of Science of Bologna University. And in America, I am a member of the American Philosophical Society of Philadelphia, and of the New York Electrical Society. I am also a fellow of the Royal Society of Sweden."

Concerning other awards, Marconi responded, "I have received The Albert Gold Medal and a Silver Medal from the Royal Society of Arts; a Gold Medal from the Italian Scientific Society of Rome; and a Gold Medal from institutions and societies from the towns of Bologna, Florence, Venice, New York, Madrid, Lisbon and others. I was awarded the Nobel Prize for Physics in 1909, consisting of a Gold Medal, a diploma and a sum of about £4000."

Marconi rattled off his honorary doctorates, which included degrees from the Engineering College of Bologna University, the honorary degree of Doctor of Science from Oxford University, and other degrees from the Universities of Glasgow, Aberdeen, and Liverpool, and also the University of Pennsylvania.

Tesla's work, Betts argued, refers to "the utilization of an electromotive force of hundreds of thousands or millions of volts" for *power* distribution. Marconi's system is set up to transmit *messages*. Since Tesla had never demonstrated the sending or receiving of wireless messages and Marconi had been conducting business in this arena for twenty years, what better proof was there as to who was the true developer of the wireless system?[10]

For the other side, ready to testify were Nobel Prize winner Karl Ferdinand Braun, who shared the prize with Marconi for his work in wireless; Jonathan Zenneck, author of the standard textbook on wireless communication; John Stone Stone, another well-known wireless inventor and president of the Institute of Radio Engineers; and Nikola Tesla. When Zenneck testified, he pointed out that Marconi's original spark-gap design "prevented the possibility of there being any large number of oscillations because of the consumption of energy as heat in the spark." Zenneck was stating that the only way to efficiently transmit wireless messages was to use Tesla oscillators, which were set up precisely for that purpose, and Marconi actually had admitted using Tesla's equipment in his 1901 article in *Electrical Review* titled "Syntonic Wireless Telegraphy" (quoted in the opening epigraph to chapter 9). Zenneck concluded, "The language of the Marconi patent therefore gives no warrant for the interpretation which the court below puts upon the term *persistent oscillator.*" In other words, where Marconi's original 1896 design used damped waves produced by spark-gap apparatuses, Tesla's 1892 lecture portrayed undamped or continuous frequencies produced by his own high-frequency oscillators, the kind of oscillators that became standard fare in all wireless systems of the day, and these were used by such operators as Fessenden, DeForest, Goldschmidt, and Marconi.

TESLA'S TESTIMONY

> In the system devised by me a connection to the Earth, either directly or through a condenser is essential. The receiver in the [Marconi system] is affected only by rays transmitted through the air, conduction being excluded; in [my system] there is no appreciable radiation and the receiver is energized through the Earth while an equivalent electrical displacement occurs in the atmosphere.[11]
>
> —Nikola Tesla

When Tesla took the stand for Atlantic, he came with his attorney, Drury W. Cooper of Kerr, Page & Cooper. Much like Marconi, Tesla also had an impressive résumé. Schooled at universities in Graz and Prague and elected vice president of the American Institute of Electrical Engineers from 1892 to 1894, Tesla had an honorary master's degree from Yale University and honorary doctorates from Columbia and the Universities of Belgrade, Bucharest, Grenoble, and Sofia, and he was a winner of Franklin Institute's Elliott Cresson Medal. Unlike Pupin, who could only state abstractly that he was the original inventor, Tesla proceeded to explain in clear fashion all of his work from the years 1891 to 1899. He documented his assertions with transcripts from published articles, from the Martin text, from public lectures such as his well-known wireless demonstration, which he had presented to the public at Columbia College in 1891 and before royal societies in London and Paris in 1892, and from well-attended lectures in Philadelphia and St. Louis and at the Chicago World's Fair, all in 1893. The inventor also brought along copies of his various requisite patents that he had created while working at his Houston Street lab during the years 1896 to 1899, and photos of his groundbreaking demonstrations in wireless transmission from Colorado Springs, which were conducted in 1899.

The key date for Marconi was 1896, the date of his patent. Aside from Tesla's actual wireless demonstrations and lectures from 1891–1893, even though his laboratory burned down in 1895, there was also an article from

Scientific American in 1895 where "the wizard" told the reporter Mr. Stephenson that he planned "to send a message from an ocean steamer to a city, however distant without the use of any wire." And then there was also a major Sunday magazine article in *The World* in March of 1896, before Marconi's patent was granted, where Tesla's initial wireless experiments in Colorado Springs were reported, where he used the principle of resonance and auto-harps to transmit the tune "Ben Bolt" four miles through Pike's Peak. This was perhaps the first instance of transmitting a song by means of wireless.[12]

In preparation for his lawsuit, Tesla had set the stage by renewing on December 1, 1914, his wireless patent #1,119,732, which dealt with the transmission of energy (electrical power) by means of wireless. From Tesla's point of view, that patent usurped the mere transmission of information, which was nevertheless spelled out and referred to therein by listing his original classic radio patents #645,576 and #649,621, which were originally filed on September 2, 1897. Tesla also used the press to plead his case, announcing his intention to sue Marconi to the *New York Sun* on August 22, 1914, stating that "the patents recently granted to Marconi are covered in detail in my own patents," as listed above, and reporting to the *New York Times*, Sunday edition, October 3, 1915 that "he had received a patent on an invention which would not only eliminate static interference, the present bugaboo of wireless telephony, but would enable thousands of persons to talk at once between wireless stations and make it possible for those talking to see one another by wireless, regardless of distance separating them."

The following dialog is taken from Tesla's 1916 deposition.

COUNSEL: What were the [greatest] distances between the transmitting and receiving stations?

TESLA: . . . From the Houston laboratory to West Point, that is, I think, a distance of about thirty miles.

COUNSEL: Was that prior to 1901?

TESLA: . . . Yes, in or about the year 1897. . . .

COUNSEL: Was there anything hidden about [the uses of your equipment], or were they open so that anyone could use them?

TESLA: There were thousands of people, distinguished men of all kinds, from kings and greatest artists and scientists

> in the world down to old chums of mine, mechanics,
> to whom my laboratory was always open. I showed it
> to everybody; I talked freely about it.

As virtually no one knew about Tesla's West Point experiments, this statement was somewhat deceptive, although it was literally true that thousands of people had witnessed Tesla's other wireless experiments through his published lectures throughout the early 1890s in Europe and the States. Referring explicitly to Marconi's system, having brought the Italian's patent along with his own, the inventor concluded:

> **Tesla**: If you [examine these two diagrams] . . . you will find
> that absolutely *not a vestige* of that apparatus of Mar-
> coni remains, and that in all the present systems there
> is nothing but my four-tuned circuits.[13]

Another jolt to Marconi came from John Stone Stone. Having traveled with his father, a general in the Union Army, throughout Egypt and the Mediterranean as a boy, Stone was educated as a physicist at Columbia University and Johns Hopkins University, where he graduated in 1890. A research scientist for Bell Labs in Boston for many years, Stone had set up his own wireless concern in 1899. The following year, he filed for a fundamental patent on tuning, which was allowed by the U.S. Patent Office over a year before Marconi's. Stone, who never considered himself the original inventor of the radio, as president of the Institute of Radio Engineers and owner of a wireless enterprise, had put together a dossier of inventor priorities in "continuous-wave radio frequency apparatus." Stone wanted to determine for himself the etiology of the invention. Adorned in a formal suit, silk ascot, high starched collar, and pince-nez attached by a ribbon to his neck, the worldly aristocrat took the stand.

> Marconi's British patent preserves in every efficient form a
> commercial apparatus in all systems used today. It is equally
> true that many of the elements in it were not of his inven-
> tion. . . . Marconi, receiving his inspiration from Hertz and
> Righi . . . impressed with the *electric radiation aspect of the sub-
> ject* . . . and it was a long time before he seemed to appreciate
> the real role of the Earth . . . , though he early recognized
> that the connection of his oscillator to the Earth was very

material value. . . . Tesla's electric Earth waves explanation was the more serviceable in that it explained [how] . . . the waves were enabled to travel over and around hills and were not obstructed by the sphericity of the Earth's surface, while Marconi's view led many to place an altogether too limited scope to the possible range of transmission. . . . With the removal of the spark gap from the antenna, the development of earthed antenna, and the gradual enlargement of the size of stations . . . greater range could be obtained with larger power used at lower frequencies, [and] the art returned to the state to which Tesla developed it.[14]

Attributing the opposition, and alas, even himself, to having been afflicted with "intellectual myopia," Stone concluded that although he had been designing wireless equipment and running wireless companies since the turn of the century, it wasn't until he "commenced with this study" that he really understood Tesla's "trail blazing" contribution to the development of the field. "I think we all misunderstood Tesla," Stone concluded. "He was so far ahead of his time that the best of us mistook him for a dreamer."

Here was a portentous and definitive legal case. All the giants in the field were there to testify, including two Nobel Prize winners, the author of the most widely accepted textbook on the topic of wireless, the head of the electrical society, and a professor of electrical engineering from Columbia University. Also partaking were Dr. Karl George Frank, head of Telefunken's company in America, Atlantic Communications; Franklin D. Roosevelt, assistant secretary of the Navy; and Roosevelt's star witness against Marconi, Nikola Tesla. But the world was at war, and so the cards were stacked. Because of Marconi's departure and the request by Attorney Betts to postpone a decision because of the war, the case was put on hold, backroom deals were made, and this momentous decision was left unresolved.[15]

In 1900, John Seymour, the commissioner of patents who had protected Tesla against the demands of Michael Pupin for an AC claim at this same time, disqualified Marconi's first attempts at achieving a patent because of prior claims of Lodge and Braun and particularly Tesla. "Marconi's pretended ignorance of the nature of a 'Tesla oscillator' [is] little short of absurd," wrote the commissioner. "Ever since Tesla's famous [1891–1893 lectures] . . . widely published in all languages, the term 'Tesla oscillator' has become a household word on both continents." The Patent Office also cited quotations from Marconi himself admitting use of a Tesla oscillator.

Two years later, in 1902, Stone was granted his patent on tuning that the government cited as anticipating Marconi, and two years after that, after Seymour retired, Marconi was granted his infamous 1904 patent.

EDWIN ARMSTRONG

Another ardent supporter of Tesla was Edwin Armstrong, who hailed the Serb's 1894 compendium *The Inventions, Researches, and Writings of Nikola Tesla* as a veritable bible. Having just graduated from Columbia University, Armstrong had invented a feedback amplifier, which further developed Lee DeForest's audion tube. The device was so sensitive that it could pick up signals from Hawaii, five thousand miles away. Armstrong struck a deal with Telefunken, which placed his "ultra-audion" at their plant at Sayville and paid the young entrepreneur a royalty of $100 per month. Destined to invent AM and FM radio and a nonconformist by nature, this new wizard had come upon his 1912 discovery because he had rejected the inferior Marconi spark-gap apparatus that most of his wireless buddies were still playing with and, like Stone, had plunged himself into the continuous wave technology developed by Tesla.[16]

With Marconi's highly visible early success, large-scale wireless enterprise, and Nobel Prize on his side of the balance scale, it became much easier to credit him with the discovery. The ongoing Great War served to further cloud the issue, and the important legal battle between Atlantic Communications Company (Telefunken) and Marconi Wireless was abandoned before it was resolved.

Due to the dangers that existed on the high seas and the rumors that the Germans were out for Marconi's head, the Senatore did not sail back on the *Lusitania*, but rather returned on the *St. Paul* in a disguised identity under an assumed name.

As Marconi set sail, the new head of the Institute of Radio Engineers, John Stone Stone, was honored at a dinner at Luchow's German Restaurant attended by a potpourri of leaders in the industry. Guests included Lee DeForest, who was about to receive a quarter of a million dollars for sale of his patents to AT&T; J. A. White, editor of *Wireless Magazine*; David Sarnoff, on the verge of launching his radio and TV empire; Rudolf Goldschmidt, the force behind the Tuckerton wireless plant; Karl Ferdinand Braun, who shared the 1909 Nobel Prize in Physics for his work in wireless with Marconi; A. E. Kennelly, a former Edison man often credited with Oliver Heaviside as the discoverer of the ionosphere; Fritz Lowenstein, who was about to earn $150,000 from AT&T for one of his inventions; Jonathan Zenneck, head engineer at the Sayville wireless plant; Ernest Alexanderson, inventor of a long-wave radio transmitter; Nikola Tesla, who stood between DeForest and Lowenstein for an official photograph; and Tesla's boss at Atlantic Communications, Dr. Karl George Frank.[17] A happy occasion, the event would be soon overshadowed by America's entrance into the Great War.

10

The Fifth Column

May 21, 1916

My dear Doctor Frank,

I have no doubt that [Fritz Lowenstein] will be able to help you out of the little difficulty. However, should he be unsuccessful, I shall be glad to assist you in every way possible.

I will be delighted to make the personal acquaintance of [Privy Councilor Heinrich Friedrich] Albert and also to see Prof. Zenneck again. Would you do me the favor of being my guests some evening at the Waldorf at mutual convenience.

Yours very sincerely,
N. Tesla[1]

In May 1916, two days before meeting the paymaster of the German secret service, Privy Councilor Heinrich Friedrich Albert (1874–1960), Dr. Karl George Frank wrote Tesla "to discuss with you today a very important matter. . . . Please call me immediately after receiving this letter."[2]

Almost a year to the day earlier, on May 7, 1915, a German submarine had torpedoed the *Lusitania*, killing 1,134 individuals. The sinking, in lieu of the alternative procedure of boarding unarmed passenger ships by military vessels, was unheard of. Quite possibly, Marconi could have been a target; however, the Germans used as their reason the alleged cargo of armaments on board headed for Great Britain. With only 750 survivors, this "turkey shoot" took almost as many lives as the tragedy of the *Titanic*. According to Lloyd Scott of the Navy Consulting Board, "Press reports stated that the Germans seemed to revel in this crime, Medals were being struck to commemorate the sinking, and holidays were given to school children." No longer neutral, former president Teddy Roosevelt decried the event as "murder on the open seas."

This huge loss of life, however, did not stop George Sylvester Viereck from supporting the German position. Having travelled by zeppelin above Berlin during the war, Viereck stated in the *New York Times* that had the weapons made it to England, "more Germans would have been killed than died in the [boat] attack." This callous argument inflamed the populace against Viereck, and the formerly renowned poet was now hailed as "a venom-bloated toad of treason."

Due to Viereck's callous and potentially treasonous attitude, the Secret Service placed a double tail on the poet. As it turned out, Viereck's newspaper, *The Fatherland*, was being secretly funded by the propaganda mill of the German government through payments made by Albert, the very man Tesla was to meet at a dinner with Frank.

One day around the time of the sinking of the *Lusitania*, Viereck stopped by the Privy Councilor's office at 111 Broadway, probably to receive more funding, and then the two stepped onto the street to go to their respective destinations. Eschewing the use of taxis, both took different el trains, so one agent followed Viereck and the other tailed Albert. Although Albert, as a commercial attaché to German Ambassador Johann Heinrich von Bernstorff, was also "chief fiscal officer for German espionage and sabotage operations in the United States," the American government was not unaware of this.[3] Like his boss, von Bernstorff, and his cohorts in the spy game, Army Captain Franz von Papen and Captain Karl Boy-Ed of the Navy, Albert was blasé when it came to security. The Americans, he assumed, as "babes in the woods," were asleep. Quite comfortable on the train, Albert nodded off, and when he reached his stop, he nonchalantly left the railcar without his attaché case. Recognizing the advantage, Frank Burke, the Secret Service agent, made off with it.

Albert ran after his tail, who had jumped onto a street tram, but was unsuccessful in catching him. In that way, many of Germany's secret-op plans for funding union strikes, poisoning horses being sent to the front, sabotaging factories, and spreading propaganda made their way to the White House, yet, astonishingly, President Wilson did nothing about it! And Albert was apparently not reprimanded by the German government. Since Wilson continued to hedge, perhaps in part because he may have been enamored of the fact that he was communicating directly with Kaiser Wilhelm via the Tuckerton and Sayville wireless stations, his advisor, Colonel Edward M. House (who, in fact, had no military service), took matters into his own hands and released the dossier to Frank Cobb, editor-in-chief of the *World*, who published the material on August 15, 1915, on the front page in an article titled "How Germany Has Worked in U.S. to Shape Opinion, Block the Allies and Get Munitions for Herself, Told in Secret Agents' Letters!"[4] This news was picked up by many other news outlets, but for reasons difficult to comprehend, the Privy Councilor astonishingly, was able to ride out the embarrassing situation without being expelled from the country.

Now it was a year later. Telefunken had Tesla install a wireless transmitter atop their Broadway offices, and it was on the blink. When Frank called because of an "important matter," it is plausible to consider that he had received a directive from the high command in Germany via Albert to get the equipment fixed as soon as possible, because the transmitter was being used to help coordinate clandestine operations.

Tesla enjoyed a pleasant dinner at the Waldorf, conversing in German with Zenneck, Frank, and Albert, and shortly thereafter met up with Lowenstein to fix Telefunken's New York City wireless transmitter. Since he was working for them, via Atlantic Communications, and since Atlantic had a contract with the U.S. Navy to install additional German-made transmitters on Navy ships, Tesla's connection to the German company, now at war with France and Great Britain, was completely aboveboard, even with the blatant skullduggery of the sinking of the *Lusitania* still a fresh and horrific occurrence. However, given that Albert had been outed as a spy a year earlier, it remains a puzzlement to this author how Tesla would be so naïve as to look forward to meeting with him. Yet simultaneously, Franklin Roosevelt, then assistant secretary of the Navy, was conferring with Telefunken and coordinating efforts with them and Tesla to fight Marconi's lawsuit, which was against all three. Keep in mind that American ships were also being sunk by German U-boats; various acts of sabotage were occurring throughout the country on a routine basis; and within two months of Tesla's meeting with Albert, a gigantic munitions dump off the coast of New Jersey known as Black Tom would be detonated by the Germans. The site, just a few hundred yards from the Statue of Liberty, caused an explosion that was so terrific that it registered 5.5 on the Richter scale. Killing five individuals, including a baby who was jolted out of her crib, the blast also knocked out hundreds of windows in lower Manhattan and was so loud that it was heard as far away as Philadelphia!

The fifth column had emerged. German spies were everywhere. Reports started filtering in that a secret submarine base on an island off the coast of Maine was being constructed and that the broadcasting station out at Sayville was not merely sending neutral dispatches to Berlin but also coded messages to battleships and submarines. The sighting of a periscope off the coast of Rhode Island further fueled the paranoia. It was even alleged by *New York Sun* reporter John Price Jones that somehow Sayville had jammed the normal wireless broadcast to the *Lusitania* from a British military ship in waters near Liverpool and had substituted a counterfeit message that drove the ship off course and past the very German U-boat that sank it![5] And that watershed event was only part of Germany's ongoing sneak attacks against the United States and also Canada.

In 1913, Germany passed the Delbruck Law, which allowed émigrés to retain their German citizenship in other lands as long as they registered with their respective local German consulates. Unlike the policy of other countries, what Germany had instituted, long before Adolf Hitler's move to Aryanize Europe, was a plan to seed the planet with their own breed, as the émigrés maintained their true allegiance to the Fatherland.

By the time the Great War had broken out, Germany, through its wireless company Telefunken, had not only set up broadcasting stations in America, but it also had stations in "Australia, Shanghai, Bangkok, Africa, China, the Dutch East Indies, the Pacific Islands, Havana, Central America, Peru, Greece, Vienna, Petrograd, Copenhagen, Tokyo, Rio de Janeiro, Buenos Aires, Mexico and Johannesburg." Perhaps the most illustrious head of secret ops was the charismatic Franz von Papen, the military attaché based in Washington, DC. Known as "a master of many disguises," and later to become the very chancellor of Germany and vice chancellor under Hitler, von Papen had "traveled extensively and worked for the [German] secret service."[6]

"Gradually and quietly, year by year, Germany spread her system of wireless communication over Central and South America, preparing her machinery for war. . . . Amateur stations [were constructed by] German reservists who had been peaceful farmers, shopkeepers or waiters all over the United States . . . who made their [secret] reports regularly to the Embassy. There, the messages were sorted by [head of operations] Count von Bernstorff."[7] As a former ambassador to St. Petersburg, London, and Cairo after marrying a wealthy daughter of a New York silk merchant, von Bernstorff was transferred to America, where he became the flamboyant German ambassador in Washington.

Shortly after the Great War began, von Bernstorff was called back to Germany, where he met with the stalwart Walter Nicolai, head of Abteilung IIIb, (Department 3B), which was the code name for German Military Intelligence, essentially the German counterpart to Great Britain's MI-6 and, later, America's OSS. Previously stationed in Russia, Nicolai rose to become chief of military intelligence. Now master of clandestine ops and soon to enlist Mata Hari as Germany's spy in France, Nicolai instructed von Bernstorff to oversee Army and Navy Captains Franz von Papen and Karl Boy-Ed in a complex three-pronged scheme against an unsuspecting and naïve United States. Funded by a staggering sum in the tens of millions of dollars, the first prong involved pro-German propaganda, setting up plausible deniability disseminated in social settings via high officials and in the press. The second and third prongs were more diabolic. They included the creation of a vast spy network covering much of the country and corresponding acts of sabotage.[8]

According to *New York Sun* reporter John Price Jones, von Bernstorff "was in possession of both the French and British secret admiralty codes . . . [resulting in] the sinking of [a number of ships such as] the *Monmouth* and the *Good Hope*."[9]

Using a cipher involving innocuous extra spaces, substituted words, and diacritics embedded into dispatches from Berlin, coded messages could be sent to Captains von Papen or Boy-Ed to coordinate continuing subterfuge. At the same time, a massive propaganda campaign was waged in the press using such "journalists" as George Sylvester Viereck writing in *The Fatherland* and wealthy German entrepreneurs and German-leaning government officials who could disavow any knowledge of fifth column activities in social settings, talks on Capitol Hill, and interviews and letters to the editor in the major papers.

> In this connection, dear Mr. Editor . . . it is known to me from reliable sources that I am to be accused of having caused the last strike in Chicago, that I am preparing to instigate labor troubles with the working-classes here, that I have planned the well-known fire on la Touraine because of the presence of an invention destined for England or France, that I am erecting a submarine base off the islands of the coast of Maine and that I am busy buying newspapers to poison the public opinion in the United States. I am only waiting to hear that it has been reported that I am putting the Giants in bed by pulling off a crooked baseball deal.
>
> —Captain Karl Boy-Ed[10]

This flagrant use of reverse psychology by the German Navy's attaché was part of an all-encompassing subversive campaign funded through payments distributed by Privy Councilor Albert. With hundreds of agents scattered around the country, an Aryan think tank came to fashion a series of subversive acts precisely along the lines denied so brazenly by Boy-Ed. The list of operations included blocking military exports; the poisoning of thousands of horses set to be exported for the French and British cavalry; acts of destruction of factories and workshops; inciting of labor troubles; monkey-wrenching critical equipment such as engine parts; setting aflame stockpiles of raw materials; planting bombs on any ship that would be used to support Germany's enemies; creating power outages at industrial centers; bribing congressmen to gain legislation friendly to Germany; coercing and intimidating employees at military establishments and munitions factories; and amassing stockpiles of armaments, including the construction of explosive devices.[11]

One major goal, left uncompleted, was the dynamiting of the Welland Canal, a major waterway parallel to Niagara Falls that connected Canada to the United States. Successful operations included the destruction of the Dupont California Powder Works in Pinole in August of 1915; the derailing of the Canadian Pacific Railroad in 1916; the demolition of hundreds of cannons at a Bethlehem Steel plant; the immolation of a warehouse in Providence, Rhode Island; arson on the battleship *Alabama* as it was docked at the Navy Yard in Philadelphia; an attack at a chemical plant in Lenhart, New York; the poisoning of the stockyards in Chicago; and the destruction of airplanes at a Curtiss factory in Depew, New York, all coordinated by Captains von Papen and Boy-Ed. This duo also planned other sorties that were never carried out onto key bridges and tunnels, such as into New York City with the idea of loading automobiles or trains with dynamite and detonating them at the proper time.[12]

With munitions demolitions and the continuing success of other acts of sabotage, the German fifth column turned their attention to targeted assassination. On July 2, 1915, Washington, DC, was rocked by a huge explosion in a Senate anteroom of the Capitol building, and the following day, the man who planted the bomb, German émigré Frank Holt, walked into J. P. Morgan Jr.'s Long Island mansion, threaded his way to the back office, and held Morgan and his wife at gunpoint with demands for the Wall Street magnate to stop shipping arms to Germany's enemies. Enraged, Morgan charged forward, but not before Holt got off several rounds, hitting the financier twice in the groin before he was wrestled to the ground by Morgan and his wife.

Holt was arrested at Morgan's home and apparently committed suicide a few days later in his Mineola jail cell when it was revealed that he was really Erich Muenter, a former Harvard University professor of German, wanted several years earlier for poisoning his wife to death with arsenic.[13]

With Tesla, just a few months earlier, boasting to Morgan that he was working for the Germans and the *Times* reporting on their front page "Grand Admiral von Tirpitz contemplat[ing] a more vigorous campaign against freight ships . . . [and planning] a secret base on this side of the Atlantic," it is quite possible that the inventor himself became tainted with a smattering of "venom-bloated toad's blood." Ever the gentleman, Tesla sent Morgan a get-well card delivered to his hospital bed. Even though he was shot twice, the wounds were not serious, Morgan healed rapidly, and there is no indication that his relationship with Tesla suffered in any way due to Tesla's partnership with German concerns. As with Anne Morgan, J. P.'s sister, Tesla continued his relationship with the Wall Street magnate for the next twenty years in uninterrupted fashion.

On Tesla's fifty-ninth birthday, the *Times* reported not only that the Germans were dropping bombs over London from zeppelins but also that they were "controlling air torpedoes" by means of radio-dynamics. Fired from airships, the supposed "German aerial torpedo[es] can theoretically remain in the air three hours, and can be controlled from a distance of two miles. . . . Undoubtedly, this is the secret invention of which we have heard so many whispers that the Germans have held in reserve for the British fleet." Although it seemed as if Tesla's devil automata had finally come into being, as the wizard had predicted nearly two decades before, Tesla himself announced in the press that "the news of these magic bombs cannot be accepted as true, [though] they reveal just so many startling possibilities."[14]

"Aghast at the pernicious existing regime of the Germans," Tesla accused Germany of being "an unfeeling automaton, a diabolic contrivance for scientific, pitiless, wholesale destruction the like of which was not dreamed of before. . . . Such is the formidable engine Germany has perfected for the protection of her Kultur and conquest of the globe." Predicting the ultimate defeat of the Fatherland, the Serb, whose former countrymen were fighting for their own survival against the kaiser, stopped doing business with von Tirpitz, although he continued his relationship with Zenneck, whom Tesla perceived as being trapped by the insanity of war.

Tesla's solution to war was twofold: a better defense, through an electronic

Star Wars–type shield he was working on that the mad scientist bragged could take down airplanes and "wipe out whole populations by pressure of a key," yet on the opposite hand, "the eradication from our hearts of nationalism." If blind patriotism could be replaced with "love of nature and scientific ideal . . . permanent peace [could] be established."[15]

The period from late 1914 to the date of the United States' entry into the war in 1917 was marked by continuing acts of subterfuge. The wireless transmitter on the roof of 111 Broadway, the building that housed Telefunken's offices, which Tesla himself had helped install, was accused of sending coded messages to the enemy. This was confirmed by Charles Apgar, who began intercepting late-night messages from the Sayville plant that had eluded Naval Intelligence. What the Germans had done was send the messages at a very high frequency so that it simply sounded like a static buzz. Apgar, however, had begun recording these nightly communiqués on Edison phonograph cylinders. With this physical recording, Apgar could slow down the vibration, and in this way, the coded messages were revealed. Working with the military, Apgar delivered his recordings to the Secret Service, and now the U.S. government had solid proof of Sayville's treachery.[16] Shortly after America's entrance into the war, Tesla informed his secretary that "the Monocle," Lieutenant Mayer, "who ran the Tuckerton operation . . . [had been placed] in a Detention Camp in Georgia," suspected of spying. Thus, the monthly stipend from Telefunken came to an abrupt end.[17]

The secretary of the Navy, whose job became to take over all wireless stations, was Josephus Daniels, with Roosevelt as his assistant. In the summer of 1915, at the verge of the United States entering the war, Daniels called upon Tom Edison with the idea of creating an advisory board of inventors. Working with Roosevelt, Edison appointed numerous inventors to this "Naval Consultant Board," including Gano Dunn, Reginald Fessenden, Benjamin Lamme, Irving Langmuir, R. A. Millikan, Michael Pupin, Charles S. Scott, Elmer Sperry, Frank Sprague, and Elihu Thomson. Waldemar Kaempffert was also put on board for his writing prowess.

It is possible that Tesla's link to Telefunken was the reason his name was not on the list, although many other inventors were also excluded, such as Hammond, Stone, and DeForest. Tesla also was not one to work again for Edison. Tesla's inventions, however, were vital to the government. President Wilson allowed his adviser, Colonel Edward M. House, to set up a secret fund to advance Tesla's developments in radio-guidance systems through the inventor's former partner, John Hays Hammond Jr. Tesla, of course, never saw a dime of the payments. However, it would also take a great deal of arm bending by Hammond's powerful father and nearly ten more years of annual pleas to be reimbursed. In 1924, or thereabouts, Hammond Jr. received a check from the U.S. government for a half million dollars, an amount that probably barely covered his development costs.

NOBEL PRIZE AFFAIR

Dear Tesla,

When that Nobel Prize comes, remember that I am holding on to my house by the skin of my teeth and desperately in need of cash!

No apology for mentioning the matter.

Yours faithfully,
RUJ[18]

On November 6, 1915, the *New York Times* reported on their front page that Tesla and Edison were to share the Nobel Prize in physics. The source for the report was "the Copenhagen correspondent of the [London] *Daily Telegraph.*"

November 29, 1915

Dear Mr. Tesla,

From the [newspapers] I have been informed of the Nobel Prize with Edison and I have expressed to you my congratulations. You have earned it by your [innovative] investigations and experiments.

With best regards,
K. G. Frank[19]

1 December, 1915

My dear Dr. Frank,

Accept my profound thanks for your friendly expression by this hypothetical opportunity. I am also convinced that I deserve the prize that I have gotten.

With best greetings,
Your loyal,
N. Tesla[20]

Although Tesla himself forwarded original copies of the announcement (also carried in a number of other journals) to J. P. Morgan Jr.; and he accepted congratulations from such notables as Dr. Karl George Frank; Jonathan Zenneck; and Karl Ferdinand Braun, who himself shared an earlier Nobel Prize with Marconi, neither Tesla nor Edison ever received the Nobel Prize. In Braun's case, he wrote a letter to "My dear and honored Mr. Tesla! . . . You cannot imagine how much joy filled my heart on hearing the news and so I would like to congratulate you heartily." [21]

Tesla replied in kind to "my dear and greatly honored Prof. Braun," expressing "a deep appreciation of my humble contribution to the engineering art and science, from such an extraordinary and celebrated man like you." Tesla noted, "In itself [your congratulations are] of greater value than the Nobel Prize that is but a conceited aspiration. But your friendly missive is especially reassuring and desirable now that I'm experiencing great suffering caused by the fact that due to some inexplicable natural degeneration, our brothers are killing each other [in this dreadful war]." [22]

As soon became apparent, the Nobel Prize was not awarded to Tesla or Edison that year (or any other year, for that matter), but rather it was awarded to W. H. Bragg and his son for using X-rays to uncover the molecular structure of crystals. As with a number of other Nobel Prize winners, Bragg happened to be a fan of Tesla's, writing to him on the wizard's seventy-fifth birthday in 1931, "I remember vividly the eagerness and fascination with which I read your account of the high tension experiments more than forty years ago. They were most original and daring; they opened up new vision for exploration by thought and experiment. Since that time the electron, X-rays, radioactivity, the quantum theory and other startling additions to our knowledge of Nature have in turn impressed their wonder upon us. But I shall never forget the effect of your experiments which used first to dazzle and amaze us with their beauty and interest." [23]

Curiously, this same *Times* article listed four other people for Nobel Prizes in literature and chemistry who also did not receive the award that year, although three of them eventually obtained it. The fourth, Troels Lund, like Tesla and Edison, never obtained the honor.

Although the announcement came in November 1915, the nomination process actually was concluded nine months earlier. There were nineteen scientists on the physics committee, each allowed two bids. Out of the thirty-eight possible bids, two were made for inventors in wireless, E. Branly and A. Righi; two were for the quantum physicist Max Planck; Edison received one bid; and the Braggs took four. According to the Academy's records,

Nikola Tesla was *not* nominated that year; however, two bids, numbers 33 and 34, were missing from their files. A week after the *Times* announcement, on November 14, Stockholm announced that Professor William H. Bragg and his son would share the award in physics.

The man who nominated Edison, Henry Fairfield Osborn, president of Columbia University (who twenty years earlier had awarded Tesla with an honorary doctorate), apologized to the committee for offering up Edison's name. "Although somewhat out of the line of previous nominations," Osborn wrote, qualifying his decision, "I would [like to] suggest the name of Mr. Thomas A. Edison . . . who through his inventions, is one of the great benefactors of mankind." One wonders why anyone would have to apologize for offering up Thomas Edison's name. Tesla eventually would be nominated in 1937 by Felix Ehrenhaft, a theoretical physicist from Vienna, the same individual who had previously nominated Albert Einstein.

Certainly, both Tesla and Edison deserved such an award, and it is nothing short of astounding that (1) neither of them ever received it, and (2) no one at the time discovered the reason behind this curious quirk of history.

O'Neill, having interviewed Tesla on the subject, stated that Tesla "made a definite distinction between the inventor, who refined preexisting technology, and the discoverer who created new principles. . . . Tesla declared himself a discoverer and Edison an inventor; and he held the view that placing the two in the same category would completely destroy all sense of the relative value of the two accomplishments."

Support for this interpretation can be found in a letter Tesla wrote to the U.S. Lighthouse Board in Washington from his Colorado Springs experimental station in 1899. The Navy had written Tesla to state that they would "prefer" to give their impending wireless contract to an American, rather than to Marconi.

"Gentlemen," Tesla responded curtly, "Much as I value your advances I am compelled to say, in justice to myself, that I would never accept a preference on any ground . . . as I would be competing against some of those who are following in my path. . . . Any pecuniary advantage which I might derive by availing myself of the privilege is a matter of the most absolute indifference to me."[24] If no one else would recognize his genius, Tesla certainly did. He would not think twice about giving up mere cash when faced with the prospect of being *compared*, in this case, to Marconi.

Note also that in his letter to Frank, Tesla refers to the Nobel Prize as a "hypothetical opportunity" and mentions that he "deserve[s] the prize that

I have gotten," which may refer to the patents that establish his priority in the fields of both AC power transmission and wireless communication.

The following letter to Johnson, which the inventor took the time to rewrite in a careful hand, was penned just four days after the announcement and four days *before* Sweden's decision to give the award to the Braggs.

> My dear Luka,
>
> Thank you for your congratulations. . . . To a man of your con-suming ambition such a distinction means much. In a thou-sand years there will be many thousand recipients of the Nobel Prize. But I have not less than *four dozen of my creations identified with my name* in technical literature. These are honors real and permanent which are bestowed not by a few who are apt to err, but by the whole world which seldom makes a mistake, and for any of these I would give all the Nobel Prizes which will be distributed during the next thousand years.[25]

In the *New York Times* interview on the day following the Nobel commit-tee's announcement, Tesla stated that Edison was "worthy of a dozen Nobel Prizes." The various Tesla biographers assumed that this was a public state-ment congratulating Edison, when, in fact, it was a piquant snub to the committee, as Tesla was stating between the lines that they recognized only small accomplishments, rather than truly original conceptualizations.

"A man puts in here [in my Tesla coil] a kind of gap—he gets a Nobel Prize for doing it. . . . I cannot stop it. . . . The inventive effort involved is about the same as that of which a 30-year-old mule is capable." Thus, Edi-son's many "better mousetraps" could all be honored, but none of them, in Tesla's opinion, concerned the *creation of new principles*. They were simply refinements of existing apparatuses.[26]

Edison would probably have agreed with Tesla on this point, for most of his inventions were actually further developments of other people's work. However, Edison did have a number of original discoveries and creations. In his own opinion, his most important contribution was the phonograph, patented in 1877, which certainly was a work of genius even if it was based, to some extent, on Édouard-Léon Scott de Martinville's "phonautograph," which recorded sound on lampblack. Scott de Martinville presented his phonautograph before the French Academy of Science in 1857 and even got a patent on the device.[27] Nevertheless, Edison's work in that realm was cer-tainly worthy of a Nobel Prize, as was Tesla's invention of the telautomaton, or remote-controlled robot. Further, Edison's unparalleled success in bringing promising creations to fruition was exactly Tesla's failure, and that, too, was a gift placing Edison in a category all by himself.

Quite possibly a letter much like the one sent to Johnson or the U.S. Lighthouse Board could also have been wired to the Nobel committee. If this hypothesis is correct, a prejudice would have persisted against Tesla and Edison, and this would explain the indefensible position of the Swedish Academy to never honor either of these two great scientists.

WIZARD SWAMPED BY DEBTS

Inventor Testifies He Owes the Waldorf Hasn't a Cent in Bank [28]

As 1915 was drawing to a close, Tesla began to find himself in deeper financial troubles. Although an efficient water fountain that he designed that year for Louis Tiffany was received favorably, his overhead was still too high. Expenses included outlays for the turbine work at Edison's Waterside Station; his office space at the Woolworth Building; salaries to his assistants and Mrs. Skerritt, his new secretary; past debts to such people as the Johnsons and George Scherff; maintenance costs for Wardenclyffe; legal expenses on wireless litigation; and his accommodations at the Waldorf-Astoria.

Some of the costs were deferred, particularly to the hotel, but the patience of the hotel's manager, George C. Boldt, had reached its limit. Tesla's uncanny elusiveness and noble air had worn thin. Rumors began circulating of peculiar odors and cackling sounds emanating from the inventor's suite. The maids were complaining that there was an inordinate amount of pigeon excrement on the windowsills. Boldt sent Tesla a bill for the total rent due, nearly $19,000. Simultaneously, Tesla was hit by the city with a suit for $935 for taxes still owed on Wardenclyffe.

Tesla signed over the Wardenclyffe property to Boldt, just as he was called into the state Supreme Court. Before Justice Finch, the inventor revealed that "he possessed no real estate or stocks and that his belongings, all told, were negligible." Under oath, Tesla revealed that he lived at the prestigious Waldorf "mostly on credit," that his company "had no assets, but is receiving enough royalties on patents to pay expenses," and that most of his patents were sold or assigned to other companies. When asked if he owned an automobile or horses, the inventor responded no.

"Well, haven't you got any jewelry?"

"No jewelry; I abhor it."

This embarrassing article was published in the *World* for all to see. Yet, as was his custom with any article about himself, the inventor had his secretary paste the mea culpa in the latest volume of press clippings. Looking much like a multivolume encyclopedia, this text, along with his other records and correspondence, would provide for posterity, through over fifty volumes, an accurate account of the inventor's rich and complicated life. The inventor had chosen his words carefully when speaking under oath to the judge. As much as he loathed being in the debtor's position, he wanted the Morgans, Marconis, Roosevelts, and Wilsons to know of his plight, for in the final analysis, this shame would be theirs as much as his. Even his editor, T. C. Martin, had turned against him, writing petty letters to Elihu Thomson at this time complaining how Tesla chiseled money out of him for the opus he had created of the inventor's collected works a generation earlier.

Attempting to raise funds in a variety of ways, Tesla continued to seek a market for his tachometer, pushed to get monies from American firms for the bladeless turbines, and tried to collect royalty payments from Lowenstein and Telefunken for the Tuckerton and Sayville plants. With the sinking of the *Lusitania* in May of 1915, the strain between the United States and Germany had reached new heights. In December, Tesla wrote Frank that "I am in the position to perfectly overcome static disturbances and materially increase the efficiency of wireless plants. . . . I greatly regret the circumstances which have prevented me from installing these improvements at your plant, but hope that this dreadful state of things will not continue much longer." Ending the letter by asking Frank "to kindly remember me to Professors Braun and Zenneck," Tesla hoped that the war would end.[29]

With the publication of his wretched state in the public forum and the transfer of Wardenclyffe to another party came also a deep sense of anger and corresponding shame, for now the world had officially branded him a dud. If success is measured in a material way, it was clear that Tesla was the ultimate failure.

On the exterior, the inventor kept up appearances, but this transfer of the property to the Waldorf would mark the turning point in his life. Now he began the slow but steady turning away from society. Simultaneously, he traveled to live in other states, in part to conduct business in a fresh atmosphere and in part to remove himself from a hostile environment. He wrote a letter to Henry Ford in Detroit, hoping, finally, that the auto magnate would recognize the great advantages of his steam engine.

"I can tell, any day, that Ford is going to contact me, and take me out of all my worries," Tesla confidently predicted to his assistant Julius Czito, who was the son of an earlier trusted assistant, Coleman Czito, and who was now working for him. "Sure enough, one fine morning a body of engineers from the Ford Motor Company presented themselves with the request of discussing with me an important project," Tesla revealed a few years later.

"Didn't I tell you!" the prophet remarked triumphantly.

"You are amazing, Mr. Tesla," Julius responded. "Everything is coming out just as you predicted."

"As soon as these hardheaded men were seated," Tesla continued, "I, of course, immediately began to extol the wonderful features of my turbine, when the spokesmen interrupted me and said, 'We know all about this, but we are on a special errand. We have formed a psychological society for the investigation of psychic phenomena and we want you to join us in this undertaking.'" Flabbergasted, Tesla contained his indignation long enough to escort the wayward explorers to the street.[30]

THE EDISON MEDAL

> Were we to seize and to eliminate from our industrial world the results of Mr. Tesla's work, the wheels of industry would cease to turn, our towns would be dark, our mills would be dead and idle. Ye[s], so far reaching is this work, that it has become the warp and woof of industry.
>
> —B. A. BEHREND, 1917

That Tesla would be nominated by an organization dwelling under the banner of the Edison name was shocking enough to the brooding Serb. Edison, himself, must have allowed the presentation to take place. Having just turned seventy, it does not appear that Edison was plagued by the reciprocal feeling of competitiveness that Tesla exhibited. It is more likely that Edison got a real kick out of the thought of giving Tesla the medal.

"If [Edison] had a needle to find in a haystack," Tesla wrote in a rather famous quote a number of years later, "he would not stop to reason where it was most likely to be, but would proceed at once with the feverish diligence of a bee, to examine straw after straw until he found the object of his search. . . . Just a little theory and calculation would have saved him ninety percent of his labor."

Tesla certainly respected Edison, but underneath it all, he remained irked that this somewhat primitive trial-and-error fellow continued to ride a huge wave of adoration, while his star had long ago dipped. According to the inventor's autobiography, written just two years after this event, Tesla suggests that when he had been working for Edison when he first immigrated to the United States in 1884, the Wizard of Menlo Park had, according to Tesla, reneged on paying him $50,000 for redesigning Edison's prevailing DC equipment. Not only had Edison completely dismissed Tesla's revolutionary and highly efficient way to harness AC, which resulted in the invention of the induction motor, the workhorse of the world, he had also embarrassed Tesla by dismissing his pecuniary promise by stating that the naïve Serb simply did not understand American humor.

Accepting an Edison Medal did not sit well with the Serbian aristocrat. And thus, his first reaction was abhorrence, and he flatly rejected the offer, but Bernard Behrend, the electrical engineer who nominated him, persisted. Here was an opportunity to recognize a worthy recipient alone for his singular contributions. "Who do you want remembered as the author of your power system?" Behrend inquired. "Ferraris, Shallenberger, Stillwell or Steinmetz?" Tesla reluctantly capitulated. To certain friends and for public consumption, Tesla was often dismissive of Edison, but the truth of the matter was that Edison had come to Tesla's rescue on more than one occasion.

As far back as 1895, when Tesla's laboratory burned to the ground, Edison offered Tesla the use of one of his metropolitan factories until the Serb could find another space. A decade and a half later, Tesla set up shop in Edison's Waterside Station in New York City to work on his turbines, and he sent Edison "hearty congratulations and good wishes, with respectful regards" on Edison's sixty-fifth birthday, and Edison reciprocated by sending back a thank-you note.[31]

Two years later, in 1914, Edison sent Tesla a formal announcement of the marriage of Edison's daughter Madeleine to the airplane manufacturer John Eyre Sloan, and Tesla sent "hearty and respectful congratulations. I hope sincerely that the marriage of your charming daughter will prove the beginning of a new life of undisturbed happiness. Trusting that you are carrying on your valuable work in the full enjoyment of mental and bodily vigor, I remain as ever, Yours faithfully, Nikola Tesla."[32]

Just several months later, like Tesla's before him, Edison's laboratory was lost in a fire, and Tesla sent his regrets but also congratulated Edison that his "valuable reports had been saved through the fore-sight of Mrs. Edison. With new and much improved facilities you will do far better and the temporary setback will prove a gain in the end." Tesla ended the letter "with best wishes for your health and success." [33]

> December 16, 1914
> Mr. Nikola Tesla
> Metropolitan Building
> New York City
>
> My dear Tesla,
>
> Allow me to express my appreciation of your kind message of sympathy in regard to the recent fire at my plant and to thank you for your good wishes.
>
> We are doing some tail hustling around here and I will be back in the game again within 60 days.
>
> Yours very truly,
> Thomas A. Edison[34]

The presentation of the Edison Medal took place on May 18, 1917, just two months before Tesla found out by telephone that vandals had broken into his Wardenclyffe laboratory and wrecked equipment valued at $68,000, and that "the Tower [was] to be destroyed by dynamite." Perhaps the greatest loss was the demise of Tesla's telautomaton, or remote-controlled robotic boat, which was indeed the work of sheer genius, comprising within its construction such inventions and future possibilities as remote control, wireless communication, cell phone technology (including the ability to create an unlimited number of wireless channels), garage door openers, TV remotes, drones, artificial intelligence, the use of a binary code to create complex commands, and robotics.

Many familiar faces dotted the crowd. The Johnsons and Marguerite Merington attended, as did Westinghouse man Charles Scott, now a professor at Yale, and Edward Dean Adams, the man most responsible for recognizing Tesla for the Niagara Falls enterprise.

The opening speech was delivered by Professor A. E. Kennelly, the former Edison crony, who was now teaching at Harvard. Long an adversary of Tesla, having been active in executing animals with AC current during the heated Battle of the Currents in the early 1890s, Kennelly spoke for fifteen minutes. During this time, the good professor managed to not mention Tesla's name even once.

"We may look forward to a time, say a thousand years hence, when like this evening the one thousand and seventh recipient will receive the

Edison Medal, and once again Edison's achievements will be honored," Kennelly droned on.

As legend has it, Tesla disappeared from the room. Panic-stricken, Behrend ran out of the building to look for him, while Charles Terry, a prominent executive from the Westinghouse Corporation, reviewed Tesla's numerous accomplishments. Finding the lonely inventor across the street in Bryant Park by the library feeding his precious pigeons, Behrend coaxed the reluctant wizard back to center stage.

During Behrend's introduction, he stated, perhaps to counter Kennelly, "The name of Tesla runs no more risk of oblivion than does that of Faraday or Edison. What can a man desire more than this? It occurs to me to paraphrase Pope describing Newton: 'Nature and Nature's laws lay hid in night.' God said, 'Let Tesla be, and all was light.'"

Behrend brought the reluctant iconoclast back to the stage, and in a rare moment, the Serbian wizard reflected on the possibility that a higher intelligence was guiding humanity. "I am deeply religious at heart, and give myself to the constant enjoyment of believing that the greatest mysteries of our being are still to be fathomed. Evidence to the contrary notwithstanding, death itself may not be the termination of the wonderful metamorphosis we witness. In this way," Tesla revealed, "I manage to maintain an undisturbed peace of mind, to make myself proof against adversity, and to achieve contentment and happiness to a point of extracting some satisfaction even from the darker side of life, the trials and tribulations of existence.

"Ladies and gentleman," Tesla continued, "I wish to thank you heartily for your kind appreciation. I am not deceiving myself in the fact of which you must be aware that the speakers have greatly magnified my modest achievements. Inspired with the hope and conviction that this is just a beginning, a forerunner of still greater accomplishments, I am determined to continue developing my plans and undertake new endeavors."

The electrical savant would go on to review much of his life, including an anecdote from his childhood about a gander who almost pulled out his umbilical cord, his early meetings with Edison and work with Westinghouse, lectures in Europe, success at Niagara, and future plans in wireless.

"I have fame and untold wealth, more than this," the inventor concluded, "and yet—how many articles have been written in which I was declared to be an impractical unsuccessful man, and how many poor, struggling writers have called me a visionary. Such is the folly and shortsightedness of the world!" [35]

INVISIBLE AUDIENCE

Tesla was aghast that Boldt had not protected Wardenclyffe adequately, as it was valued at a minimum of at least $150,000. Even though he had signed it over to the hotel, he had done so, according to his understanding, so as to honor his debt "until [his] plans matured." As the property, when completed, would yield $20,000 or $30,000 *a day*, Tesla was simply flabbergasted that Boldt would allow the property to be vandalized and then move to destroy it.

Photos of Tesla's office at this time show numerous papers scattered all over the floor. One can only guess about the valuable documents that were stolen or destroyed. Boldt, and/or "the Hotel Management," saw Wardenclyffe now as theirs, free and clear, even though Tesla offered as proof "a chattel mortgage" on the machinery that the inventor had placed at his own expense. The hotel's insurance was only $5,000, whereas Tesla's covered the machinery, valued at $68,000. Why would Tesla independently seek to protect the property if he didn't still have an interest in it? Tesla saw the contract as "a security pledge," but the paper he had signed did not specify any such contingency. The hotel's lawyer, Frank Hutchins of Baldwin and Hutchins, stated callously, "It was [a] bill of sale with the deed duly recorded two years ago. We fail to see what interest you have." [36]

Storming into the lawyers' offices on Pine Street, Tesla demanded to find out firsthand what was to happen. Tesla's equipment, specialized radio and lighting tubes, and valuable papers and probably blueprints were now missing, and he wanted answers.

"You will have to ask Smiley Steel Company. They are the ones in charge of salvage operations."

J. B. Smiley informed Tesla that, indeed, the tower was to be taken down, its parts sold to cover outstanding debts. "A great wrong has been done," the inventor wrote in reply, "but I am confident that justice will prevail."

"Pay no attention to Tesla whatsoever, but proceed immediately with wrecking as contracted," Smiley told his demolition crew after conferring with Hutchins.

> Waldorf-Astoria Hotel Company
> July 12, 1917
>
> Gentlemen:
>
> I have received reports which have completely dumbfounded me all the more so as I am now doing important work for the

Government with a view of putting the plant to a special use of great moment. . . .

I trust that you will appreciate the seriousness of the situation and will see that the property is taken good care of and that all apparatus is carefully preserved.

Very truly yours,
N. Tesla[37]

With the country still at war and Germany still on a track towards "world domination,"[38] the wizard decided that the only way to save Wardenclyffe was to extol its virtues as a potential defensive weapon for the protection of the country. Capitalizing on the excellent Nobel Prize publicity, the inventor once again strained the reader's credulity with another startling vision.

TESLA'S NEW DEVICE LIKE BOLTS OF THOR

He Seeks to Patent Wireless Engine For Destroying Navies by Pulling a Lever

To Shatter Armies Also

Nikola Tesla, the inventor, winner of the 1915 Nobel Physics Prize, has filed patent applications on the essential parts of a machine the possibilities of which test a layman's imagination and promise a parallel to Thor's shooting thunderbolts from the sky to punish those who had angered the gods. Dr. Tesla insists there is nothing sensational about it. . . .

"It is perfectly practicable to transmit electrical energy without wires and produce destructive effects at a distance. I have already constructed a wireless transmitter which makes this possible."

"Ten miles or a thousand miles, it will be all the same to the machine," the inventor says. Straight to the point, on land or on sea, it will be able to go with precision, delivering a blow that will paralyze or kill, as it is desired. A man in a tower on Long Island could shield New York against ships or army by working a lever, if the inventor's anticipations become realizations.

—*New York Times*[39]

Tesla would not draw up an official paper on the particle beam weapon, or "death ray," for another twenty years, yet it is clear that he had conceived the machine by this time, probably creating prototypes as far back as 1896, when he was bombarding targets with roentgen rays. Tesla "recall[ed] an incident that occurred often enough when he was experimenting with a cathode tube. Then, sometimes, a particle larger than an electron, but still very tiny, would break off from the cathode, pass out of the tube and hit him. He said he could feel a sharp, stinging pain where it entered his body, and again at the place where it passed out. The particles in the beam of force, ammunition which the operators of the generating machine will have to supply, will travel far faster than such particles as broke off from the cathode, and they will travel in concentrations, he said."[40]

Tesla's hope was that by revealing to the world what would come to be called the death ray, the military might come to his aid and help save his precious tower. As he waited hopefully for positive news, he wrote to former investors such as William Crawford of the Simpson-Crawford Department Store to make his case, but like the others, Crawford decided to pass.[41] In "a serious plight," with nowhere else to turn, the inventor contacted Morgan once again to ask for assistance. This was his last chance to protect his wireless patents and save the tower. "Words cannot express how much I have deplored the cruel necessity which compelled me to appeal to you again," the inventor explained, but it was to no avail.[42] He still owed Jack $25,000 plus interest on the turbine work. The financier ignored the entreaty and quietly placed Tesla's account in a bad debt file.

In February of 1917, at the height of the war, the United States broke off all relations with Germany. With the U.S. Navy having surreptitiously placed sensors on Sayville's wireless equipment, U.S. Attorney General Mitchell Palmer was able to confirm that Sayville was a highly successful commercial enterprise doing, in his words "enormous business." However, at the same time, as he learned, it was illegally owned by the enemy and through clandestine means had indeed become an "instrument of warfare on the allies . . . [disseminating through wireless means] propaganda and sabotage instructions . . . to German agents and spies," and thus Palmer seized it.[43] "Thirty German employees of the German-owned station were suddenly forced to leave, and enlisted men of the American Navy have filled their places." Guards were placed around the plant as the high command decided what to do with the remaining broadcasting stations lying along the coast.

With Palmer confirming that the enemy had also stealthily and improperly obtained large tracts of U.S. lands, he moved to seize these properties and place Thomas J. Tunney, head of the Bomb Squad for the New York Police Department, as the lead investigator following the trail of the fifth column throughout the metropolitan area. Creating a task force to uncover spy rings and track down terrorists, Tunney scored an unprecedented coup when he tied would-be assassin Frank Holt (Eric Muenter) to Abteilung IIIb and arrested Paul Koenig, a "detective superintendent of the Atlas Line," a shipping subsidiary for the kaiser.

Running the docks much like a Prussian warlord, Koenig had been recruited by Captains von Papen and Boy-Ed with the official start of the Great War in August of 1914. Early on, Koenig was placed in charge of a New York State network of spies and saboteurs. Main objectives included the planting of bombs aboard ships bound for England; the destruction of tunnels and bridges, particularly those into New York City; the demolition of the Welland Canal; and the use of germ warfare, which in this instance involved inoculating cavalry horses bound for the war with anthrax.

Thinking he was invincible, Koenig was caught off guard with his arrest, and that is how Tunney obtained the German agent's secret codebook. A treasure trove of how and where to send coded messages, Koenig's black book also contained a list of aliases for von Papen, Boy-Ed, and numerous other agents; evidence of a connection between Holt and the detonation of a bomb aboard a ship bound for Great Britain; and a list of operations, many of which fortunately never came to fruition.[44] The key here is that if Holt was being funded by Abteilung IIIb, it meant that the German high command had moved its operations from dirty tricks to targeted assassinations, in this instance, the American megafinancier J. P. Morgan Jr.

19 MORE TAKEN AS GERMAN SPIES

Dr. Karl George Frank, Former Head of Sayville Wireless Among Those Detained[45]

In February, 1917, the American cargo ship the *Housatonic* was sunk off the coast of England, although the crew was allowed to get into lifeboats, and then in April, President Wilson issued a proclamation: "Seiz[ing] all radio stations." Enforcement of the order was delegated to Secretary of the Navy Josephus Daniels.[46] Clearly, an overt decision had to be made about the fate of Wardenclyffe. Tesla's expertise was well known to Daniels and Assistant Secretary Franklin Roosevelt, as they were actively using the inventor's scientific legacy as ammunition against Marconi in his patent suit against them. Coupled with the inventor's astonishing proclamation that his tower could provide an electronic aegis against potential invasions, Wardenclyffe might have been placed in a special category. However, there were two glaring strikes against it. The first was that Tesla had turned the property over to Boldt to cover his debt at the Waldorf, and the second was the transmitter's record of accomplishment: nonexistent. What better indication of the folly of Tesla's dream could there be than the tower's own perpetual state of repose? To many, Wardenclyffe was merely a torpid monument to the bombastic prognostications of a not very original mind gone astray. From the point of view of the Navy, Tesla may have been the original inventor of the radio, but he was clearly not the one who made the apparatus work.

On July 6, 1917, Zenneck was arrested by Deputy Marshall Linford Denny of the Department of Justice. It was just three months since America's entrance into the war. With his advanced knowledge of wireless apparatuses and the fact that he had been part of the campaign that stormed through Belgium, Zenneck was portrayed as "one of the most dangerous German subjects in this country." Along with other dignitaries such as Nobel Prize winner Karl Ferdinand Braun (who would die on American soil at the war's end in 1918) and suspected spies such as Carl Heynen, "a German munitions expert," and Dr. Karl George Frank, the head of Atlantic Communications who had hired Tesla to work at Sayville, Zenneck was handcuffed and taken to Ellis Island and placed behind bars.[47] It had taken a number of years, but finally, the U.S. government had woken up. They uncovered a vast ring of spies and saboteurs and now had the incontrovertible evidence to deport all the German so-called dignitaries who were, indeed, the brains behind all the chicanery. In this manner, Count von Bernstorff and his two henchmen, Captains von Papen and Boy-Ed, were finally kicked out of the country. In actuality, they deserved to be arrested, but they had diplomatic immunity and were thus untouchable.[48]

There is no doubt that many of the people of German descent who were also arrested were indeed part of the spy and sabotage network, and there is no way to know if Zenneck was included in that clique. However, it seems that Nikola Tesla didn't think so, as he made it his business to visit Zenneck while he was interned at Ellis Island, before the suspected spy was shipped to the prisoner of war camp at Fort Oglethorpe in Georgia, where many of the others ended up.[49]

Years later, Zenneck would refer obliquely to this time by stating that Nikola Tesla was one of the nicest gentlemen he had ever known, writing this endorsement two years before being appointed to head the Deutsches Museum, a comprehensive technology museum in Munich. The year was 1933. Now a member of the Nazi party, Zenneck carried out the new civil service statute by firing two Jewish employees at the museum.

While interned at Oglethorpe and under oath, the paymaster of the German spy organization, F. A. Burgemeister, revealed that the kaiser had budgeted literally tens of millions of dollars for covert operations in America. Through Burgemeister, these funds were transferred to von Bernstorff, who dispersed them through Albert to von Papen and Boy-Ed as well as to other spies and German propagandists, such as George Sylvester Viereck.[50]

After the Great War, Tesla resumed his friendship with Frank, who after his release became a "consulting engineer" with offices on Liberty Street not far from Tesla's very first laboratory. Frank tested Tesla's tachometer but remained doubtful that he could help the inventor sell the invention to the German manufacturers whom he was still in touch with. Decades later, on Tesla's seventy-ninth birthday, Frank wished Tesla "heartiest congratulations . . . confident you will see your hundredth birthday as predicted,"[51] and Tesla reciprocated, his "confidence of passing the century mile post still unshaken."[52]

In April of 1917, when the U.S. Navy took over all wireless stations, the key ruling concerning the true identity of the inventor of the radio became neatly sidestepped by the war powers act proclaimed by President Wilson, because it called for the suspension of all patent litigation during the time of war. France had already recognized Tesla's priority via their high court, and Germany recognized him through Slaby's affirmations and Telefunken's decision to pay him a consulting fee, but in America, the land of Tesla's home, the government backed off and literally prevented the courts from sustaining a decision. The Marconi syndicate, in touch with kings from two countries and with equipment instituted on six continents, was simply too powerful.

If we had to identify an American culprit, Secretary of the Navy Joseph Daniels stands out. A lifelong newspaper editor from North Carolina, Daniels was appointed to the position of Navy secretary by President Wilson in 1913. A military leader with pacifist tendencies, as war with Germany became inevitable, Daniels dragged his heels so much that important naval warships were not ready for battle until 1919, which was a year after the war ended. A white supremacist who saw African Americans as intellectually inferior and therefore not worthy of the right to vote,[53] Daniels, who apparently lacked objective abilities, purposely denied compensating Tesla for his contract with the Tuckerton wireless enterprise when the Navy took over the site, even though the Navy continued to reimburse Homag for Rudolf Goldschmidt's alternator.

Perhaps one could say there was too much confusion as to who really was the inventor of the wireless, but that argument falls apart for several reasons. When Homag, the company that ran the Tuckerton plant, asked for reimbursements for monies paid out to Goldschmidt for his alternator, they also asked for reimbursement for Armstrong for his ultra-audion and Tesla for his contract, which stipulated 5 percent of the gross revenues of the company for use of his entire wireless scheme. This worked out to about $15,000 for Tesla on gross revenues of about $300,000 for Homag in 1916. Daniels refused to honor payments, not only to Tesla but to Armstrong as well.[54]

Since Armstrong had specific patents involved in the deal, (patents #1,113,149 and #807,338), Daniels's decision on that score simply doesn't make sense. Homag argued that the Armstrong patents increased reception, particularly in the summer months, and allowed them to earn an additional $2,800 per month. Therefore, paying Armstrong $250 per month for that right made perfect business sense.

In Tesla's case, it is more complicated. Homag, as a German company, had constructed Tuckerton to be sold to the French, and the French Supreme Court had ruled in favor of Tesla over Marconi. Rather than agree to the validity of specific Tesla patents, Homag had decided to instead avoid litigation with the Serbian inventor and simply give him what he wanted, namely 5 percent royalties on gross revenues. Homag had decided that to battle Tesla in the courts would be too costly. They surmised it was simply a "more prudent business decision" to pay him the royalty. However, clearly, they understood that the odds of winning in court against Tesla were unlikely because of the French ruling and that, further, Tesla was also actively suing Marconi at the time of the deal. Since they were a German company, they also knew that Adolf Slaby, one of the founders of the German wireless conglomerate, fully backed Tesla, as did Jonathan Zenneck and Nobel Prize winner Karl Ferdinand Braun, who were both testifying to that specific concern in litigation between Telefunken and Marconi at the very time of this arrangement.

Daniels, of course, knew all this, particularly when it was specifically pointed out to him by Homag's attorney, Crammond Kennedy. And to add to the muddle, Marconi was, at that very time, also suing the U.S. Navy for patent infringement for using Telefunken's equipment, and one of the key reasons that the Navy thought they would prevail was Tesla's priority over

Marconi! This was established by Daniels's assistant secretary of the Navy, Franklin Roosevelt, who stated overtly in a watershed document dated September 14, 1916, that Tesla's work predated Marconi's and, further, that these findings "should be used in forthcoming litigation in which the Government is involved."[55] So Daniels literally wanted it both ways. He would use Tesla's priority to bolster the Navy's claim that the use of Telefunken's wireless equipment on their ships did not infringe on Marconi's patents, yet at the same time, he chastised Homag for paying Tesla a 5 percent royalty when their contract did not specify which particular Tesla patents they were using. One could argue that Tesla's decision to yield to Homag on that point gave Daniels an out, but that argument simply falls apart because Daniels refused to reimburse Homag for their payments to Edwin Armstrong as well, when his patents *were* specified.

With the suspension of all patent litigation and the country in the midst of a world war, Roosevelt, as assistant secretary of the Navy, penned the famous Farragut Letter. This document allowed such major companies as AT&T, General Electric, Westinghouse, and American Marconi the right to pool together to produce each other's equipment without concern for compensating rightful inventors. Further, it "assured contractors that the Government would assume liability in infringement suits."

On July 1, 1918, Congress passed a law making the United States financially responsible for any use of "an invention described in and covered by a patent of the United States." By 1921, the United States Government had spent $40 million on wireless equipment, a far cry from Secretary of the Navy John Davis Long's policy of refusing to pay a few thousand dollars for Tesla's equipment eighteen years earlier. Thus, the Interdepartmental Radio Board met to decide various claims against it, and nearly $3 million in claims were paid out. The big winners were Marconi Wireless, which received $1.2 million for equipment and installations taken over (but not for their patents); International Radio Telegraph received $700,000; AT&T, $600,000; and Edwin Armstrong, $89,000. Tesla received nothing for his own patent #1,119,732 "for the transmission of electrical energy without wires," which had been renewed in 1914; however, he did receive a minuscule portion through Lowenstein, who was awarded $23,000. Under Roosevelt's direction, the government also transferred the Tuckerton wireless property to the French, as they were the ones for which the plant was originally intended. Since the U.S. Navy didn't pay Tesla for use of his patents, most likely, the French didn't either. Disgusted with this wholesale backing of Marconi by the U.S. government, having transferred Wardenclyffe to the Waldorf-Astoria, and now essentially out of funds, Tesla was forced to let the whole thing go.

The United States government, through Roosevelt, *knew* that Marconi had infringed on Tesla's fundamental patents. In fact, it was Tesla's proven declaration that was the basis and central argument that the government had against Marconi when Marconi sued in the first place. However, rather than deal in the midst of war with the truth and with a difficult genius whose present work appeared to be in a realm above and beyond the operation of simple radio telephones and wireless transmitters, Roosevelt, Daniels, President Wilson, and the U.S. Navy took no interest in protecting Tesla's tower.[56]

In July of 1917, Tesla packed his bags and said good-bye to the Waldorf-Astoria. Having lived there for two decades, he talked George Boldt Jr. into allowing him to keep a large percentage of his personal effects in the basement until he found a suitable place for transferring them. "I was sorry to hear about your father," Tesla told the new manager, whose father, George Boldt Sr., had died just a few months before.

Preparing to move to Chicago to work on his bladeless turbines, Tesla was invited to the Johnsons' for a farewell dinner. Soon to become the ambassador to Italy, Robert was now directing the affairs of the American Academy of Arts and Letters, an organization that counted among its ranks Daniel Chester French, Charles Dana Gibson, Winslow Homer, Henry James and his brother William, Charles McKim, Henry Cabot Lodge, Teddy Roosevelt, and Woodrow Wilson. Taking a weekend train, Tesla moved into the Blackstone Hotel, alongside the University of Chicago, as Wardenclyffe waited silently for its ignoble end.

It was an unusual winter, with both rainstorms and snowstorms and a long cold snap that froze the ponds and also created a sheet of ice atop the cupola at Wardenclyffe. Local kids had been sneaking onto the property for years, challenging each other to climb to the top, which rose eighteen stories above the flat land that was Long Island.

On a day for the ages, Dave Madison called on his girlfriend and told her to bring her ice skates.

"Where we going?" Dorie wanted to know.

"It's a surprise," Dave said. Sneaking onto the property, he took her to the base of the tower. She looked way up and then agreed to accompany him on the climb.

When they hit the tenth "floor," Dorie had had enough. She wanted down. It was scary up there, but Dave egged her on. "Trust me," was all he said, an amazing gleam in his eye.

Eight "floors" later, they reached the top, and there it was, a perfect ice skating rink, nearly twenty stories up, high above the landscape, a full fifty-eight feet across.

The young couple put their skates on and undertook this breathtaking experience, one unparalleled in the annals of ice skating lore. From atop this soaring edifice, they could look down to Tesla's mysterious laboratory, still filled with electrical paraphernalia including lathes and broken parts of motors; they could look east or west to the occasional church or hamlet and the hundreds of acres of potato fields, or they could look north, out over the Sound all the way to the Connecticut shores. It was a magical moment skating up there, one they would remember for the rest of their days. Dave wanted to take her again the following weekend, but Dorie's legs were still wobbly, and then it was too late to return.[57]

On Monday morning, the inventor hired a limousine to drop him at the headquarters of Pyle National Corporation. Having already shipped prototypes to Chicago to give workers a head start, now Tesla could work at an intense pace in an entirely new setting, his goal being the perfection of his revolutionary bladeless turbines.

At night, the elder wizard liked to walk across the street from his hotel to the Museum of Arts and Sciences, the only building remaining from the World's Fair of 1893. There, he could stand by the great columns and think back to a time when, daily, hundreds of thousands would stream into this enchanted city powered by his vision. One Saturday, in the heat of summer, he took the mile walk along Lake Michigan, past the Midway to a series of small lakes and a park where once stood the Court of Honor. There, at the entranceway to a spot near Lake Michigan, Tesla learned that the Statue of the Republic, which had been destroyed in a fire, was to be replaced. Although the replica would be nowhere near as high, it was still gratifying to know that the great Chicago Fair would get more of the recognition it deserved. With him was a letter from George Scherff.

August 20, 1917

Dear Mr. Tesla,

I was deeply grieved and shocked when I read the enclosed, but I have the supreme confidence that more glorious work will arise from the ruins.

I trust that your work in Chicago is progressing to your satisfaction.

Yours respectfully,
George Scherff[58]

It was during the height of the world conflagration, when the Smiley Steel Company's explosives expert circled the gargantuan transmitter to place a charge around each major strut, and nail the coffin shut on Tesla's dream. With the Associated Press recording the event and military personnel apparently present, the magnifying transmitter was leveled, the explosion alarming many of the Shoreham residents.

And with the death of the World Telegraphy Center came the birth of the Radio Broadcasting Corporation, a unique conglomerate of private concerns under the auspices of the U.S. government. Meetings were held behind closed doors in Washington between President Wilson, who wanted America to gain "radio supremacy;" Navy Secretary Daniels; his assistant Franklin Roosevelt; and representatives from General Electric, American Marconi, AT&T, and the Westinghouse Corporation. With J. P. Morgan and Company on the board of directors and the Marconi patents as the backbone of the organization, the Radio Corporation of America (RCA) was formed. It would combine resources from these megacorporations, all who had cross-licensing agreements with each other and all who co-owned the company. Here was another entente cordiale reminiscent of the AC polyphase days, which was not so for the originator of the invention. It was a second major time that Tesla would be carved from his creation, with a secret deal probably concocted that absolved the government from paying any licensing fee to Marconi in lieu of them burying their Tesla archives. David Sarnoff, as managing director, would soon take over the reins of the entire operation and morph it, over the next twenty years, into NBC, the National Broadcasting Radio and TV Company. The *New York Sun* inaccurately reported:

U.S. BLOWS UP TESLA RADIO TOWER

Suspecting that German spies were using the big wireless tower erected at Shoreham, L.I., about twenty years ago by Nikola Tesla, the Federal Government ordered the tower destroyed and it was recently demolished with dynamite. During the past month several strangers had been seen lurking about the place. The destruction of Nikola Tesla's famous tower ... shows forcibly the great precautions being taken at this time to prevent any news of military importance of getting to the enemy.[59]

At the end of the war, President Wilson returned all remaining confiscated radio stations to their rightful owners. American Marconi, now RCA, of course, was the big beneficiary. Seeing an opportunity, France decided to give up their wireless plant at Tuckerton and sold it to RCA for a quick profit.

After the true demise of the Wardenclyffe dream, Tesla made several efforts over the next few years to manufacture equipment that he hoped to sell to generate the capital he needed to resurrect the operation. He had a dozen or more other projects and a revenue stream from foreign patents, but in terms of generating the kind of money he needed to rejuvenate his precious tower, ultimately, he failed. At the same time, because the place had been ransacked, the aging wizard lost a large percentage of his records and equipment that had been on site.

In 1920, the Westinghouse Corporation was granted the right to "manufacture, use and sell apparatus covered by the [Marconi] patents." Westinghouse also formed an independent radio station that became as prominent as RCA. At the end of the year, Tesla wrote a letter to E. M Herr, president of the company, a man he had known for a quarter century, about the new developments in the radio industry. "It will probably be news to you that, with the exception of non-essentials, nothing is employed in the art except devices of my invention. But the plants are ridiculously inefficient owing to the fact that the experts are completely misled by the Hertz-wave theory. . . . If you are desirous to inaugurate a wireless system a century ahead of that in use at present, I can put you in a position to do so, provided that your company is willing to come to an understanding with me on terms decidedly more generous than those under which they acquired my system of power transmission thirty years ago."[60]

Unfortunately, Herr wrote back with "regret that under the present circumstances we cannot proceed further with any developments of your activities. I thought it best to let you know this frankly so that you would not be under a misapprehension of our position," signing the letter, "Yours very truly, E.M. Herr, President."[61]

However, this did not stop the Westinghouse Corporation, a few months later, from requesting the lone inventor to "speak to our 'invisible audience' some Thursday night in the near future [over our] radiotelephone broadcasting station."[62]

November 30, 1921

Gentleman,

Twenty-one years ago I promised a friend, the late J. Pierpont
Morgan, that my world-system, then under construction . . .
would enable the voice of a telephone subscriber to be transmitted to any point of the globe. . . .

    I prefer to wait until my project is completed before address-
ing an invisible audience and beg you to excuse me.

Very truly yours,
N. Tesla [63]

11

Tesla's Mysterious 1931 Pierce-Arrow

The transmission of power by wireless will do away with the present necessity for carrying fuel on the airplane or airship. Th[eir] motors ... will be energized by the transmitted power, and there will be no such thing as limitation of the radius of action, since they can pick up power at any point on the globe ... just as trains on tracks are now supplied with electrical energy through rails on wires.[1]

—NIKOLA TESLA,
Reconstruction,
July 1919

During the Roaring Twenties and right up until the time of the Great Depression, Tesla began work on several top secret devices. One of the most controversial was an electric car, a Pierce-Arrow that supposedly derived its power from a distant source.

Tesla's interest in electric cars went all the way back to the 1890s, when he studied Jean Jacques Heilman's steam-driven electric vehicles. Heilman was moving in the direction of creating locomotives for railroads, but the same principle would also apply to automobiles.

In 1904, Tesla wrote a portentous letter to the editor for *Manufacturer's Record* wherein he noted that, except for Heilman, steam-driven locomotives and automobiles "are still being propelled by the direct application of steam power to shafts or axles." In other words, steam is created to power the axle and that causes the wheels to turn. What Tesla suggested was that the steam generated should be used to power a dynamo instead and that this dynamo, in turn, would be hooked up to an induction motor to power the car electrically, which would result in a gain in efficiency of somewhere between 50 to 100 percent. "It is difficult," he concluded, "why a fact so plain and obvious is not receiving more attention from engineers."[2]

Almost one hundred years to the day of this letter, Martin Eberhard and his partner Marc Tarpenning started Tesla Motors, and their inspiration was precisely the substance of this letter to the editor, namely that cars could be run by electric motors instead of gasoline. The big difference between Eberhard and Tarpenning's roadster and the substance of Tesla's letter is the source of power. In Eberhard and Tarpenning's case, their roadsters would run on lithium batteries, whereas Tesla is suggesting here to run the car on steam. In honor of this modern electric car, Eberhard and Tarpenning named the company Tesla Motors, and shortly thereafter they sold the enterprise to Elon Musk, who inherited and decided to keep the name.

At this juncture, Tesla was talking about a steam-driven electric car; however, he was well aware that electric cars could also be powered by batteries. One of the first battery-powered electric vehicles can be traced back to the Austrian inventor Franz Kravogl, who displayed a two-wheeled contraption at the World Exposition in Paris in 1867. By the early 1900s, there were many battery-operated electric automobiles, and Thomas Edison even drove one as late as 1913. However, there were two main problems: limitation in range and the lack of recharging stations. So over time, electric vehicles tended to be restricted to gated communities and golf courses. The great efficiency and attractive price of Henry Ford's Model T changed the dynamics considerably, particularly when the Machiavellian iconoclast bought up and pulverized the mass-transit railways, as he did in Los Angeles. Having perfected mass production, low-cost gasoline-powered automobiles became the mainstay.

However, Tesla never abandoned the idea of using an induction motor to propel a car, and he talked frequently of the idea of running such vehicles by means of wireless power derived from broadcasting towers much like Wardenclyffe. In 1927, Tesla traveled up to Niagara Falls to confer with Francis Fitzgerald of the Niagara Power Commission to discuss with him the creation of a wireless transmission tower to be set up at the Falls with the backing of the Canadian Power Commission.[3] According to legend, three years later, Tesla met with Heinz Jerbens, director of the Inventor's House from Hamburg, Germany, on the director's trip to New York. Jerbens had taken an ocean liner to the States to meet with Edison. "On board ship. Heinz Jerbens met a former Serbian-Austrian officer Peter Savo, who introduced himself as a relative of Nikola Tesla. During a dinner at a restaurant on the ship, Peter Savo insisted that the German director should meet Nikola Tesla, claiming he was the greatest inventor in America. Thus, after meeting with Edison, Jerbens apparently hooked up with Tesla on November 26, 1930 [in the lobby of the] Waldorf-Astoria hotel in New York."[4]

Sworn to secrecy, Jerbens was taken by Tesla to Buffalo to see his top secret vehicle, a standard Pierce-Arrow. From an email from Jerbens's son to the author, we learn that the original motor was removed and, instead, the car was "installed with an 80 HP-electric motor" that replaced the "fuel engine."[5] The new motor possessed a six-foot-tall aerial that somehow drew power as the car moved along. Tesla attached a converter box comprising a dozen radio tubes, "24 resistors and diversified cables," which, it must be assumed, derived its energy from a secret transmitter he had supposedly set up near Niagara Falls. After "inserting two rods into the engine," the car was ready to roll, achieving, according to Jerbens, speeds in excess of ninety miles per hour.

If the story is to be believed, it would mean that Tesla had to have installed a wireless transmitter, most likely close to the power station by the Falls, and the transmitter would have had to be turned on during the demonstration and, of course, turned off afterwards. Also, the vehicle would have had to be housed in some nearby location and a driver hired to take them out on a test run.

There are two key problems with the story. Assuming Jerbens really came to America in 1930, it would have been impossible for him to meet Tesla in the lobby of the Waldorf-Astoria, as the building was torn down in 1929 to make way for the Empire State Building, and the new hotel was constructed in 1931. There was no Waldorf-Astoria in 1930. It's certainly possible that if indeed Jerbens met Tesla at a posh hotel, through the years, the actual hotel could have morphed into the Waldorf. However, this is a clear aspect to the story that could not be true. The second problem concerns the actual car. According to the story, Tesla would have had to replace the combustion engine with a powerful induction motor along with some type of receiving device that would pick up the power from the transmitter. It was certainly within Tesla's wheelhouse to undertake this venture; however, Buffalo is quite a distance from New York City and thus it would have undoubtedly taken several trips to the Niagara Falls region for Tesla to construct the transmitter and install the equipment necessary inside the automobile. So far, no evidence has arisen that suggests that Tesla commuted to the Buffalo region at this time. In addition, since this vehicle has never been located and since no one in the Buffalo region has ever come forward to confirm the story, tremendous doubt remains as to whether or not this wireless experiment with the automobile ever took place.

12

Telephotography

1931

Greetings to Professor Tesla,

*Sending you my . . . heartiest congratulations for the great achieve-
ments due to you in the fields of physics and electrotechnics. I hope
you will be pleased by my pointing out anew how Tesla currents were
useful in the first stage of phototelegraphy. An evacuated tube, nowadays neon tubes [was] made luminescent by Tesla currents sent as rays
through a small window on the receiving photographic paper, and
the Tesla currents were modulated by the signals arriving from the
transmitting station.*

*The first photo ever sent over a telegraphic line (Munchen-
Nurnberg-Munchen) by the aid of a photoelectric cell in the transmitting station was received in 1904 in this manner.*

This was the beginning of modern phototelegraphy.

Yours very truly,
Arthur Korn[1]

Aside from the idea of wirelessly sending faxes of typed copy, such
as the reproduction of a newspaper to ocean liners, Tesla also
planned to transmit photographs and, it seems, even color reproductions.
Drawing inspiration from the structure of the eye and his work from the
mid-1890s in projecting X-ray images of skeletal tissue, Tesla described in
1901 the basis for transmitting faxed images and the precursor to television. Tesla's idea was to create two similar plates, one for transmitting and
one for receiving the image.

On May 30, 1901, Tesla wrote, "When an image is projected upon . . .
these cells, the corresponding transmitter will . . . produce variations of
density in the oscillations proportionate to the action on the cells. The

image must resemble the picture projected . . . because the receiver plate is composed of elements arranged in the same manner. Effect of color . . . is [also] certainly practicable."[2]

As he was a student of history, there is little doubt that Tesla was aware of James Clerk Maxwell's circa 1855 ideas on color photography based on the principles of color vision. In fact, the ability to create color photography had been around for several decades by the turn of the century. Tesla had also studied the concepts behind the transmission of pictures over telegraph lines.

In 1880, having studied the work of Londoner A. C. Brown on transmitting speech by way of light rays, Alexander Graham Bell duplicated and improved upon this invention so that he could invent and patent a photophone which transmitted spoken sounds by means of a beam of light several hundred feet to a receiving device. A forerunner of fiber optic telephone lines, in a recent Zoom conversation with Vint Cerf held in his honor by the Tesla Science Center at Wardenclyffe, Vint pointed out that Bell also invented a way to improve upon Edison's quadraplex design, which expanded the transmission of four conversations along the same transmission lines to ten or more channels. Tesla, of course, was well aware of Bell's wireless photophone and Bell's and Elisha Gray's ability to disseminate multiple conversations along the same transmission line, and these inventions played a significant role in helping him develop his own creations in wireless transmission, which he expanded to virtually unlimited channels for transmitting voice and pictures by multiplying frequencies, laying the groundwork for what became radio, TV and cell phone technology.[3]

In 1905, the *Severance News* reported on page 1 of their July 14 issue that "the picture telegraph of Dr. Korn of the University of Munich has been so perfected that in ten to twenty minutes a photograph 4 × 7 inches in size can be sent through a resistance corresponding to one thousand miles. The portrait or design to be transmitted is on a transparent film, which is wound around a glass cylinder and upon which a lens focuses a point of light that passes through the film to a selenium cell in the cylinder. The bright and dark portions . . . [cause] the ray of light to vary the resistance of the selenium cell to an electric current passing through it, and this variation produces a corresponding instantaneous brightening [or] darkening of the glow in a Tesla vacuum tube at the receiving end of the wire . . . giving a new photograph accurate in minute detail."

In Tesla's 1920 article on the history of telephotography, he traces "the original idea . . . to Alexander Bain, a Scotch mechanician, who secured a

British patent disclosing the invention in 1843. His plan contemplated the transmission of printed letters, drawings and pictures" by creating a grid of pixels on both the transmitting end and the receiving end, where a chemical paper lying on a metal plate would respond to each pixel being electrified. This led to efficient fax machines that appeared in Electricity Hall at the Chicago World's Fair of 1893. By combining this idea with Tesla's brush radio tube that could be moved by changes in the electromagnetic field at many times per second, the first major hurdle was overcome in the development of the wireless transmission of still and moving pictures. Using Tesla currents, a variation of a Tesla brush tube, and the proper grid and selenium cells, the first step was taken when photographs were transmitted along transmission lines by Arthur Korn in 1904. Korn, a German scientist, was quite forthright in stating that his success was based on the use of Tesla continuous wave currents.[4]

Through the years, Tesla extended his telephotography system to a realm that is usually associated with science fiction, conceiving of a typewriter that would be "electrically operated by the human voice,"[5] and eighteen years later, of a machine that would read a person's thinking process.

An article from the September 10, 1933, issue of the *Kansas City Journal-Post* reported the following:

> "I expect to photograph thoughts," announced Mr. Tesla in the same tone of voice that a person occupied with some trivial things in the scheme of life might announce that it was going to rain. . . .
>
> Mr. Tesla points out that the ideas of television and radio and airplane were scoffed at in their infancy. . . .
>
> "In 1893, while engaged in certain investigations, I became convinced that a definite image formed in thought must, by reflex action, produce a corresponding image on the retina, which might possibly be read by suitable apparatus. This brought me to my system of television, which I announced at that time.
>
> "My idea was to employ an artificial retina receiving the image of the object seen, an 'optic nerve' and another such retina at the place of reproduction. These two retinas were to be constructed after the fashion of a checkerboard with many separate little sections, and the so-called optic nerve was nothing more than a part of the earth.

"An invention of mine enables me to transmit simultaneously, and without any interference whatsoever, hundreds of thousands of distinct impulses through the ground just as though I had so many separate wires . . . [and] a scanning apparatus or a cathodic ray, which is a sort of moving device, the use of which I suggested in one of my lectures.

"Now if it be true that a thought reflects an image on the retina, it is a mere question of illuminating the same property and taking photographs, and then using the ordinary methods which are available to project the image on a screen.

"If this can be done successfully, then the objects imagined by a person would be clearly reflected on the screen as they are formed, and in this way every thought of the individual could be read. Our minds would then, indeed, be like open books."[6]

Coincident with Tesla's seemingly radical idea of photographing thought, in the present day, Arnav Kapur, a graduate student at MIT's Media Lab, has actually achieved this amazing goal by inventing a device he calls AlterEgo. The problem with Tesla's idea was the extreme difficulty, if not impossibility, of hooking up some type of receiving device to the retina. Kapur solved this problem by placing his apparatus along the neck. "The device intercepts electrical signals which the brain normally sends to vocal cords and sends that information to a computer." At the same time, Kapur has also rigged this setup to a transmitter attached to his inner ear so that he has both output and input as well as the ability to produce visually the product of his thoughts and the answers that come from the hyperlink onto a big computer screen.

On the April 22, 2018, edition of *60 Minutes*, reporter Scott Pelley asked Kapur a complex mathematical problem that would be next to impossible to compute in one's head. Kapur thought out the problem mentally. In the process of envisioning the words corresponding to the numbers involved, this signal was sent to his vocal cords/voice box then intercepted by his invention, and the question was then sent silently to an Internet search engine, most likely Google. Kapur's genius lay in his ability to create an interceptor that could translate the "thought" of the words of each number, e.g., twenty-seven times three thousand and forty-two, and this question was then sent to a search engine. The answer to the mathematical question was retrieved by the computer, and the answer was then converted as a verbal response to a hearing aid attached to his inner ear, within a matter of seconds giving Kapur the answer—which he correctly relayed.

Pelley then asked Kapur to name the capital of Bulgaria and report the population there. Kapur then *thought the words* in his brain and then the signal was sent along his neck en route to his vocal cords where it was again intercepted and sent to a search engine. Kapur almost immediately received the answer, which was sent directly to his inner ear via his unobtrusive microphone, and Kapur replied, "Sofia, population 1.1 million," and he was correct.[7]

This brilliant, potentially Nobel Prize–worthy invention is conceptually similar to Tesla's idea. The main difference was that Tesla wanted to intercept signals to the retina, whereby Kapur came up with a much more doable contraption that intercepted the signals of the words being sent to the vocal cords, and *voila!,* with a direct connection to the Internet and an input to his ear, after simply *thinking* the question, the answer was covertly and elegantly given to him. It seems to me that as this device is perfected, it could become more and more miniaturized and proficient over time, and if introduced to children, we would literally spawn a new race of individuals who would have immediate access to a vast amount of information in a way that would seem almost miraculous.

Aside from being able to find almost immediately the answer to just about any mathematical question or other problem, a person could also almost seamlessly converse with someone who speaks an entirely different language even though the person wearing AlterEgo might not know or even recognize a single word from that other language.

PART IV

Death Ray

By scientific application, we can project destructive energy in thread-like beams as far as a telescope can discern an object. The range of the beams is only limited by the curvature of the Earth. . . . Should you, say, send in 10,000 planes or an army of a million, the planes would be brought down instantly and the army destroyed. The plane is thus absolutely eliminated as a weapon, it is confined to commerce, and a country's whole frontier can be protected by one of these plants producing these beams every 200 miles.

—Nikola Tesla quoted in
"Beam to Kill Army at 200 Miles,
Tesla's Claim on 78th Birthday,"
by Joseph Alsop,
New York Herald Tribune,
July 11, 1934

13

The Day Tesla Died

"I've been feeding pigeons, thousands of them for years. . . . But there was one, a beautiful [female] bird, pure white with light grey tip[ped] wings; that one was different. . . . I would know that pigeon anywhere. I had only to wish and call her and she would come flying to me. She understood me, and I understood her.

"I loved that pigeon . . . as a man loved a woman, and she loved me. When she was ill, I nursed her back to health. . . . The joy of my life, if she needed me, nothing else mattered. As long as I had her, there was purpose to my life."

—NIKOLA TESLA,
circa 1942

This remarkable story was told by Tesla to his biographer and *Herald Tribune* reporter John O'Neill, who informed the reader that if he had not been accompanied by his associate, *New York Times* reporter William Laurence, "I would have convinced myself that it was nothing more than a dream." Having known Tesla for roughly forty years, O'Neill commented that although Tesla planned to "engineer the complete elimination of love," en route to becoming a "self-made superman," Tesla left "unguarded . . . a channel" that revealed the old wizard's true need for love. Now resembling a walking skeleton, Tesla concluded his story as follows:

"Then one night as I was lying in the dark, solving problems as usual, she flew in through the open window and stood on my desk. . . . I knew she wanted to tell . . . [that] she was dying. . . . And as I got her message, there came a dazzling light . . . from her eyes . . . more intense than any I had ever produced in my laboratory. When that pigeon died, something went out of my life. . . . I knew my life's work was finished."[1]

In December of 1939, the first Canadian Infantry Division commanded by General Andrew McNaughton was shipped over to England to help defend the coast from a possible German invasion. Earlier that year, Germany had invaded Poland, and the year before that, Hitler had taken Austria and the Sudetenland. After Canadian authorities conferred with the newly elected British prime minister, Winston Churchill, the Second Canadian Infantry Division also arrived. The day Churchill came to power, May 10, 1940, was the very same day Germany launched a highly successful invasion of Belgium and France, marching triumphantly into Paris just one month later.

The following year, the Nazis went east and invaded the Soviet Union, and six months after that, on December 6, 1941, Pearl Harbor was attacked. Along with Japan, Germany declared war on the United States, upping the European conflict into a full-blown world war, and a few weeks later, on January 19, 1942, a German submarine sank a Canadian ocean liner, the RMS *Lady Hawkins*, off the American coast at Cape Hatteras, killing more than 250 people, with 71 surviving.

Just two months later, at the height of the conflagration, McNaughton, now a close ally of Churchill's, traveled to Washington to confer with President Roosevelt.[2] McNaughton, who had made the cover of both *Life* and *Time* magazines, had been in extensive top secret discussions for several years with Nikola Tesla concerning the construction of his particle beam weapon as a means for protecting the British Empire. Whether or not the topic of Tesla and his weapon came up in his discussions with the president is not known; however, just a few months later, on January 2, 1943, Roosevelt directed his minions to contact the head of Columbia University "to get the low-down on Dr. Nikola Tesla," ostensibly to set up a meeting. His longtime appointments secretary, Marvin H. McIntyre, located Tesla's extensive biography in *Who's Who* and contacted the secretary to the president of Columbia, Nicholas Butler, who stated that "he was quite familiar with the case of Dr. Tesla." According to the secretary, "The old gentleman had never taught at Columbia, and . . . confidentially, they looked upon him much more prominently in newspapers and magazines than in scientific circles." McIntyre, who thirty-five years earlier had been editor of the *Washington Post*, an aide to Franklin Roosevelt when he was assistant secretary of the Navy during the Great War, and then campaign manager when Roosevelt began his quest to become president, questioned the president. "Frankly," McIntyre stated bluntly, "I can't see any reason or justification for your [probe]."[3]

The reason was most likely, information provided to President Roosevelt from his wife, the first lady. On January 2, Roosevelt sends a memo to Columbia University to get the "low down" on Nikola Tesla, most likely because Roosevelt was interested in discussing with Tesla his particle beam weapon. Five days later, Tesla died and two days after that, Roosevelt sent the following memorandum to Mrs. Roosevelt: "I was having this looked into but the paper, yesterday, carried the story that Dr. Tesla died. Therefore, I am returning the enclosures herewith." What all of this suggests is that it was Eleanor's prompting that caused the president to look into the military implications of the inventions of Nikola Tesla.[4]

Just two weeks prior to Roosevelt's inquiry about Tesla, on December 16, Roosevelt had obtained an ominous memo from Vannevar Bush, head of the Office of Scientific Research and Development, which was responsible for new weapons development, including the Manhattan Project. Bush reported that his head physicist, Leo Szilard, feared that "Germany must surely lead in the [atomic] bomb race . . . [and thus] it is quite possible that [they] may well be able to produce bombs sooner than we can."[5] The implication was that the White House was now seriously considering learning more about Tesla's electronic shield between nations, and it is likely that this information was so secret that even McIntyre was kept out of the loop.

Simultaneously, Abram Spanel, CEO of the Latex Corporation, was making inroads to contact Vice President Henry Wallace about Tesla, attempting to arrange his own tête-à-tête, while also trying to set up a meeting between the old wizard and one of his greatest admirers, electrical engineer Bloyce Fitzgerald. Now a private in the U.S. Army, working and soon to graduate from "the Detroit College of Applied Science with a degree in Radio Mechanical Engineering," Fitzgerald's service record lists him as being "a consulting engineer for special weapon armament projects, rocket guns, hydraulics and gun turrets" under the auspices of the NDRC (National Defense Research Committee).[6] With this background, Fitzgerald was able to attend classes at MIT and meet with Military Intelligence in Washington, and higher-ups at Wright-Patterson Air Force Base in Ohio, with his goal being to develop the particle beam weapon based on designs provided by Tesla.[7] Little did all concerned know that Tesla was losing ground rapidly, continuing to mourn the death of his beloved White Pigeon, hallucinating daily, and calling on messengers to send hundred-dollar bills to Mark Twain, even though the venerated author had been dead over thirty years.

> The President and I are deeply sorry to hear of the death
> of Mr. Nikola Tesla. We are grateful for his contribution
> to science and industry and to this country.
>
> —ELEANOR ROOSEVELT,
> January 12, 1943

Five days after the Roosevelt memorandum, Alice Monaghan, a floor maid
for The New Yorker Hotel, went past the "Do Not Disturb" sign on Tesla's
door and entered. Except for a pair of leggings, the inventor was completely
nude, lying dead on his bed. He was eighty-six years old. Dr. Palmer, who
filled out Tesla's death certificate, wrote in the coroner's report that Tesla
died "suddenly . . . while asleep," and further, that there was "nothing
suspicious." [8]

Described as having been in failing health the last two years of his life,
the inventor had displayed a cadaverous appearance throughout most of
the previous year, due, most likely, to a restricted diet of vegetables and a
thin soup prepared specially for him by a hotel chef. The help at the hotel
described the inventor as a peculiar man who constantly washed his hands
and would not allow any of them to come within three feet of him. [9]

Rumors persisted that Tesla might have been murdered by Nazi saboteurs posing as OSS agents who were after the inventor's particle weaponry
papers. Names were even dropped, such as Otto Skorzeny, an SS comman-
dant and Hitler's "go-to guy" for helping Mussolini escape and other clandestine ops, including a failed attempt to assassinate the Big Three: Stalin,
Churchill and Roosevelt. [10] However, anyone who had seen the emaciated
inventor in the last few weeks and days of his life would have realized the
simple truth: Tesla died of old age.

> The greatest inventor who ever lived on this Earth died here in
> New York a week ago last night. That is the opinion of William
> Leonard Laurence, the *Science News* editor of the *New York
> Times*. In fact, Mr. Laurence told me that no other inventor in
> all history could even hold a candle to this man I am talking
> about. His name is Tesla . . . Nikola Tesla; and not one person
> in forty thousand ever heard of him; yet . . . few other men
> in all history have done so much to change civilization. His
> inventions ushered in the electrical age.
>
> —Radio broadcast on station WJZ, New York,
> by Dale Carnegie, January 27, 1943

Although Tesla was often portrayed as having slipped into obscurity, this generally accepted portrayal was contradicted by the fact that over two thousand people attended his funeral. The service, conducted in Serbian by prominent Greek Orthodox priests at the Cathedral of St. John the Divine, was attended by "inventors, Nobel Prize winners, leaders in the electrical arts, high officials of New York and the Yugoslav government, and men and woman who attained distinction in many other fields." Honorary pallbearers included "Newhold Morris, President of the City Council; Dr. Ernest F. W. Alexanderson of the General Electric Company, inventor of the Alexanderson alternator; Professor Edwin H. Armstrong of Columbia University, inventor of . . . many important radio devices [e.g., FM and AM radio]; Dr. Harvey C. Rentachler, Director of Research Labs for Westinghouse; Gano Dunn, President of the J. G. White Engineering Corporation [who had helped Tesla display some of his first experiments in wireless at Columbia University back in 1891]; Colonel Henry Breckenridge [former assistant secretary of war under Woodrow Wilson, fencing champion, and FDR opponent in a presidential campaign]; David Sarnoff [president of RCA and what became NBC TV, soon to be named reserve brigadier general of the Signal Corps]; D. M. Stanoyevitch, Counsel General Prime Minister of Yugoslavia; and William H. Barton, Curator of the Hayden Planetarium." [11]

WESTERN UNION TELEGRAM:

Dr. Tesla's great inventions have placed the whole world in his debt. Physicist[s] in particular. Will deeply regret his irreparable loss.

—JAMES FRANCK,
University of Chicago

A floral wreath from King Peter II of Yugoslavia was presented along with telegrams of condolence from such dignitaries as Vice President Henry Wallace, Eleanor Roosevelt on behalf of the president, and the Nobel Prize winners Robert Millikan and three members of the top secret Manhattan Project: James Franck, Arthur Compton, and their leader, Ernest Lawrence, who wrote, "I along with scientific colleagues the world over show a feeling of loss in the passing of Nikola Tesla who contributed so greatly to electrical science and technology." The service was topped in spectacular fashion with a radio broadcast from the inimitable Mayor Fiorello LaGuardia, who read over the airwaves in his iconic lilting and impassioned voice a eulogy written in flowery fashion by one of Tesla's greatest admirers, Louis

Adamic, a Croatian-born poetic writer, champion for celebrating diversity in America, and winner of a Guggenheim Fellowship.

> On last Thursday night, here in our city of New York, a man who was 87 years of age died in his humble hotel room. His name was Nikola Tesla. He died in poverty, but he was one of the most successful men who ever lived. His achievements were great and are becoming greater as time goes on. Nikola Tesla could have amassed hundreds of millions of dollars, could have become the richest man . . . in the world, if he wished for riches. He did not. He did not . . . have time for anything that spells success for too many people. . . . A great humanitarian, a pure scientific genius, a poet in science, he did extraordinary, amazing, miraculous things during his life amongst us. He did them simply to serve mankind, and for his services he did not charge anything. Money—he did not care for it. Honor—who was anybody to honor anybody else? That was his attitude. Gratitude—he did not expect or demand it. Nikola Tesla did not care to be paid for anything he did for the human race. . . .
>
> But Tesla is not . . . really dead. Only his poor, wasted body has been stilled. The real, important part of Tesla lives in his achievement, which is great, almost beyond calculation and an integral part of our civilization, our daily lives, our current war effort.[12]

Based on a look at Tesla's pallbearers, an interesting window into Tesla's later life emerges. Clearly, he must have spent time at the Hayden Planetarium, looking up at the stars. Tesla also liked to peruse Grand Central Terminal because, much like the planetarium, the ceiling of that great rotunda was adorned with the constellations mapped out as such mythological gods as Orion, Taurus, Cassiopeia, Gemini, and so on. Simultaneously, he must have also conferred with the planetarium's curator, William Barton and, no doubt, talked about the concept of the plurality of worlds.

The fact that Colonel Henry Breckenridge was also a pallbearer supports the hypothesis that like Breckenridge, who actually beat Roosevelt in a New Jersey primary, and like another longtime friend, William Randolph Hearst,[13] the inventor too was vehemently opposed to Roosevelt's New Deal. Very much in antithesis to the image portrayed by Louis Adamic, Tesla was an elitist who eschewed the common man. In 1934, in a letter to J. P. Morgan Jr., who himself detested the "class traitor" so much that aides cut Roosevelt's photo out from the morning newspaper "for fear that it would spike his blood pressure,"[14] Tesla alludes to an impending deal he had with the Russians to sell them a weapon that would help protect them from a Japanese invasion. "Some years ago," he wrote, "Lenin made [me] a tempting offer to come to Russia, but I could not tear myself away from my laboratory work."

Writing in part, on the one hand, to try to "square [my] account with you," that is, pay back the money Morgan's father had loaned Tesla to create Wardenclyffe, the inventor was also seeking funds to help in his various endeavors. "Mr. Morgan," Tesla wrote, "you are still able to help an undying cause, but how long will you be in this privileged position? We are in the clutches of a political party which caters openly and brazenly to the mob and believes that by pouring out billions of public money, still unequalled, it can remain in power indefinitely. The democratic principles are forsaken and individual liberty and incentives are made a joke. The 'New Deal' is a perpetual motion scheme which can never work, but is given a semblance of operativeness by unceasing supply of the people's capital. Most of the measures adopted are a bid for votes and some are destructive to established industries and decidedly socialistic. The next step might be the [re]distribution of wealth by excessive taxing if not conscription."[15] Breckenridge, also a strong opponent of the New Deal, could not have said it better. So there is little doubt that Tesla and Breckenridge must have spent time together, as elitists do, ruminating about the problem that Roosevelt was still in power.

Obtaining a true portrait of Tesla is not an easy task. Like many individuals, he was a man of contradictions that throw a monkey wrench into the rose-colored portrait painted by Adamic.

Under the heading "Unknown Subjects, Equipment, Experiments and Research of Nikola Tesla, Deceased. Espionage," Special FBI agent Percy Foxworth, chief of Special Intelligence and liaison to the British Security Office, wrote a detailed communiqué on the events surrounding the death of Nikola Tesla, drafted just three days before Foxworth himself was killed in an airplane crash on a secret mission to North Africa at the special request of General Dwight D. Eisenhower.[16] He was just thirty-seven years old.

The FBI memorandum reads, in part:

> On the night of January eight, Sava Kosanovic, George Clark and Kenneth Swezey visited Tesla's hotel with a representative of Shaw Walker Co., in order to open the safe in the room of

Tesla . . . to search for a will . . . in the presence of three assistant managers of Hotel New Yorker as well as representatives of the Yugoslavian Consulate, identities not yet known.

After the safe was opened, Swezey took . . . a book containing testimonials sent to Tesla on the occasion of his 75th birthday . . . arranged by Swezey. Kosanovic took . . . three photographs. . . .

According to managers . . . nothing else was removed. . . . The safe was then closed under a new combination . . . now in possession of Kosanovic.

On January nine, Gorsuch and Fitzgerald of Alien Property Control went to Hotel and seized all the property of Tesla consisting of about two truckloads . . . and transferred [the material] to the Manhattan Storage & Warehouse Co., New York where [there was also] approximately thirty barrels and bundles belonging to Tesla which had been there since about 1934. These have also been sealed and are now under the orders of Alien Property Custodian. . . .

Alleged by informant Fitzgerald . . . [Tesla had a] working model of an invention in a safe deposit box in Governor Clinton Hotel in New York. Inquiry shows that this was placed there by Tesla in 1932 as security for $400 owed Hotel. This bill is still owed and Hotel appears unwilling to release this property . . . until debt is paid. . . .

Concerning Tesla, Hotel managers report he was very eccentric if not mentally deranged during past ten years, and it is doubtful that he has created anything of value during that time, although prior to that he probably was a very brilliant inventor. Therefore any notes of value were probably those made prior to that time. . . . In view of the fact that the notes and other material would be highly technical . . . [they] could not be reviewed except by a trained person from the Office of Scientific Research Development [who] might be interested.[17]

In August of 2017, I toured the Hotel New Yorker as part of a segment for *The Tesla Files* TV show, accompanied by physicist and rocket scientist Dr. Travis Taylor, investigative reporter and ex-military man Jason Stapleton, and former hotel maintenance manager Joe Kenny. Some interesting facts emerged from the investigation. Kenny repeated the rumor that members of Military Intelligence and perhaps the FBI may have had their own apartments and were monitoring Tesla just several rooms away.[18] Further, as a unique feature, the hotel had in its basement its own generator, the largest private electrical generator on the planet, and thus was able to power its entire forty-one floors, encompassing 2,500 rooms, independent of the power grid. During the Great Blackout of 1965, although the entire Northeast was electronically wiped out, the Hotel New Yorker kept humming. With 35 chefs, 150 laundry workers, 95 switchboard operators, and the largest barber shop in the city, the Hotel New Yorker was a thriving metropolis, its own self-contained world. Kenny mentioned that a photograph of J. Edgar Hoover was used in an advertisement promoting the hotel. At the time it was opened, throughout the 1930s and 1940s and even into the modern era, the Hotel New Yorker remained one of the posh "in" places to be. Numerous celebrities, such as baseball player Joe DiMaggio; boxer Muhammad Ali; actors William Powell, Spencer Tracy, and Joan Crawford; and numerous politicians including President Kennedy stayed there. Benny Goodman and other big band leaders played in its ballroom during their heyday, and with a million square feet of space, the hotel also boasted an ice rink and its own theater.

Tesla had a two-room suite on the thirty-third floor. Since the quirky inventor often did things in threes, such as circling a block three times before entering a building, various Tesla experts have made the case that Tesla chose his room, 3327, because it was divisible by three. Kenny pointed out in our elevator ride up to Tesla's room that $3^3 = 27$. Tesla's bed, desk, and bath were in 3327, and his filing cabinets, safe, and probably his pigeon coops were in 3328.

What impressed me about his main room was the magnificent view he had of both the Chrysler and Empire State Buildings out his east window. Since the Empire State Building was constructed on the site of the original Waldorf-Astoria, where Tesla lived for nearly twenty years, one can only imagine the thoughts and memories that were most likely triggered whenever the aging wizard gazed out his window.

Tesla was living at the Hotel New Yorker in the run-up to and during the height of World War II, and it was apparent that he was residing in a very modern building that would without doubt be attractive to military personnel. If there had been a blackout or some type of attack on New York and the power grid went out, what better place would a military unit want to be than in a building able to withstand such a catastrophe? Taylor speculated that Tesla may have had a work space in the building, and Kenny intimated that the penultimate forty-first floor, which was closed off, may have offered such a possibility. There beyond a complex room typical for telephone operators (now, of course, out of service for decades), was indeed a sizeable empty corner room three or four steps away from access to the roof which certainly could have been a place for Tesla to run his experiments.

By coincidence, at this very same time, I located a newspaper article published in 1935 about a pigeon that flew into the fortieth or forty-first floor, and when the management arrived they noticed that it was an International Federation homing pigeon with a leg-band reading "I.F. 34 N. U. 1283." Nikola Tesla was called on, and he took the pigeon to his rooms and cared for it.[19] This led to the speculation that Tesla could have been working clandestinely with the War Department all along and that his interest in pigeons may have served a hidden purpose as well, namely to pass messages to a military liaison or perhaps raise and care for homing pigeons for the military.

Another interesting feature of the building was the classy Art Deco–tiled tunnel two levels below the main floor that led to the New Yorker's own subway stop, which suggested that, undetected, Tesla could easily drop down to this bottom floor and have direct access to Penn Station, the Empire State Building, or any number of other hotels where the wizard kept a variety of trunks and safety deposit boxes.

When I wrote *Wizard*, I was well aware that a government dossier of Tesla's effects known as the Trump Report was completed just two weeks after Tesla's death. Written by MIT professor John G. Trump, the dossier stated overtly that Tesla had sold the details of his particle beam weapon to the Soviet Union via the Amtorg Trading Corporation for $25,000 and that Tesla also attempted to sell the weapon to the British.

I also knew that by 1942, Tesla was indeed working with the U.S. military, because I had interviewed a somewhat mysterious World War II veteran, Ralph Bergstresser, who told me outright that the aging inventor had lent him various documents, including the particle beam weapon paper, so that Bergstresser could make copies to help the country in the war effort.

Bergstresser stated that he knew Tesla only in the last six months of the aging wizard's life. Claiming that the wizard's mind was sharp to the very end, he felt that Tesla "stepped over" to the other side at death. During our conversation, Bergstresser recalled he would go to Tesla's apartment, borrow the papers, make copies, and return them several days later. He also claimed to have been the last person to see the inventor alive. This would therefore date his connection to Tesla from about June of 1942 until the inventor's death in early January 1943.[20] Several years earlier, on July 12, 1940, Tesla told the *Baltimore Sun* in a detailed article on his particle beam weapon that "all my inventions are at the service of the United States Government." This offer was repeated in several other newspapers. "But," Tesla icily added, "there must be no interference by experts."

This 1940 piece gained more prominence when another rendition of it was published in the *New York Times*. Written by the esteemed science reporter William Laurence, who just four months earlier had penned a front-page shocker on the discovery at Columbia University that one pound of uranium-235 contained the explosive power of five million pounds of coal,[21] the death ray piece was, at the time, arguably comparable. Prompted by the article, FBI special agent J. J. Keating was moved to alert the higher-ups and have Tesla interviewed.[22] Since the FBI files give scant information for this period, that is, from 1940 until Tesla's death in January of 1943, articles on the web speculate that Tesla was under surveillance for that period and that these files have been removed.

On October 24, 1947, the assistant director of the Office of Alien Property wrote to the Air Technical Service Command at what was then Wright Field in Dayton, Ohio, requesting the return of Tesla's particle beam weapon paper, which had been sent to Colonel T. B. Holliday at their equipment laboratory. Yet, when I repeatedly wrote to Wright-Patterson under the Freedom of Information Act, their response covering a span of thirty years was that they had no files on Tesla.[23]

Given that there are numerous FBI and OAP memos from colonels and brigadier generals from Wright-Patterson discussing Tesla's files, their constant denials and the simple fact that the OAP has never released the particle beam weapon paper, even though their documents state overtly that this paper was in their possession, supports the contention that aspects of Tesla's work, even to this day, are still top secret.

Tesla's published statements on his particle beam weapon trace back to World War I, but Tesla himself dates its origin to the mid-1890s, when he was experimenting with X-ray transmissions based on the new discoveries of Wilhelm Roentgen.

> Dramatically, Tesla describes how this titanic voltage would hurl into space billions of microscopic electrical particles of matter that would bring down invading airplanes as insects are dropped by a spray gun.[24]

In January of 1916, W. H. Ballou, a writer for the *Washington Post*, reported that Tesla had invented an electrical machine that could potentially be placed on airplanes to shoot down hostile craft, sending out a "death dealing blow at the speed of 300 miles per second. No [person] or corporation should be allowed to possess such an invention. The Tesla designs are in the patent office, and should be seized and appropriated by the government for its own uses and defense, or for ending the European war at will. . . . A demonstration of its power to destroy a distant fortress or warship would be ample [proof] to end the present and all future wars. Mr. Tesla announced that his machine would travel around the world in one and a half minutes—under perfect control. Not a second should be lost in seizing and appropriating it." [25]

This piece was in reference to the *New York Times*'s December 8, 1915, article, "Tesla's New Device Like Bolts of Thor," wherein Tesla told the reporter that "a man in a tower on Long Island could shield New York against ships or army by working a lever, if the inventor's anticipations become realizations." [26] Thus, we can therefore date Tesla's first formal announcement of his death ray to specifically December of 1915.

A number of key points emerge from these World War I articles, the first being that Tesla basically kept this research a secret for many years. The most likely reason Tesla decided to reveal the existence of the weapon was his last ditch effort to portray Wardenclyffe as an essential asset because of the Great War. If he could prove the military importance of his broadcasting station, perhaps the government and maybe even President Wilson himself would come to his aid and save his precious tower.

Clearly, Tesla could work on top secret experiments and tell no one. It is quite possible that the details of his work in capturing and harnessing cosmic rays fell into this category. Coincidentally, both World War I articles mention that Tesla submitted the plans for this death ray to the patent office. Whether this occurred or not is unknown. However, the patent office does have a secret division for classified patents, and John Hays Hammond Jr., who at one time worked at the patent office, had patents sequestered into this category.

Without doubt, Tesla was working for or with the government throughout at least the last half of 1942 until he died. Two years earlier, when the William Laurence article appeared, Tesla was questioned by the FBI. Yet, if one reads his FBI dossier, there are glaring omissions, which J. Edgar Hoover himself inadvertently reveals. In one of the first FBI files released through the Freedom of Information Act, dated two days after Tesla passed away, J. Edgar Hoover states, "A review of the Bureau files reveals considerable information concerning Nikola Tesla and his inventions and it should be noted that one Nikola Tesla ... made a speech at the Grange Hall, Springfield, Massachusetts, on June 4, 1922 under the auspices of the Friends of Soviet Russia."[27] Where is this "considerable information"? Except for one document from 1940, which noted fear that Tesla could be kidnapped because of his special knowledge, there do not seem to be any files on Tesla released for any period before his death. However, there are scores of FBI files for Tesla *after* he died.

From a letter from the White House dated while Tesla was still alive, it is clear that the inventor became a person of interest in the Oval Office, and that led Roosevelt to request in writing more information on the inventor. All of this suggests that a meeting with the wizard was being organized.[28] But just a few days later, Tesla was dead. So the question remains, in the run-up to and early part of World War II, was the man I interviewed, Ralph Bergstresser, the only government official working with Nikola Tesla?

Born in 1912, in Pueblo Colorado, Ralph Bergstresser was raised in Wichita, Kansas, ostensibly with his high school buddy, Tesla acolyte Bloyce Fitzgerald, who was four years his junior. Born in Alberta, Canada, but having moved to Wichita as a boy, Fitzgerald had known Tesla only through telephone calls and letters to the inventor since his freshman year in college. On the other hand, Bergstresser conferred directly with Tesla at his apartment in the last six months of the inventor's life, while Bergstresser was working for Military Intelligence before being deployed to Southeast Asia.

After returning from the front in Burma and China, right after the war, in October of 1945, Bergstresser approached Frederick E. Cornels, special agent in charge, to alert the FBI that Tesla's secret weaponry papers could pass to the Communists if they were released to Tesla's heir, his nephew, Sava Kosanović, ambassador from Yugoslavia.

In a detailed brief to the FBI, Bergstresser stated that rather than allow Tesla's estate to be passed behind the Iron Curtain, he was working with Fitzgerald to support a plan to keep the motherlode in the States. Fitzgerald was working at Wright Field Air Force Base (which in 1945 would combine with nearby Patterson Field to become Wright-Patterson Air Force Base) "with a group of young Army scientists" and also with professors at MIT, endeavoring to develop a working particle beam weapon, as they both felt that this was the best defense against the new curse of the atom bomb. "Their ultimate goal," Bergstresser went on, is to create a "memorial foundation ... for the protection of Tesla's experiments and for the preservation of the inventor's memory . . . [and to create] a sort of 'idea factory.' . . . [Thus] they intended to contact Henry Ford Sr., to solicit his aid in this regard." [29]

During the war, Bergstresser had been stationed throughout Southeast Asia as a participant in Roosevelt's new Office of War Information, which was a sanitized name for an office of psychological warfare. On June 13, 1942, through an executive order, Franklin Roosevelt had created the new office, which was, in fact "America's official propaganda agency," whose goal was to control information coming from the front lines and "sell the war to the American people." In 1942 and 1943, Bergstresser worked with the OSS for General Joseph Stilwell in Burma, involved in a disinformation campaign to denigrate the Japanese in the eyes of the locals. Other propaganda crusades involved the dropping of leaflets to "deceive, mislead, puzzle and frighten the enemy." [30]

Upon his return from the horrors of war, having witnessed the death of some of his colleagues, Bergstresser "had driven General Stilwell's car from Washington to the West Coast as a personal favor to the General's wife." After making his way back East, he set out to uncover whatever he could about the elusive inventor and his secret weaponry papers. [31]

Bergstresser's first stop was to interview Tesla's closest associates, starting with *Herald Tribune* writer John O'Neill. O'Neill had just published his Tesla biography, *Prodigal Genius*, and he had known his subject intimately for four decades. He described the inventor as extremely fastidious. "Hotel employees," O'Neill related, "said it was common to see Tesla transfixed, and when they were cleaning his room, they had to work around him. If they took too long," Tesla would say, "You will leave me now." He would continue cogitating or feed his pigeons. O'Neill also noted that Tesla would go down to the barbershop three times a week for half-hour scalp massages to stimulate his brain. When it came to Tesla's trunks, O'Neill said he thought Tesla claimed to have eighty of them, mostly located in a storage facility, but that some trunks might still be in the basements of the various hotels where Tesla had lived. [32]

Kenneth Swezey (1905–1972), a writer for *Popular Science* and the *New York Sun*, was Bergstresser's next stop. One of Tesla's closest confidants, Swezey, who wrote several science books, including one on Einstein, had helped organize several anniversary celebrations for Tesla. He started the Tesla Society, and during the last years of the aging wizard's life fielded numerous phone calls from the old codger, some at three or four in the morning.

Bergstresser also spoke to Tesla's secretaries, Muriel Arbus, Dorothy

Skerritt, and Charlotte Muzar, who related that Tesla loved going to the movies, mostly comedies, and that he did some of his best thinking in movie houses. From these discussions, Bergstresser's list expanded. He interviewed Tesla's draftsman, Leon Kirsch, who, according to Skerritt, had important information on the construction of some of Tesla's most secret inventions;[33] Clifford Denton, science reporter for the *Daily News*; Mr. Crosby of the Cramps Shipbuilding Company in Philadelphia, who expressed interest in Tesla's remote-controlled boat; Julius Czito, (Coleman Czito's son), Tesla's lab assistant from 1916 to 1929, who showed Bergstresser a photograph of many of Tesla's workers; and Hugo Gernsback, who had featured the inventor in many articles, including Tesla's autobiography, which was serialized in *Electrical Experimenter* in 1919.

Bergstresser queried Tesla's lawyers, Parker Page and John Kerr, and the inventor's closest friends, including Lady Ribblesdale, formerly Mrs. John Jacob Astor; Robert and Katharine Johnson's daughter, Agnes Holden, who had inherited Tesla's extensive correspondence to her parents;[34] and Tesla's speculated love interest, playwright Marguerite Merington, who remembered from the late 1890s seeing Tesla operate his remote-controlled boat. From this work, Bergstresser got the name of John Hays Hammond Jr., who had his own invention factory in his castle in Gloucester, Massachusetts. Journeying to the castle, this dogged researcher spent a long afternoon with Hammond, who told Bergstresser that Tesla had been a veritable genius, the man most responsible for the foundation of the radio-guidance systems that Hammond had become the leading expert in. This investigation led Bergstresser back to New York, where he met with Emil Lowenstein, brother of the deceased Fritz Lowenstein, one of Tesla's most trusted workers and former associate of Hammond, who had passed away in 1922.

Bergstresser went on to interview Sava Kosanović, Tesla's nephew; William Laurence, the science writer for the *New York Times* who wrote a major article on Tesla and his particle beam weapon and who had accompanied Bergstresser to Tesla's funeral two years earlier; Jordan Mott, who, through his family, funded some of Tesla's inventions; Mrs. George Scherff, the wife of Tesla's longtime secretary; and Lieutenant Denmark of the U.S. Army, who had information on Tesla's particle beam weapon, having been involved in the sealing of Tesla's estate. The information from all of these individuals was compiled into a growing dossier that Bergstresser would come to give to his FBI contact and his boss at the Office of War Information.

Leaving no stone unturned, Bergstresser continued to uncover additional Tesla contacts with the hope of discovering the location of the inventor's most secret inventions. He interviewed William Dubilier, president of the Cornell Dubilier Company, which manufactured condensers, and who told Bergstresser that he also possessed "original notes of the 1899 experiments in Colorado Springs"; E. G. Gage, a New York City engineer who stated that one of his friends, J. S. Leach, now deceased but "formerly of Redbank, N.Y., made electrical parts for Tesla's laboratory"; employees of the Commercial Photostat Company, located on the eleventh floor of the Woolworth Building, who had photocopied Tesla's various patents when Tesla had offices there; and Francis Fitzgerald of the Niagara Power Commission, who told Bergstresser of Tesla's 1927 plans to set up a wireless transmission tower up at Niagara Falls with the backing of the Canadian Power Commission.

Continuing on his quest, Bergstresser then located Harry Secor, a "model craftsman" and writer who had authored a front-page article in *Electrical Experimenter* on Tesla's invention of radar stemming back to World War I; and A. J. Shirk, who owned the Inventor's Model Shop at 70 West 100th Street in New York City. Bergstresser noted, "Mr. Shirk has met Tesla on three occasions and appears to know quite a bit about his model work and laboratory developments, particularly in recent years." Shirk and one Charles Baumgarten, whom Bergstresser met at Tesla's funeral, both had information on Tesla's secret laboratory, which has never been located but was apparently associated with Coleman Czito's machine shop, alongside the Queensborough Bridge at Fifty-Seventh Street and Third Avenue.[35]

Bergstresser passed all of this information to FBI Agent F. E. Cornels,[36] but what he didn't pass on was the pièce de résistance, a copy of Tesla's particle beam weaponry paper that Bergstresser had obtained and microfilmed from Tesla himself. Most likely, Bergstresser made duplicates that he shared with Military Intelligence, Bloyce Fitzgerald and his higher-ups at the Office of War Information. And then forty years later he most likely shared the paper with Bob Beck of the Psychotronic Society, who passed the paper to Andrija Puharich, who shared it, in 1984 via the International Tesla Society Meeting, with the world.

Although Fitzgerald had corresponded with Tesla since 1935, the budding engineer had only met Tesla during the last two weeks of his life, if he physically met him at all, whereas Bergstresser had been meeting with the wizard for several months. Bergstresser, however, was concerned that Fitzgerald had befriended a communist sympathizer, one Abram Spanel, a Russian-born immigrant and the CEO of the Latex Corporation (now Playtex), who had also met with Tesla. Bergstresser feared Spanel would pass Tesla's secrets to the Soviets.

Because of these concerns, Bergstresser penned an impassioned communiqué on Spanel to that dogged Communist-hunter, Senator Joseph McCarthy. Many years later, in the early 1980s, in a phone conversation with me, Bergstresser revealed that he also thought that the *New York Times* journalist, William Laurence, was also a communist sympathizer. However, I think a case could be made that Bergstresser, battle weary from his forays into China and Burma, was paranoid in this regard because Laurence was not a communist sympathizer; in fact, he was the only journalist to cover live the dropping of an atomic bomb in Japan, and that suggests he had top secret clearance. And in Spanel's case, Bergstresser also could not have been more in error.

As an immigrant, Spanel was so proud of being an American that "having invented floating pontoon stretchers for soldiers wounded in amphibious landings, he turned back his million dollar profits to the U.S. government to support the war effort." Two years ahead of Bergstresser, Spanel became so alarmed that the enemy could obtain Tesla's secret weaponry papers that the day Tesla died, he rushed to Washington to alert Mr. Bopkin of the Department of Justice to call on J. Edgar Hoover, and he also contacted Dr. D. Lozado, advisor to Henry Wallace, vice president of the United States. His main concern was his fear that Kosanović, Tesla's nephew, whom Spanel characterized as someone Tesla did not really like, would pass the inventor's secrets to the enemy.

> Describing Tesla as ... one of the greatest inventors that has ever lived ... Spanel stated that Tesla had an invention in which he was able to direct electrical current without the means of a conductor. ... For this reason [Spanel] felt that Tesla's inventions and patents should be put into the hands of proper Government officials. ... The day before Tesla died he had tried to get in touch with [the] War department ... [to] make available ... patents and inventions ... he had developed. ... Spanel also stated that he believed Bloyce Fitzgerald ... secretary to Tesla, has been contacted by Special Agent Cornels and that if the New York Office wishes to contact Spanel he may be reached at his home.[37]

Born in North Alberta, Canada, in 1916 from "not-so-rich parents," Bloyce Fitzgerald immigrated with his mother to the United States at age seven, taking odd jobs as he went to school, ending up in Wichita, Kansas, where he gained employment in the flour and oil industries and apparently met

Ralph Bergstresser, four years his senior, who also had an engineering background. At the age of twelve, he built his first Tesla coil and probably also constructed crystal radio sets. At age eighteen, he attended the University of Michigan in Ann Arbor, where he enrolled in engineering classes and "conduct[ed] studies on the structure of the sodium molecule and electronic effects at high pressure." He also attended classes to study cosmic rays with Nobel laureate and Tesla acolyte Robert Millikan. It was at this time, circa 1935, that Fitzgerald began sending birthday greetings to Tesla with the hope of meeting him.[38]

At age twenty-three, Fitzgerald traveled to New York on related business and summoned the courage to phone Tesla. But Tesla declined to meet with him. The year was 1939. Searching for a place to live, Fitzgerald eventually found an apartment in Brooklyn. He either rented a car or took a train out to Wardenclyffe to "visit the site of your previous experiments."[39] Earlier, on March 9, Fitzgerald had written the inventor from Wichita, Kansas, alerting him of his sojourn to New York.

"I am a consulting electrical engineer of this city and I have followed for a long time . . . your exploits and researches." Having been employed at one time by General Electric and also Westinghouse Electric, Fitzgerald decided to go on his own. "Interested in high frequency and high potential circuits [I] have read practically every published book on that subject, among them *The Inventions, Researches and Writings of Nikola Tesla. . . .* Your photograph with the Tesla high frequency coil [from] your Colorado Springs plant of many years ago adorns my desk."

The young engineer then went on to describe a plethora of inventions that he was developing. These included a forty-foot helicopter that he planned to test fly in Pasadena, California; an air conditioning system for cars; and "a special Rochelle crystal microphone which responds to low frequencies" that Fitzgerald was attaching to an oscilloscope and recording device to monitor and amplify "heart sounds." Claiming to have constructed twenty-six complete sets that he had installed in hospitals throughout the States, Fitzgerald hoped for a return note and asked if they could meet in July.[40]

Two years later came the attack at Pearl Harbor, and Fitzgerald volunteered for military service, becoming a private in the U.S. Army. Keenly interested in Tesla's particle beam weapon, with the war in high gear, Fitzgerald wangled a position with the NDRC and made his way to MIT, bringing his knowledge of Tesla's work to Professors Keenan, Woodruff, and Kaye "in the solution of certain problems regarding the dissipation of energy from rapid-fire weapons."[41]

In December of 1942, Fitzgerald telephoned Tesla to wish him a Happy Christmas and also try once again to finally meet the ailing inventor. "My entire time and effort has been towards assisting our war effort [with my work at MIT] so that I may later help to create a better world in which to live," the young engineer related, as he made yet another effort "to meet . . . and discuss my radiation problems with you. Your vast experience and contributions to the scientific world have been an inspiration to me for many years."

As of December 20, 1942, Fitzgerald had yet to meet Tesla.[42] By this time, however, Fitzgerald had formed a friendship with Spanel, who tried to find the youngster a job and also help sell an antitank gun Fitzgerald had designed to the Remington Arms Company, but the deal fell through. According to Spanel, he had himself met with Tesla, and a slim possibility exists that he and Fitzgerald, together, may have indeed seen the inventor a week or so before Tesla died. Photographs of Tesla at that time show him to be in a truly horrific state. However, Bergstresser, who apparently passed the particle beam weapon papers to Fitzgerald, claimed he was the last person to see Tesla alive.

The night Tesla died, Fitzgerald, now twenty-six years old, moved into high gear. Forming an association with Colonel Ralph Doty of the War Department, Fitzgerald contacted FBI headquarters in New York City to set up a meeting with Assistant Director Percy Foxworth. By this time, a number of Tesla's closest associates had been to Tesla's two rooms at the Hotel New Yorker. Tesla's nephew, Sava Kosanović; science writer Kenneth Swezey; and RCA museum director George Clark had been there to crack open his safe and change the combination by using a locksmith. Where Kosanović removed several photographs, Swezey took the 1931 compendium of letters from Nobel laureates and other dignitaries wishing Tesla a happy seventy-fifth birthday. Alerted, sci-fi magazine editor Hugo Gernsback moved swiftly to make a death mask, and discussions ensued about having Clark work to set up a museum in America in Tesla's honor.

This activity was done under the eyes of hotel management and also Military Intelligence, as FBI reports reveal that both Fitzgerald and Doty were present at some, if not all of these events. One report labeled them as hotel managers. Whether this was a mistake or disinformation, one way or another, Fitzgerald met with Swezey and convinced him to loan the 1931 book of testimonials so that Fitzgerald could bring it to his parley with Foxworth. But Foxworth had been called away by Eisenhower for a secret assignment in the war effort, and so Special Agent F. E. Cornels took the meeting.

According to Cornels, "Bloyce Fitzgerald, [a young] electrical engineer, had been quite close to Tesla during his lifetime and in fact was the protégé of Tesla." Fitzgerald recounted Tesla's accomplishments, showed Cornels various newspaper clippings and the book of testimonials, and told the assistant director the somewhat erroneous information that Tesla had been a consultant to Marconi and Edison and had sold his patents on the hydroelectric power system to George Westinghouse "for $2,500,000."

In fact, Tesla had worked for Edison back in 1884–85. He had, of course, never been a consultant to Marconi. According to a January 6, 1899 letter he wrote to John Jacob Astor and other supporting documents, he probably received about $500,000 from Westinghouse for the sale of his induction motor and AC polyphase system, about half of which he most likely had to split with his initial backers.

Cornels's account went on:

> Fitzgerald stated that of late years, he had become a confidant of Tesla and knew that Tesla . . . was carrying on extensive experiments [for a revolutionary type of torpedo, and] for transmitting electrical power by wireless and with propelling electrical rays possessing sufficient power to destroy implements of warfare, such as airplanes and submarines. Within the past month Tesla told Fitzgerald that his experiments in this connection had been completed and perfected . . . [and] that the complete plans . . . of the basic theories of these things are some place in the personal effects of Tesla. . . . A working model . . . of the so-called "death ray" . . . which cost more than $10,000 to build [is apparently] in a safety deposit box . . . at the Governor Clinton. . . .
>
> In past conversations, Tesla has told Fitzgerald that he has some 80 trunks in different places containing manuscripts and plans having to do with experiments conducted by him . . . stored . . . in a warehouse some place in New York City. . . .
>
> When asked if Fitzgerald was primarily interested in getting possession of Tesla's papers, he stated . . . what he wanted was for them to be made available to the government . . . for use in the present war. Because he stated, the so called Death Ray would be a deadly weapon in the hands of any power, and

the use of the wireless transmission of electrical power would make possible the most efficient airplanes conceivable.

Fitzgerald's concern was aroused . . . over the presence of the two nephews in whom he has no confidence as to their loyalty and patriotism to the Allied Nations, and also because of the pension Tesla received from Czechoslovakia [actually Yugoslavia] he feels that that government may feel that it is entitled to any revolutionary ideas Tesla may have had and indicated on paper.

Fitzgerald informed Cornels that he had worked with MIT professors on the death ray "until three months earlier, and that presently he was working as a designing engineer with the Ordinance Department of the U.S. Army, both in Washington, DC and New York City." He concluded the meeting with a plea to make sure that the FBI secured the safety deposit box at the Governor Clinton Hotel and to also make sure that "other technical papers and designs of Tesla be safeguarded so that they cannot be removed from the country or made available to . . . other nations."[43]

The government had certainly been aware of Nikola Tesla, his particle beam weapon, and his state of health for many years. In 1940, when the William Laurence *New York Times* article on Tesla's weapon was published, J. Edgar Hoover himself wrote Special Agent J. J. Keating, "Your courtesy and interest in bringing this information to my attention are indeed appreciated, and you may be assured your letter will receive appropriate consideration." Since Tesla had ties to foreign countries, including the Soviet Union, it seems likely that he would certainly be under tight scrutiny.

Add to the mix the timing of this eye-catching 1940 announcement and Laurence's close ties to individuals secretly involved in the Manhattan Project. The young chemist Glenn Seaborg, an émigré from Sweden and later a Nobel Prize winner who as early as 1940 was bombarding uranium with deutrons to create neptunium, was well aware that enough information was now available, as Laurence also reported, for it to be just a matter of time for "the Germans, like us . . . to prepare a bomb assuming they know about [plutonium]." Seaborg conveyed this concern to other members of the Manhattan Project team, particularly Arthur Compton and Eugene Wigner, namely that the Nazi scientists could develop a bomb theoretically as early as 1942. The Americans, he figured, could at the earliest develop one two years later.[44] That was why Fitzgerald was pushing so hard at MIT and Wright-Patterson to perfect Tesla's death ray, as he felt it might be the best and possibly only way to protect the homeland from such an awful and truly terrifying possibility.[45]

Within twenty-four hours of Tesla's death, the Oxford University-educated lawyer Lawrence M. C. "Sam" Smith (1902–1975), head of the National Recovery Administration for Roosevelt's administration, voiced his own concern of "the possibility of enemy agents confiscating some of the trunks of Tesla. [Smith] understands that the War Department was interested in this matter and that apparently the Alien Property Custodian's office was taking some action. He desired to know whether the Bureau would take some steps to refrain relatives of Tesla from taking the contents of these trunks and whether the Bureau would seize possession of the trunks."[46]

Hoover had his own concerns. On January 21, 1943, the seemingly omnipresent head of the FBI wrote an extensive memo to his Special Agents Little, Donegan, and Ostholthoff on the situation. Under a banner titled "ESPIONAGE," Hoover voiced his concern that Kosanović, whom he described as "a member of the Yugoslavian Government in exile," had apparently been in the country for several years, arriving by a British ship in September of 1941.[47]

In writing *Wizard*, I received the original Tesla FBI files with redacted portions. Recently, these files have been unredacted. The unredacted document reveals that Hoover noted that Kosanović was not simply a diplomat but rather "Chairman of the Board of Minister of State for Yugoslavia." Kosanović, Hoover suspected, was most likely a Communist, and in that capacity, he was funding various Serbian American newspapers. Further, he was also on a planning board for a post–World War II Europe. Hoover therefore considered having Tesla's nephew arrested for unlawfully breaking into his uncle's safe because he feared that Kosanović "might possibly make certain material available to the enemy." However, since Kosanović, as Tesla's nephew, was legally entitled to Tesla's estate, Hoover's idea was quickly scrapped.

NAZI CONNECTION

T. J. Donegan, the FBI agent in charge of dealing with Kosanović, noted this situation potentially resembled the Ludwig case. In the spring of 1941, a taxicab ran over a man purported to be a Spanish courier, Don Julio Lopez Lido. The odd thing was that the person who was with Lido, instead of attending to his fallen colleague, grabbed the fallen man's briefcase and disappeared into the crowd. As it turned out, this deceased individual was not Spanish at all; rather, he turned out to be a German spy, Captain Ulrich von der Osten, and the man who ran off, later captured, was Kurt Frederick Ludwig, an American-born German individual in charge of a sizeable ring of German spies who were spread throughout the metropolitan area.[48]

It was well known that Nikola Tesla had also been struck by a taxicab, back in 1937, and not too long after that his apartment had been ransacked. Although Tesla claimed that nothing of value was taken, it was further known that in the run-up to the Second World War, the inventor had been approached by officials of the new Nazi regime. Further, Tesla was closely associated with George Sylvester Viereck, a city poet of Prussian descent and a German propagandist from the First World War. It thus seems likely that Viereck, who was covertly working for the Nazis, had contacted them to arrange a meeting with the old wizard.[49]

Tesla saw Viereck as his "many sided friend," famous for having conversed with individuals ranging from Theodore Roosevelt, George Bernard Shaw, Aleister Crowley, Albert Einstein, and Sigmund Freud to Kaiser Wilhelm and Adolf Hitler.[50] Agreeing during the height of the Great Depression to co-write an article, Tesla sent his friend a note, quoting from Viereck's gloomy poem *The Cynic's Credo*, editing out the more erotic sections: "'We are two storm-tossed sailors stranded on the selfsame reef.' And a little of the filthy lucre would be soothing to this plight."[51] Happy to share the writing fee, Viereck thereupon penned with Tesla a piece published in *Liberty Magazine* in 1935 titled "A Machine to End War," which discussed not only Tesla's particle beam weapon or defensive shield between nations but also such topics as the difference between religion and science, the advent of television, a coming age where automatons will do the work of men, and eugenics—that is, the systematic plan to prohibit inferior individuals from reproducing.[52]

Six years later, during the height of WWII, which would be two years before Tesla died, Viereck was arrested for his traitorous activity in his continuing decision to spread Nazi propaganda through the printed media. Having traveled to Germany in 1939, Viereck met, on a regular basis, Hermann Goering, Adolf Hitler, and the former chancellor of Germany, Franz von Papen, whom Tesla may himself have known when von Papen was stationed in America during the first world war.

Born in Germany in 1884 and rumored to be the illegitimate son of the kaiser, Viereck had immigrated to the United States in 1897. By the time of his arrest, he had two sons at Harvard. Surveilled by the FBI for many months, his bail was set at $15,000. FBI files reveal that Viereck was paid $5,000 per year by the German Tourist Information Office to ostensibly write publicity and travel ads, but in actuality, his real post was to spread Nazi propaganda.

Having met with Hitler in the early 1920s and again in 1933 and 1934, Viereck returned to America to speak before a crowd of twenty thousand "Friends of the New Germany," all scarily in unison giving the "Heil Hitler" salute at Madison Square Garden, where, incredibly, he drew parallels between Hitler and Roosevelt! Telling his audience to "sympathize with National Socialism without being anti-Semites," Viereck ended up losing many of his Jewish contacts, most notably Sigmund Freud, who stated that Viereck on the one hand was a "lion hunter," seeking to meet the greats in society, yet on the other hand was a "narcissist who had delusions of persecution with a fixation on the Fatherland," and thus Freud severed all ties.[53]

Throughout much of the 1930s, Viereck was planning strategies with the ultimate counterintelligence agent, Franz von Papen, who was one of the master architects of the fifth column in America, not only during the First World War, when von Papen was stationed in America coordinating the vast network of German spies and saboteurs that finally caused his ouster from the country, but also, through Viereck and other American minions, in the run-up to the Second World War. The same subversive article written by both men would appear under Viereck's name in America and under von Papen's name in Germany.[54] This tack is clearly evident in the Tesla article, wherein eugenics is discussed. Although it is certainly possible that Tesla believed in population control, what is more likely is that Viereck directed the conversation to this topic because it espoused Nazi beliefs, and thus this kind of indoctrination could make its way into a mainstream article as part of a concerted plan to influence public policies in a foreign land in a subversive way.

Tesla researcher Branimir Jovanovic, director of the Tesla Museum in Belgrade, notes that "two days after the publication of the article, Tesla was approached in writing by G. von Haeften, the vice-consul of the German mission in New York." A protégé of von Papen and an emissary of the burgeoning Nazi regime, von Haeften sought a meeting with Tesla, expressing interest in his defensive weapons. However, Tesla declined the engagement and let von Haeften know that "he was not willing to make any kind of agreement with representatives of the new German state, although he replied politely in the letter."[55]

One can't know precisely Tesla's attitude towards Germany during this period, the time of the Great Depression. However, what is evident is that Tesla was still conversing with Dr. Karl George Frank, former head of the Telefunken wireless company Atlantic Communications from the First World War, who had been arrested at that time as a spy. Tesla was also working with the architect Titus de Bobula in designing a replica Wardenclyffe tower. A virulent anti-Semite, de Bobula was also an arms merchant, anarchist, and Nazi sympathizer who had joined a pro-Hitler group as early as the mid-1920s.[56] In his letter writing, Tesla occasionally disparaged the Jewish people, even though quite a number of Jewish individuals played major roles in Tesla's life. For instance, in a letter to Katharine Johnson in the late 1890s, which Tesla atypically began, "My dear Kate," the Waldorf-Astoria dandy invites the Johnsons to dinner at the hotel so that they could be separated from "plebians, drummers, grocery men, Jews and other social trilobites."[57] On a separate occasion, while writing to J. Pierpont Morgan in yet another plea for additional funds, Tesla rails, "Will you now let me go from door to door to humiliate myself to solicit funds from some jew or promoter and have him participate in that gratitude which I have for you? I am tired of speaking to pusillanimous people who become scared when I ask them to invest $5,000 and get the diarrhea when I call for ten."[58]

It is imperative when undertaking historical analysis to take into account the period in history. Anti-Semitism was rampant through the early part of the twentieth century. One of Tom Edison's closest friends was Henry Ford, with whom he shared adjoining estates in Fort Myers, Florida. Ford's anti-Semitic diatribe published in the *Dearborn Independent* was so egregious that Hitler adopted sections of it in *Mein Kampf*. Charles Lindbergh, another ardent anti-Semite, had a secret family in Germany and proudly wore a Nazi insignia given to him by Hermann Goering. J. P. Morgan Jr. was also known to be anti-Semitic, in particular "out to get the Jewish press" for how they portrayed his father, although he did maintain cordial relations with Belmont and the Guggenheims. Other Tesla associates of this ilk included Long Island railroad magnate and rabid anti-Semite Austin Corbin, who set up beach clubs along the Atlantic Ocean that disallowed Jews; and African explorer and congressman William Astor Chanler, who was a strong proponent of the premise of the infamous *Protocols of the Elders of Zion*, which was a spurious document espousing the belief that the Jews were secretly trying to take over the world.[59] And the situation in England wasn't any better, with neo-Nazi marches in the streets of London in the mid-1930s and the horrible public relations problem the House of Windsor had with King Edward VIII, who had to abdicate, in part, so that the monarchy could keep his friendly meetings with Adolf Hitler out of the public eye.

At this same time, in 1939, President Roosevelt would not allow the *St. Louis*, a transatlantic oceanliner carrying more than nine hundred Jewish individuals fleeing Germany, to enter the United States. Cuba also turned them away. In the case of my Jewish uncle who was in the army during World War II, when stationed in the Carolinas, some of the soldiers wanted to see his horns. And even as late as the 1950s and 1960s, I myself was well aware that certain golf clubs on Long Island, actually not far from Wardenclyffe, did not allow Jews.

In Tesla's case, as a Serb, he must have certainly encountered prejudice against his race, both in Europe in his homeland and in America, in part because World War I was sparked by the assassination of Archduke Ferdinand by a Serb. During World War II, the Croats tended to side with the German forces, and ninety thousand Serbs were slaughtered in their own holocaust. The members of a third group in that region, often called the Bosnians but actually of Moslem heritage, were also hated by both these other tribes, in part because the Ottoman Empire had slaughtered, beheaded, raped, and pillaged Croats and Serbs for many centuries after the Battle of Kossovo of 1389. When I had speaking engagements in Croatia and Serbia in 2007, there was no direct flight between Zagreb and Belgrade because of this animosity, even though these two cities had been part of the same country for decades. Concerning anti-Semitism and Tesla's disparaging comments, we can see, even today, many instances of this prejudice appearing throughout the world.

That said, Jews played a prominent role in Tesla's life. Arthur Korn, ostensibly the first person to use Tesla technology to electronically transmit pictures, was Jewish, as was Jacob Schiff, who, in 1906, agreed to give the inventor $10,000 towards the completion of Wardenclyffe, if Tesla were to raise $100,000 "of new capital from responsible people."[60]

August Belmont, who funded the Westinghouse Corporation, was Jewish, as was Fritz Lowenstein, Tesla's closest technical assistant in both New York and Colorado Springs.[61] Hugo Gernsback, also a Jew, not only published Tesla's autobiography, he also played a key role in getting the Westinghouse Corporation to cover Tesla's rent at the Hotel New Yorker for the last ten years of his life. Given that Tesla lived in two rooms, his rent was double the normal rent. There is evidence that the induction-motor expert Bernard Behrend, who nominated Tesla for the Edison Medal, helped pay some of Tesla's debts at some of these hotels, and Behrend was probably also Jewish.

Years later, Tesla tried to interest other Jews, such as Simon Guggenheim, in a new process Tesla had developed to increase the strength and conductivity of copper.[62] Tesla also wrote to movie producer Carl Laemmle about his special effects expert Kenneth Strickfaden, who used Tesla's technology for its visual power in such movies as *Frankenstein*. "Tesla called Laemmle a genius because Laemmle, like Tesla, successfully fought the Edison clique . . . when Edison held a monopoly on key movie making patents and would not allow competitors to use them."[63] Recognizing a kindred spirit, Tesla wrote to discuss with the producer the idea of the plurality of worlds. "Undoubtedly," Tesla said, "there are myriads of planets in the universe which were, are, or will be abodes of life in its highest forms."[64] If Tesla had been a rabid anti-Semite on the order of a Henry Ford or Charles Lindbergh, it is doubtful he would have interacted with so many people of that religion. What Tesla reveals in his disparagement of the Jewish people in letters to Katharine Johnson and J. Pierpont Morgan, and later to the British War Office, is a hypocritical, sobering, and frankly disappointing aspect to his nature. A product of his times, like many of his cohorts, the Serbian inventor was an individual who bought into the generally accepted notion of anti-Semitism with exceptions to the rule. It certainly is a damning flaw in his personality.

Given the moment, the height of World War II, and the fact that the Germans did indeed have a strong fifth column in the United States and that Communism was also a threat, it is understandable why Hoover had serious concerns not only about Tesla's associations with Nazi sympathizer Viereck, who was now in prison for his traitorous activity; the anarchist Titus de Bobula, whom they had a file on; and Tesla's Yugoslavian nephew Sava Kosanović, who was known to be supporting Serbian newspapers; but also about Tesla himself, a man who shirked his debts, lived a secret existence, and had already sold top secret military weapon designs to foreign governments.

14

The Trump Report

As a result of this examination, it is my considered opinion that there exist among Dr. Tesla's papers and possessions no scientific notes, descriptions of hitherto unrevealed methods or devices, or actual apparatus which could be of significant value to this country or which would constitute a hazard in unfriendly hands. I can therefore see no technical or military reason why further custody of the property should be retained.

—JOHN G. TRUMP
Technical Aide
Division 14, NDRC[1]

Military Intelligence had moved swiftly upon hearing of the inventor's death. In conjunction with the Office of Alien Property, they transferred Tesla's estate to the Manhattan Storage Warehouse, which was located one block east of Hell's Kitchen at Fifty-Second Street and Seventh Avenue. Two weeks later, on January 26 and 27 of 1943, the official examination of Tesla's effects was undertaken by MIT professor John G. Trump along with a Navy photographer and stenographer and Willis George, a counterintelligence agent and safe cracker associated with both Naval Intelligence and the OSS. For arbitrary and even puzzling reasons, according to Trump's call, no investigation was made of material already in existence at the warehouse or of any trunks that had remained untouched in the basement of the Hotel New Yorker for ten years prior to Tesla's death.

E = MC²

In 1909, the physicist Lise Meitner heard Albert Einstein lecture on his famous equation. It took her awhile to realize the implications, namely that Einstein had created a mathematical way "to calculate the conversion of mass into energy." Three decades later, in 1938, Meitner was studying the puzzling finding that if a "large uranium nucleus split into two smaller nuclei, the smaller nuclei would weigh less in total than the common parent." When she used Einstein's equation, which involved multiplying the missing mass by the speed of light squared, Meitner found that this missing part, which was one-fifth of the proton mass, "was equivalent to 200 MeV" of energy, that is, 200 million volts![2]

This was a startling finding in 1938, gargantuan in comparison to normal chemical reactions. On the one hand, it drove home the hard-to-believe reality of the enormous energies that are locked inside of such tiny spaces, yet on the other hand, it portended the development of the atomic bomb, an idea that H. G. Wells had forecast many years earlier.

Several years before this, George Gamow, one of the founding fathers of quantum mechanics, had taken a job at George Washington University under the proviso that the school would host yearly conferences on theoretical physics. Now, in 1938, based on Meitner's finding, when Niels Bohr arrived in town, he immediately sought out Gamow, who in turn called Edward Teller. "Bohr has just come in," Gamow said. "He has gone crazy. He says a neutron can split uranium."

Gamow opened the conference with the venerated scientist, noting that Bohr's news "galvanized the room."[3] The viability of atomic warfare had just become reality.

In March and April of 1940, Einstein sent off two letters to President Roosevelt urging him to begin funding nuclear research. The fear was that if the Germans developed an atomic bomb, they could win the war literally overnight. Vannevar Bush, the designer and head of the new National Defense Research Committee (NDRC) who reported directly to the president, stated, "The show was going before that letter was ever written." When asked if Einstein could be brought aboard what would come to be called the Manhattan Project, the tall slender head of the new NDRC, described as resembling Uncle Sam without the beard, stated, "I wish very much that I could place the whole thing before him and take him into confidence, but it is utterly impossible." With "perhaps excessive self-confidence, a rare mastery of the intricacies of the Washington power structure, a shrewd, imaginative, yet wholly practical mind,"[4] and a tremendous penchant for secrecy, Bush had assessed that Einstein's philosophy, coupled with his prominence, was too much of a risk. Scientists who came into the project were not allowed to voice philosophical opinions about the bomb; they were either on board or not, and so Einstein was excluded.[5]

History was moving rapidly. In 1939, Poland was invaded by the combined forces of the Soviet Union and the rapidly expanding Nazi empire. The following year, President Roosevelt, by executive order, created the Special Defense Unit out of the Department of Justice to "coordinate all Justice Department activities relating to espionage, sabotage, and subversion." A key part of the assignment was to compile a list of all aliens or foreign nationals in the country and rate them according to what countries they were associated with and how dangerous they might be. Roosevelt appointed Lawrence M. C. Smith, "heir to the Smith-Corona typewriter fortune," as head.[6]

Two years later, as an offshoot of this unit, Roosevelt created the Office of Alien Property "to safeguard the property of enemies and their allies in this country pending the war." Using the Trading with the Enemies Act for his authority, President Roosevelt created the organization under a division of the Treasury Department to give the government official authority over foreign-owned property. He assigned Leo Crowley, a former banker and chair of the FDIC, as its head and also placed Homer Jones, a University of Chicago economist, in charge of its Division of Investigation and Research.[7]

Because of Nikola Tesla's ties to foreign nationals, he was on their radar, particularly because the aging wizard had been touting his own ultimate weapon. Billed as a defensive shield between nations since the First World War, Tesla's device had generated a big resurgence of publicity throughout the 1930s. In 1931, for his seventy-fifth birthday, Tesla made the cover of *Time*, and dignitaries from America and abroad flocked to wish him a hearty happy birthday. Among the individuals who congratulated him were Nobel Prize winners Robert Millikan, Ernest Rutherford, William Bragg, Wilhelm Roentgen, Albert Einstein, and Arthur Compton. Also included was Vannevar Bush, dean of engineering at MIT, who wrote, "Dear Dr. Tesla: I am glad to have the opportunity of sending you my personal greetings on the occasion of your seventy-fifth birthday, and I wish to join to my own tribute of admiration for your unique career the congratulations of the Massachusetts Institute of Technology, where the contribution which your original genius has made for the benefit of mankind is fully appreciated."[8]

Bush, who wrote his doctoral dissertation on oscillating electrical circuits, certainly had an intimate knowledge of Tesla's field and contributions to electric theory. A brilliant scientist who helped design electronic devices, including analog computers that made use of the binary code via on-off voltages, Bush had graduated conjointly from MIT and Harvard for his Ph.D. in 1916 and then went on to co-found the engineering company Raytheon in the early 1920s. In 1940, after a distinguished career at MIT and Carnegie Institute, Bush was able to obtain a one-on-one with President Roosevelt, accompanied by Secretary of Agriculture Henry Wallace, "not because he was [rumored to become] vice president of the United States, or because he was [slated to become] chairman of the National Defense Board, but because he was a scientist (a plant geneticist), the only one in the cabinet."[9]

According to legend, armed with a single sheet of paper outlining his plan and seeing science as "an Endless Frontier," Bush walked out of the meeting as head of the NDRC, an organization set up, in Roosevelt's words, as one "of teamwork and cooperation in coordinating scientific resolution of the technical problems paramount in war. Its work [shall be] conducted in the utmost secrecy and carried on without public recognition of any kind."[10]

Shortly after that meeting, James Conant, Bush's colleague and president of Harvard University, traveled to England as an emissary from the secretaries of war and the Navy to meet with the British War Office and report back to President Roosevelt on "the need [to create] special teams to work in England on immediate battle problems and . . . the need of an Office of Research and Development for liaison purposes with the War and Navy Department[s]." Possibly influenced by Canada's National Research Council, which had been in existence for nearly twenty years, Conant drew an organization chart placing the head of this new Office of Research and Development on a par with the secretaries of war and the Navy, who, like them, would report directly to the president. "In the enclosed diagram, I have put in Dr. Bush's name as Director, as all of us who have been in contact with him feel that he is the one man who could undertake this difficult task."[11] Roosevelt agreed and thereby bumped Bush up from head of the NDRC to weaponry czar of the ORD. This meant that Bush would now oversee literally thousands of scientists and operations being carried out at such agencies as the National Academy of Sciences, National Advisory Committee for Aeronautics (precursor to NASA), and liaison to the British War Office. Further, he would also have control over research and development at the War Department, including all military branches and also "other agencies concerned with defense research." In avoiding publicity, *Fortune Magazine* noted that the new Bush agency was "the most supersecretive," more than any of the four military branches. What this meant for Tesla was that his work became even more buried once his estate was taken over.

Moving swiftly, with the war looming, Bush gathered together an elite team of the most brilliant physicists to work on the Manhattan Project, a venture so secret that even Harry Truman, when he became vice president, was kept out of the loop.

Having been the dean of MIT throughout the 1930s, Bush chose his NDRC and ORD advisors from his closest associates. Aside from Harvard president James Conant, who recommended Bush for the position, the team also included Karl Compton, president of MIT; his brother, the Nobel Prize winner Arthur Compton; Captain Lybrand P. Smith, winner of the Legion of Merit; Major General Clarence C. Williams, chief of ordinance going back to World War I; Roger Adams, an organic chemist and expert in explosives; and several other individuals who worked on the Manhattan Project, including Frank B. Jewett, president of the IEEE, which bills itself as "The world's largest technical professional organization for the advancement of technology," and also Bell Labs; Richard C. Tolman, vice chairman of the NDRC and the astrophysicist most responsible for enticing Albert Einstein to immigrate to the United States; and Irvin Stewart, the head of Cold War University and executive secretary of the atomic bomb project.

Once Tesla died, Smith, as head of the Special Defense Unit, contacted Hoover to inform him that the Office of Alien Property would have the authority to impound Tesla's estate, and Bush directed Karl Compton to assign Trump to head up the investigation. It was Trump's task to make a detailed list of the contents of Tesla's holdings and see if there was any military significance to any inventions.

Meanwhile, in Canada, the military experts took a different approach. "I was down in Washington two weeks ago and had an interesting conference with Dr. Vannevar Bush," Dr. C. J. Mackenzie, the new head of the Canadian National Research Council, wrote the man he replaced, General Andrew McNaughton, who, in 1940, was now in charge of the Canadian armed forces in Europe. McNaughton, at this juncture, was still trying to determine the viability of Tesla's weapon, having been in extensive negotiations with the inventor for the last several years.[12]

"The most interesting and significant developments recently have to do with co-operative arrangements which we have been able to make with our American friends," Mackenzie went on. "The Americans are busy on similar work." Mackenzie was referring to new technologies and advanced weaponry systems such as radar, subsurface warfare, jet planes, high-speed projectiles, fire-proof clothing, power-operated gun turrets, and torpedoes that steer themselves. "The facilities of the large industrial research laboratories will be of great value to all of us. ... I have been to Washington several times to confer with Dr. Bush and members of the committee there. As you well know, there is a most extraordinary pro-British sentiment in such quarters. Scientifically and technically, they are at war as much as we are—and one has to remind oneself continually that they are still a neutral country."[13]

Two years later, Mackenzie informed McNaughton that "The Americans have followed the suggestion you made on your recent visit [to Washington] and now have a three-man committee operating on a very high level, that is, reporting directly to the President. It is called the Joint Committee on New Weapons."[14]

Just six months later, Tesla, who had been negotiating for several years with McNaughton, was dead, his papers were in the hands of Bush, and America now was very much enmeshed in war. Born just a mile from Bunker Hill, along the Mystic River in a suburb of Boston, the grandson of a whaler and the son of a preacher, Bush was described in a cover story for *Time* as "lean, sharp, salty . . . [and] a Yankee whose love of science began as a boy tinkering with gadgets." At age fifty-four with two sons in the military, one in the Army and one in the Air Corps, Bush, who was rarely seen without a pipe in his craw, was perceived as "a shrewd, imaginative physicist . . . [whose] job is unprecedented in U.S. history. As chairman of the Army and Navy's Joint Committee on New Weapons & Equipment, he is the first civilian technician ever to sit in the highest war councils."

With a budget exceeding $135 million per year, $50 million alone for radar development and submarine warfare, "Bush's army consists of 6,000 of the top U.S. scientists . . . their work surrounded with fantastic secrecy. Clerical workers often do not know even the name of the weapon being developed in their laboratory. A few super-secret projects are carried on in isolated walled villages which no-one is allowed to leave except on special permits."[15]

Looking over the possible individuals to study Tesla's holdings, Bush settled on someone he knew very well, the associate professor of physics at MIT John G. Trump, an excellent choice to study the wizard's holdings. Born in 1907, Trump had earned his master's degree in physics at Columbia University before transferring to MIT, where he obtained his doctorate under the watchful eye of Bush, the dean of the school. The year was 1933. Trump, who for a time had worked in transformer design for General Electric, developed for his doctoral thesis "a vacuum-insulated rotating electrostatic motor" while working with Robert Van de Graaff on "high voltage phenomena and vacuum insulation studies."

Six years Trump's senior, Van de Graaff was by this time world famous because of his generator. No doubt Tesla came to the attention of these two young physicists, not only because the Tesla coil was crucial to their experiments but also because Tesla wrote a prominent article on the Van de Graaff generator that was published as a spectacular cover story in the March 1934 issue of *Scientific American* and because Tesla was nominated for a Nobel Prize shortly after that by the man who nominated Albert Einstein, the physicist Felix Ehrenhaft.

Trump's research in this field "led to the design of compact megavolt X-ray generators for use in medical therapy [involving] penetrating radiation." By 1940, over one thousand patients had been treated for cancer and other maladies with Trump's machine, which was sold to many hospitals.

In 1942, Trump became secretary of the Microwave Committee for the NDRC, working as a technical aide to Dr. Karl Compton, president of MIT and chairman of Division D of the NDRC. Two years later, during the height of World War II, as consultant to General Carl A. Spaatz, commander of Strategic Air Forces in Europe, Trump worked on "vital war projects," including infrared experiments and the use of microwave radar as a defense against German air attacks.

From his work with Van de Graaff, Trump headed up MIT's high-voltage research lab, which in the 1930s involved creating particle accelerators that directed beams of electrons, ions, and X-rays. In 1937, before developing his medical X-ray equipment, Trump had been able to accelerate atomic particles in vacuum insulation environments and, with Van de Graaff, produce thirty megavolts, which was among the highest man-made voltages ever recorded.[16] Clearly, Trump, who would end up after his death becoming uncle to a president of the United States, was a superb choice to analyze the viability of Tesla's particle beam weapon, and Bush was keen to assign him to the task.

In 1941, Bush created the Office of Scientific Research and Development (OSRD, which Conant had called the Office of Research and Development, or ORD), and that became a giant umbrella organization that superseded the NDRC. Having very much "the ear of the president" and overseeing a plethora of new weapons, as a civilian now being depicted in the press as "a kinetic egghead charged with melding thousands of American scientists into a 'superbrain,'" Bush was often at loggerheads with the military higher-ups, whom he sometimes referred to as "technical dunderheads."[17] Nevertheless, despite ego battles and territorial differences, Bush naturally became more closely tied to the War Department, particularly when his work on the development of radar greatly helped a reluctant Navy to suddenly begin on a more regular basis to locate German U-boats.

On the domestic side, Bush was closely tied to the Special War Policies Unit, an offshoot of the Department of Justice, run by Lawrence M. C. Smith, which oversaw the legal arm that would have control over foreign property. This became the Office of Alien Property (OAP), whose head was Leo Crowley, now a cabinet member in Roosevelt's administration.

After Tesla died, Crowley contacted Irvin Stewart, who directed Walter Gorsuch, head of the New York OAP office, to take charge, and Gorsuch assigned Irving Jurow, a young attorney, to the case, and it was he who allowed Trump and his associates to study Tesla's personal effects to determine if they could be released to Tesla's heir, Sava Kosanović, who happened to also be ambassador from Yugoslavia. Tesla had another nephew, Nikola Trbojevich, William Terbo's father, but Trbojevich, an American citizen from Detroit and inventor in his own right of the hypoid gear which streamlined the automobile so that it no longer needed a running board, gave up his rights, as he agreed with other relatives in Europe with the idea of his cousin setting up a museum in Belgrade in Tesla's honor.

By this time, all of Tesla's estate had been transferred to the Manhattan Storage Warehouse, which now housed as many as 120 containers and trunks, including smaller barrels and crates, joining those that Tesla had placed there since he moved out of the Hotel Governor Clinton in 1934. About two years before his death, Tesla simply stopped paying storage fees to the warehouse, and so Manhattan Storage put a notice in the paper to auction off, at this time, the fifty or so trunks. This auction was noticed by *Herald Tribune* reporter John O'Neill, Tesla's biographer. He rushed over to Sava Kosanović's apartment with the news, and Kosanović came up with the money to cover all past debts, and he also agreed to continue to cover present and future expenses, about fifteen dollars per month from then on, thus staving off the devastating possibility of having Tesla's priceless legacy dissipated willy-nilly in an auction.

On January 26 and 27, 1943, Trump entered the warehouse with Charles Hedetniemi, chief investigator from Washington for the OAP, and three individuals from Naval Intelligence: Edward Palmer, who took photographs and made microfilm copies of as much of the artifacts and correspondence as possible; John Corbett, who served as Trump's stenographer; and Willis George, who was listed as a "civilian agent."[18]

However, Willis George was not just a "civilian agent" working for Naval Intelligence, he was also a spy for "Wild Bill" Donovan, head of the OSS. "Most people call it burglary, but when it's done for Uncle Sam, *surreptitious entry* is the preferred piece of verbal camouflage used to mask U.S. government breaking-and-entering operations," George proclaimed on the back cover of his book of clandestine operations, aptly titled with that euphemism.[19] An admitted cat burglar who stole documents from foreign diplomats for the government, there is little doubt that George was chosen because of his talent for picking locks, for—most likely—Tesla's various trunks and Tesla's safe, for reasons of national security.[20]

A decade later, when Kosanović finally obtained control of the physical estate and opened the safe, he found a set of keys missing (they were found placed in another trunk) as well as Tesla's Edison Medal, which was never recovered. The safe had definitely been broken into, and George was the most likely culprit. Did George also sneak into the Hotel Governor Clinton and steal Tesla's particle beam weapon, which was allegedly housed in a safety deposit box at the hotel? That possibility exists. Because when Trump went there to retrieve the box's contents, expecting to find the weapon, he found instead a standard electrical device worth several hundred dollars.

This was the concern of the ubiquitous Bloyce Fitzgerald. Now a private in the United States Army, Fitzgerald's military record lists him as also working for the NDRC as "a consulting engineer" on "aviation, rocket guns and gun turrets, hydraulics and special weapon armament projects," and thus, of course, on the development of Tesla's particle beam weapon. This unusual position for a private suggests that he may have been working directly for Major General Clarence C. Williams, head of Ordinance for the NDRC. One way or the other, this position enabled Fitzgerald to travel to MIT, where he worked with Professors Woodruff, Keenan, and Kaye on this device, and also spend time at Wright Field Air Force Base, where he consulted with engineers and military personnel.[21]

Tesla had told Fitzgerald that he had constructed such a weapon and repeated the story that he told to hotel management, that the device was being held at the Hotel Governor Clinton in a safety deposit box and that the key was tucked away in his safe. Fitzgerald therefore alerted the FBI and other authorities in Military Intelligence of this important piece of information. Thus, when the weapon had not surfaced, Fitzgerald kept looking for it.

Now that the device was missing, Fitzgerald sought advice from Colonel Thomas Parrott of Military Intelligence. Parrott, who would eventually go on to become a spy for the CIA, speculated that since Willis George, a hired thief, also worked for "Wild Bill" Donovan, head of the OSS, Fitzgerald should look there. A meeting was arranged with James R. Murphy, Donovan's head of Counter Espionage, and he was asked to search for any information as to the weapon's whereabouts. The substance of this encounter was passed to Colonel T. B. Holliday at Wright-Patterson, who by this time had received photostatic copies of the full particle beam weapon paper from Lloyd Shaulis of the OAP, and Holliday followed up with a letter to the OSS recounting the Fitzgerald meeting and requesting "any copies, translations of Tesla material particularly those concerning the propagation of high frequency currents" to be sent to the air base, but Murphy wrote back to say the OSS had no such material.[22]

Fitzgerald then met with Laurence Cardee Craigie, who was not only a brigadier general at Wright-Patterson in its Chief Engineering Division, he was also the first military pilot to fly a jet plane. That was accomplished while Tesla was still alive in the early 1940s, while the war was still raging. To put this into perspective, to those in the know, Craigie could be considered the Chuck Yeager or John Glenn of his day, except for the fact that the accomplishment was so secret that when his plane was merely docked, a wooden propeller was attached to the nose and tarps were placed on the air intakes to mask its real abilities.[23] Convinced of the importance of Tesla's invention, in complete contradiction to the smug decision by Trump to dismiss it, Fitzgerald had gone to the very uppermost echelon of military circles in his continuing quest to develop what he envisioned as possibly the only means to protect the country from invasion by an enemy possibly able to employ an atomic bomb. Craigie in turn wrote to Hoover, doing his best to enlist his help "for the defense of the nation," but the FBI, which checked with Bush, was also unable to locate Tesla's prototype.[24]

The elusive inventor had owed the Hotel Governor Clinton $400 in back rent. During the height of the Great Depression, when one could feed a family of five for less than a dollar, this was a great deal of money. Tesla gave the hotel a device he claimed was the particle beam weapon, having placed a value of $10,000 on it, and warned them that it was very dangerous. Here it was just about a decade later, and Tesla told Fitzgerald that the prototype was still there, locked in a safety deposit box and that he had the key. This story was corroborated by pigeon handler Charles Hausler, who remembers Tesla telling him in the early 1930s that a device Tesla had on the roof of the hotel next to the pigeon coops was "very dangerous."[25] The hotel also provided a handwritten note by Tesla establishing this claim. So the mystery remains as to whether (1) Tesla ever built a prototype of the particle beam weapon, (2) whether he gave this prototype to the hotel as collateral against what he owed them, or (3) whether the wily inventor simply scammed the hotel to escape the debt he owed. The fact that Willis George was not only a paid bandit working for Military Intelligence but also proudly boasted about this activity in a book gives credence to the possibility that he may have indeed snatched the so-called death ray before Trump got there, and if that is the case, it may still exist in some secret holding tank of Military Intelligence.

In less than two days, Trump searched through Tesla's entire estate, comprising, among other things, nearly two hundred thousand personal documents, such as letters, newspaper and technical articles, artifacts, and so on, deciding ahead of time not to open any trunk that was more than ten years old. This would suggest that Trump admitted to not opening somewhere between sixty and eighty crates and trunks, although it is also possible, even probable, that Edward Palmer, the photographer and chief yeoman for Naval Intelligence, on his own, photographed much of the correspondence from this earlier period.

The Library of Congress microfilm archives regarding Tesla, which I have studied, include personal correspondence from the 1890s and early 1900s between Tesla and such people as his secretary George Scherff, Robert and Katharine Johnson, George Westinghouse and the Westinghouse Corporation, J. Pierpont Morgan, and his son, J. P. Morgan Jr. Since Tesla's invention of a particle beam weapon began well before the arbitrary date assigned by Trump (1943 minus ten years), his decision to ignore the earlier trunks was flawed and in fact contradicted the FBI report that suggested that Tesla was less cogent during the last few years of his life, which would make his earlier writings more important than his later ones. In any event, even though time was short, Trump did an extraordinary amount of work, resulting in a dossier known as the Trump Report, which outlined in detailed fashion many of Tesla's key findings and also bombshell events. Sent to the Washington Office of Naval Intelligence, copies of the report were also relayed to the OAP and the FBI and a few years later to Wright-Patterson Air Force Base.

Not only did Trump discover the elusive highly classified particle beam weapon paper, whose details he outlined in his report, he also unearthed the startling discovery that Tesla had sold the weapon to the Soviet Union for $25,000 in 1935 and may have also sold it to the Canadians and the British War Office as well!

A year earlier, in the spring of 1934, Tesla had dined with Samuel Montgomery Kintner, assistant vice president of the Westinghouse

Corporation. Tesla certainly had mixed feelings because he had been trying to sell Westinghouse his radio patents for thirty years. However, by the time his patents had lapsed, Westinghouse was creating its own broadcasting station in direct competition with Marconi, whose patents and organization were being absorbed into RCA under the leadership of David Sarnoff, whom Tesla knew. When Sarnoff and his company received $700,000 or more from the U.S. government when RCA was formed and the government made good on the use of various patents it had usurped during the war, Tesla gained a small percentage of the $23,000 that Fritz Lowenstein received in that settlement.[26]

Because of the dominance of RCA, the Westinghouse Corporation was being maneuvered into a back-seat position. Forced to upgrade their equipment, they spent several million of their own dollars on various patents and accompanying equipment. For instance, they paid Edwin Armstrong $335,000 for his regeneration and superheterodyne patents and probably an additional $200,000 to Lee DeForest for similar patents in order to head off a legal dispute between these two inventors.[27] This could not have sat well with Tesla, whose work lay at the basis of these wireless inventions. However, here it was, well over a decade later. Tesla was in a good mood. Kintner was a colorful man who had invented a "Negro robot," which, in a way, was an imitation of Tesla's telautomaton, that is, his remote-controlled robotic boat. Kintner, in one famous photo, placed an apple on the robot's head and pretended to shoot it with a bow and arrow. They had a grand time swapping stories.

Carried away with the moment, Tesla tried to talk Kintner into purchasing his particle beam weapon, which he billed as a "Peace Ray." The idea was that if all countries had such a powerful defensive shield, war would become obsolete because it would be fruitless to invade another country, as that option would become impossible.

"I have groped for years trying to find some solution of the most pressing problem of humanity," Tesla told Kintner, "that of insuring peace, and little by little, I have been led to the ideal means to this end. . . . [My particle beam weapon] will provide perfect protection to every country . . . and the International Peace Conference will insist on its immediate and universal adoption."

Kintner passed this information on to the research and development division at Westinghouse, but they remained skeptical. Tesla insisted that "I have demonstrated all the principles involved and am going ahead with perfect confidence which all the experts in the world could not shake."

Kintner wished Tesla well, but the Westinghouse representative declined to pursue the matter further.[28] Having been turned down by a major American firm in his quest in this odd way to help bring about world peace, Tesla turned his attention to foreign lands.

Tesla was notoriously flippant when it came to money, but in the 1930s, his income was greatly diminished, and considering that he probably received at this time a check from the Soviets for $25,000 ($500,000 in today's dollars), it's hard to see where all the money went. There is evidence that he may have set up a secret laboratory near the Fifty-Ninth Street Bridge at Third Avenue in Coleman Czito's machine shop, and maybe he did indeed construct prototypes of his particle beam weapon while there. Experimenting is expensive, and Tesla was working on a number of lines of research at that time. For instance, he was bouncing beams off the Moon with Czito's son, Julius, and working to capture cosmic rays as a source of renewable energy, as he told John O'Neill. He probably was also funding this secret laboratory, helping to cover Czito's rent, and he was also working with the architect Titus de Bobula to design a more modern-looking World Telegraphy Center than Wardenclyffe, which actually greatly resembled the large Van de Graaff generators that graced the article Tesla wrote in 1934 for *Scientific American*. The inventor also gave Robert Johnson several thousand dollars in his later years. But what is troubling is that he never covered his other debts, totaling about $2,800, to the Hotels Pennsylvania, St. Regis, and Governor Clinton. If Tesla had really come into such a large sum, one would think he would be motivated to fulfill these obligations. However, this was nothing new for the metropolitan dandy. "His correspondence with [his] lawyer George Foster from 1912 to 1917 [alone] reveals 34 lawsuits, and this was not the total number. In the majority of cases, he was sued for [lack of] repayment of funds related to various liabilities." These lawsuits included claims from "the Gavin Machine Company for . . . work on a steam turbine, the Yellow Taxicab Co. for unpaid taxi bills [and] the City of New York for unpaid taxes," and that was just for one five-year period.[29] When it came to Tesla's unwillingness to pay his rent or cover these past debts, a real character flaw is revealed.

On top of that was the issue of selling this fantastic weapon to the Soviets in the first place. No doubt this was a top secret transaction for a variety of reasons, one being that Tesla's closest friend, Robert Johnson, would have railed against such an arrangement. Had Johnson known that Tesla was selling an advanced weaponry system to the Soviets, this brazen move may have fractured their incredibly close friendship. As ambassador to Italy under Woodrow Wilson's administration, Johnson had written to Wilson's secretary of state, Bainbridge Colby, to congratulate him for what was known as the "Colby Note."

In 1920, Colby had written the Italian ambassador expressing his keen antipathy to the idea of the Soviets spreading their communist ideas to Poland. This, according to Johnson, was in reaction to Lloyd George, the prime minister of England, who saw Poland as "not a nation, but only a mask. And what can you expect from a country that has a piano player as Prime-Minister?" The individual in question was the pianist Ignace Paderewski, who did indeed become prime minister of Poland and who was a good friend of both Johnson and Tesla.[30] What is interesting about this event was that even as far back as 1920, the English were adopting a policy of appeasement, this time to Stalin; the next time, of course, would be to Hitler.

If we jump ahead to the 1930s and the deal Tesla was construing with the Soviets, the election of Franklin Roosevelt in 1932 was a wake-up call for Johnson because Roosevelt had decided to open U.S. doors to the Soviets for trade. The instrument would be the Amtorg Trading Corporation.

Johnson wrote to Colby that he was "deadset" against Roosevelt's policy with the Soviets. "I am doing what I can to oppose this on the ground that the Soviets have organized ... a propaganda [campaign] against representative government here and elsewhere"[31] A year earlier, he had written to Colby, asking, "Have you a recollection of the fact that in August 1920, when the Russians were swooping down on Warsaw, I cabled the Department expressing the hope that you (or it) would see the way to follow the lead of France in friendly policy towards Poland?"[32]

What these letters tell us is that Tesla's decision to help arm the Soviets was not only a very controversial maneuver but also that it went smack in the face of the wishes of Tesla's dearest friend. Did Tesla just do this for the money? Or did he really believe that if every nation had such a device, this would mean the end of war? And if that was the case, why, at this time, did he not offer the device to the U.S. government? Although when war broke out several years later, Tesla did indeed give freely the details of his particle beam weapon to the U.S. military, at this juncture, in 1935, when he was selling the details to the Soviets, his reason for not offering it to the Americans appears to be influenced by his political leanings. Like many of the upper crust, he vigorously opposed the policies of Franklin Roosevelt.

In a letter to Frank Vanderlip, a banking giant and Morgan associate and one of the founders of the Federal Reserve, Tesla said, "There is not the slightest doubt in my mind that I could get the Government to support it if I cared to pull the wires, . . . but I don't like advancing my discovery with money provided by Fascist methods against the wishes of the population." [33]

It is hard to imagine today the latent hostility many people of the opposing political party had for the Democrats or for Roosevelt when he is so often portrayed today in a revered manner. Comparing Roosevelt's presidency to a fascist regime in the 1930s was actually commonplace and had a different connotation than it would have during and after the war. In general, what that criticism was referring to was Roosevelt's centralizing of power much the way Mussolini did in Italy. [34] And Tesla, much like many of the Republicans of that era, was clearly against that. Both Tesla and Johnson were opposed to Roosevelt's policies for almost opposite reasons, Johnson because Roosevelt opened the door for the Soviets and Tesla because of the New Deal.

Stalin had done an amazing job of presenting an idealized image of the Soviet Union to the outside world, so much so that many left-leaning intellectuals were sympathetic to the idea of sharing the wealth, whereas the truth was that he ruled the hidden interior of his country with a murderous iron fist, a reality of which Tesla and much of the West were completely ignorant. Many idealistic communist sympathizers such as the Hollywood Ten and various intellectuals bought into the rose-colored image that Stalin was able to project, and Tesla apparently bought into that image as well, which was precisely the opposite view that his good friend Johnson had.

Although it is apparent from Irving Jurow's letter to me (see chapter 1) that the U.S. government had an inkling that Tesla had sold the details of his weapon to the Russians for a princely sum, it is highly doubtful that Johnson or anyone else, for that matter, except for Tesla's secretary, would have figured out Tesla's real link to a country whose goals were so antithetical to the American way of life. [35]

In the following peculiar letter to Andrew Robertson, chairman of the Westinghouse Corporation, possibly referring to previous correspondence concerning a proposed display of wireless transmission of power for the 1939 New York World's Fair, Tesla makes reference to obtaining precisely the figure of $25,000 for the sale of one of his inventions:

New York
May 22, 1941

Westinghouse Electric & Mfg. Co.
150 Broadway, New York, NY

Attention of Andrew W. Robertson, Chairman

Gentlemen:

Permit me to remind you that recently I approached you with a specific proposal which I explained fully to your engineers. Their attitude . . . caused me a painful surprise. . . . [It concerned] an exhibition of rotating field phenomena on a large scale. . . . With a small field ring and only 100 horse power applied, delicate devices would spin in a radius of twenty feet around the ring, and with a large ring and ample power . . . the phenomena could have exhibited at great distances.

It would have been a triumph for the Westinghouse Corporation. But this avenue being closed, I got in touch with a concern which had three scientists of world repute on their staff. . . . As a business basis, they paid me twenty-five thousand dollars and promised to see me again in a couple of months.

On returning, they reported that their tests had proved satisfactory and that they were ready to go ahead. This meant much money for me but I had devoted years to raising poultry which involves several billions of dollars annually. The artificially raised poultry is unfit for the human stomach. The eggs are leathery and the meat is saturated with oils owing to which the body can not assimilate the proteins. With the help of a very able . . . chemist I have prepared a compound which I have named "Factor Auctus," signifying "Creator of Growth" which . . . given [in] daily measure . . . to the fowls and pigeons separate from their food . . . yields wonderful results. You will be grateful to me when you get the delicious eggs and meat obtained by this revolutionary process.

Believe me,
Yours very truly,
Nikola Tesla[36]

Factor Auctus became a trademark created by Tesla, using a rooster as its symbol. It was a special food made for pigeons and chickens and probably also for humans from hearts of cabbage, carrots, cauliflower, leeks, celery hearts, white potatoes, sweet potato, spinach, fresh tomato, white turnips, lettuce hearts, and tapioca.[37]

This letter is peculiar for many reasons. Robertson had undermined Tesla in a small book he wrote about George Westinghouse that essentially attributed the discovery of the rotating magnetic field, invention of the induction motor, and AC polyphase system to William Stanley! Robertson curiously also wrote about the animosity that society often feels toward the great inventor while he so blatantly disrespected Tesla in the same treatise by ignoring him for Stanley.[38]

The psychological mechanism known as projection comes to mind, whereby the writer attributes to others traits he himself possesses. For the head of the Westinghouse Corporation to attribute literally the cornerstone of their company to the wrong person is an incredibly bizarre example of the animosity that existed in certain factions of the Westinghouse camp towards the Serbian inventor. Thus, when Robertson talks about the animosity society has for the individual inventor, it seems to me he is actually talking about himself.

On first reading, Tesla's letter to Robertson reads like it is from a crazy person because of the dramatic non sequitur. The letter starts out discussing a technique of wireless transmission of energy associated with Tesla's rotating magnetic field and ends up talking about feeding chickens and pigeons a special diet. Tesla is quite old by this time, headed into his mid-eighties. He is not well, and it may be that his thinking is clouded. Or it may be that he is angry with Robertson for stripping Tesla of his iconic discovery. Robertson is chairman of a company that literally owes a huge chunk of its livelihood to Tesla's invention. So Tesla has either gone a little bit nuts or he is writing a letter to somehow get back at Robertson.

Another point, however, is that Tesla mentions receiving $25,000 for the sale of an invention, the exact figure he most likely received from the Soviets five years earlier when he turned over the details of his particle beam weapon to A. A. Vartanian of the Amtorg Trading Corporation.

This was by far the most important revelation in the Trump Report, namely that Tesla sold the details of this coveted device to an enemy nation. Keep in mind that the Soviets were using propaganda in attempts to undermine America's ideology and that shortly after the deal, they aligned with Nazi Germany when together they invaded Poland. It wasn't until Operation Barbarossa, the Nazi war plan for the takeover of Ukraine, that the Russians became our allies. The date was June 22, 1941, a date that marked the turning point in the battle because Hitler had stupidly ensured a two-front war.

Why did Hitler invade Russia? Part of the reason was paranoia. Since he could easily breach an agreement, he assumed Stalin could as well. However, the main reason had to do with his ultimate goal, namely expanding his territory into Ukraine so that he could wipe out indigenous Slavs, whom he saw as inferior, and repopulate this rich farmland with his superior Arian progeny. After studying the situation, it becomes more understandable why Hitler invaded France to begin with. He feared that if he attacked Russia first to take over Ukraine, France, since it had a treaty with Russia, would come in and he *would* be stuck in a two-front war. So he went west first to create a buffer zone so that he could then go east to engulf a large fertile area that could feed his people and allow them to expand. Since so many British soldiers had escaped when Germany stormed into France during the miracle at Dunkirk, this was not the best move because Hitler still risked a two-front war. However, the Allies were severely crippled at this juncture, with nearly two million French, British, and Polish soldiers captured with the invasion of Belgium and France, and, most importantly, America in June of 1941 was still neutral.

Had Japan not attacked Hawaii when it did, it is quite likely that America's entrance into the war would have been further delayed, and that might have given Hitler the extra time he needed to complete his takeover of Russia. The cold winters, which literally killed between three and four million German soldiers, may have still been too much of a loss to overcome. Literally millions simply froze to death. Hitler, as an arrogant and ultimately self-destructive narcissist, was just plain pig-headed not to listen to his generals and retreat with the onset of winter, but the two-front war that ensued, coupled with the United States' entrance just six months after Operation Barbarossa, laid the groundwork for the final blow. Had Japan not attacked Pearl Harbor, or had it attacked six or eight months later, it is quite possible that Hitler may have perfected his jet planes or further developed his V-2 rockets, which were planned to skim along the ionosphere so as to hit American targets like New York City, and had this happened, the United States would have been in an entirely different position, and thus may have been forced to settle for a treaty to end the war, and had that happened, Hitler would have stayed in power. The more one studies World War II, the more it becomes apparent that the Allies were fortunate to have prevailed. That

is one of the reasons why Tesla's particle beam weapon was so important
to key factions of the armed forces, who no doubt were flabbergasted by
Trump's dismal conclusion that there was nothing of military value to the
wizard's inventions. Trump put it this way:

Exhibit F—Trump Report

"The New Art of Projecting Concentrated Non-dispersive
Energy Through Natural Media." This undated document by
Tesla describes the electrostatic method of producing very high
voltages capable of very great power. This generator is used
to accelerate charged particles, presumably electrons. Such a
beam of high energy electrons passing through air is the "concentrated non-dispersive" means by which energy is transmitted through natural media. As a component of this apparatus
there is described an open-ended vacuum tube within which
the electrons are first accelerated.

The proposed scheme bears some relation to present
means for producing high-energy cathode rays by the coop-
erative use of a high-voltage electrostatic generator and
an evacuated electron acceleration tube. It is well known,
however, that such devices, while of scientific and medical
interest, are incapable of the transmission of large amounts
of power in nondispersive beams over long distances. Tesla's
disclosures in this memorandum would not enable the concentration of workable combinations of generator and tube
even of limited power, though the general elements of such a
combination are succinctly described.[39]

There are a number of ways to look at what occurred. It certainly is true
that a particle beam weapon along the precise lines that Tesla laid out in
his plans has yet to be realized. Trump also had much experience generating extremely high voltages and constructing particle accelerators. So, to
that extent, Trump was correct. However, Trump makes several fundamental errors in his assessment by suggesting that Tesla planned to "accelerate
charged particles, presumably electrons," and that the voltages he would
produce would be inadequate in any event to send such particles a hundred or more miles. Tesla's article, which appears to be a precursor to a top
secret patent application, states explicitly that he planned to emit particles
of tungsten or minute droplets of mercury from this open-ended vacuum

tube, *not* electrons. Further, Tesla notes that the production of thirty million volts would be insufficient for repelling such particles out the barrel of the gun. "By the application of my discoveries and inventions, it is possible to increase the force of repulsion more than a million times, and what was heretofore impossible, is rendered easy of accomplishment."[40]

The end result was that Trump's report dismisses Tesla's invention, and this was accepted by Homer Jones, chief of Investigation and Research for the NDRC, and by Frank Jewett, another NDRC official who happened to also be working on the Manhattan Project, but neither of these men did their homework. They accepted Trump's conclusion but did not study the actual Tesla paper, which explicitly refutes Trump's criticisms. The final report said it this way:

> With respect to the so-called death ray machine . . . Dr. Frank Jewett, President of the National Academy of Science . . . advised that if Dr. Trump is willing to stake his reputation we may be assured that there is nothing in the ideas of Tesla which would be of interest to the United States Government. Dr. Waterman has also advised . . . that Dr. Trump can be relied upon and his opinion should be considered final.[41]

Jewett, who was known to be egotistical and perhaps even petty,[42] went so far as to go public with this critique, cited in *Electrical Engineer* in August of 1943: "In the interval since Tesla did his really great creative work, his name has been associated with so many projects of questionable scientific quality that I imagine they may have raised doubts in the minds of many as to his real achievements."[43] Given that Bell Labs was the corporation to develop the Osprey helicopter-airplane based on Tesla's patents and that Jewett was the director of Bell Labs, this criticism gains a particular form of irony, but it also reflects the sentiments of a significant percentage of people in the know in terms of their opinion of Nikola Tesla.

In May of 1943, just a few months after Tesla's death, J. Edgar Hoover contacted Vannevar Bush "to seek his advice on the mysterious subject of 'death rays.' Hoover apparently was intrigued by the idea of killing people from afar with powerful beams of energy, and he wanted Bush to check it out. Since this wasn't the first inquiry about death rays—nearly three years earlier, the Army had inquired about a similar fantasy weapon,—Bush made quick work of Hoover's query, replying that 'it does not seem to me that this disclosure warrants further inquiry.'"[44]

What this passage suggests is that Bush, like Jewett and Trump, may have also dismissed completely Tesla's ideas, and if that were the case, then like his cohorts, Bush would have miscast the invention, as Tesla stated explicitly that he had *not* conceived of a death ray but rather of a weapon that shot out minute particles, which is a very different animal.

Given Bush's great intellect, if indeed he had misconstrued Tesla's actual plan, he would have certainly been responsible for pulling all funding in the development of this intriguing and potentially revolutionary creation.

Bush's biographer, G. Pascal Zachary, took Bush to task with regards to his dismissal of rocketry, in particular because Bush "never considered" rocketry pioneer Robert Goddard, who, much like Tesla, was often denigrated in ivory towered enclaves. "I don't understand how a serious scientist or engineer can play with rockets," Bush said before the war. Now, by 1944, with his country in serious jeopardy and the Germans raining death on London with their supersonic V-2 rockets, Bush realized his shortsightedness, stating that "a fully developed guided missile program might even revolutionize many aspects of warfare." Yet, still as late as December of 1945, when he testified before a Senate committee, Bush reiterated his position that a "high angle rocket . . . in my opinion, such a thing is impossible. . . . I say technically, I don't think anybody in the world knows how to do such a thing and I feel confident it will not be done for a very long period of time to come." In fact, within eight years, the United States was developing long-range missiles along this very same path, and just a few years after that, we sent a man to the Moon.[45] Bush's missteps with regards to not seeking advice from Goddard were to Zachary "puzzling, if not egregious."[46] The same might be said about his easy dismissal of Tesla.

Further, this negative attitude towards Tesla on the particle beam weapon probably carried over to the rest of his work, which included a global communication system that used ELF waves, which would have been a huge advantage to the Allies, who were desperately trying to track German U-boats, and Tesla's advances in remote control, which had direct applications for radio-guidance systems, sending encrypted messages, and also for torpedoes and drone warfare.

Bush, of course, was very interested in how to create reliable radio-guided weapons, so much so that he came to author an article with James Conant on how to train kamikaze pigeons to be placed inside of warheads to peck a screen that would keep the aerial bomb on target until it reached its destination.[47] The idea was scrapped, but it shows an odd similarity concerning an interest in the intelligence of pigeons with a man whose ideas he seemed to dismiss.

On the other hand, with Bloyce Fitzgerald leading the opposition, key professors at MIT and military personnel at Wright-Patterson Air Force Base took quite seriously the inventor's particle beam paper. Thus, there were two opposing camps when it came to studying Tesla. Some people were keenly interested in understanding the inventor's entire scheme, so there is a good chance that, through the years, numerous highly classified prototypes were constructed that may very well have helped provide the basis for what President Reagan, in the 1980s, would come to call Star Wars technology.

15

The Russian Connection

Exhibit Q—Trump Report

An agreement dated April 20, 1935 between Nikola Tesla and the Amtorg Trading Corporation, in which Tesla agreed to supply plans, specifications, and complete information on a method and apparatus for producing high voltages up to fifty million volts, for producing very small particles in a tube open in air, for increasing the charge of the particles to the full voltage of the high potential terminal, and for projecting the particles to distances of a hundred miles or more. The maximum speed of the particles was specified as not less than 350 miles per second. The receipt of $25,000 fee for this disclosure was acknowledged in this agreement, which was signed by Nikola Tesla and by A. [Vartanian] of the Amtorg Trading Corporation. The method referred to in this agreement [i.e., Tesla's article "The New Art of Projecting Concentrated Non-dispersive Energy Through Natural Media"] is an effort by Tesla to clear up the questions raised by Soviet engineers after the subject disclosure had been made. . . . [But] it should . . . be expected that this disclosure subsequently proved unworkable.[1]

B y studying recently declassified Russian documents and Tesla's correspondence with the Soviets at the time of the Great Depression, some important aspects of their relationship come to new light.

In July of 1934, after obtaining an incredible amount of publicity about his so-called death ray, Tesla walked into the Soviet Consulate to meet with Consul General Leonid Tolokonsky. A generation earlier, in 1911, Tesla had gained the ear of Czar Nicholas II, who was eager to obtain Tesla's technology for the wireless transmission of power because of the czar's wish to modernize his country. This desire, in fact, continued after the czar was deposed, when Vladimir Lenin seized control, sending his own communiqué concerning an invitation for the inventor to come to the Soviet Union to help implement his AC hydroelectric power system there.

In the 1920s, it was common practice for sympathetic Americans to travel to Russia to help build the new Soviet country, and in fact, I myself had an uncle, Philip Steigman, an architect who made such a trip to help in that capacity.

Tesla told Tolokonsky about his invention, which could direct electronic rays to bring down airplanes and stop tanks. When questioned why Tesla was coming forth with such an offer, the Serbian spoke of their common Slavic heritage, Tesla's sympathy for the Russian people and "as a Jugoslav," Tesla's appreciation of the fact that "the Russians went to war on our behalf" in the Great War. His weapon, Tesla argued counterintuitively, would actually be an instrument of peace. It was billed as a defensive weapon because it was so devastating; if all countries had such devices, war would become obsolete.

Tesla's fee for providing the details of what Tolokonsky would soon learn was not really a death ray but rather a particle beam weapon was $25,000. Needing more information, the consul general contacted Commissar Nikolai Sinyavsky, chief of the Red Army's Signal Division, who looked over the proposal and assigned G. P. Brailo, Representative of People's Commissariat of Heavy Industries from Amtorg Trading Corporation, and two electrical engineers/party members stationed in America to meet with Tesla, T. Smolentsev and A. A. Vartanian,[2] the latter of whom invited the fellow Slav to his apartment, where he could meet the Soviet engineer's wife and children.[3]

Tesla's friendship with Vartanian, which began in 1934, immediately blossomed. Chief of the Engineering Department and involved with the sales and purchase of machinery and inventions for Amtorg Trading Corporation, located at 261 Fifth Avenue, Vartanian maintained their friendship until Tesla's passing, exchanging Christmas cards and other correspondence throughout the years. In March of 1935, Tesla forwarded a magazine to Vartanian that was most likely the *Scientific American* issue that contained both his cover story on Van de Graaff and a stylized photograph of Tesla fashioned by Napoleon Sarony. Shortly thereafter, a legal document with a fancy green cover titled "Agreement between Nikola Tesla and Amtorg Trading Corp, representative of the U.S.S.R." was drawn up, whereby "the party of the first part [Nikola Tesla] provided the party of the second part information and data necessary to construct a complete installation embodying and utilizing said [particle beam weapon and related] inventions with all accessories required for . . . effectively prevent[ing] aerial raids and other forms of attack." The document, which was signed by both parties, further stated that Tesla will provide all "such technical advice . . . necessary for completion of same."[4]

Amtorg was formed in 1924 by combining a company created by Armand Hammer with a Russian "products-exchange corporation." This enterprise, which started out as part American, became the purchasing and trading agent for the Soviet Union. With over one hundred employees, many in America, Amtorg began trading with such American corporations as International Harvester, Ford Motor, RCA, DuPont, General Electric, and "more than a hundred other companies during the First Five-Year Plan." No doubt, Amtorg also served as a front for Russian spies, but it remained a legitimate trading arm to the Soviet Union for the next fifty years.[5]

In short order, Tesla revealed to the Soviets the following characteristics of his invention:

1. A method for "crushing matter [such as mercury or tungsten] into tiny particles approaching the size of a molecule."
2. The ability to electrify the particles to a specific charge.
3. The ability to create a tremendous repelling force with the capability to hurl such particles out the barrel of a gun to a distance of hundreds if not thousands of miles.

> As Tesla explained to the Soviets, "The underlying idea of my process is to impart to a given mass the greatest possible kinetic energy. This is accomplished by subdividing the mass into innumerable minute parts which can then be given a prodigious velocity and then consolidated in their effect at great distances. In this manner, the kinetic energy of a projectile, as the bullet of a rifle, can be made millions of times greater than when the mass is projected in the usual way."[6]

Vartanian reported to Sinyavsky that these tiny projectiles could be directed in a straight line at almost the speed of light to pierce and destroy any object that would lie in their path, so then a dossier was compiled and forwarded to Moscow.

On April 4, 1935, Staff Officer M. Galaktionov of the USSR Defense Commissariat conveyed this information to Joseph Stalin, chairman of the

Politburo, and his two most trusted henchmen, Lazar Kaganovich and Kliment Voroshilov.[7] The date is very important because the following year, Stalin began his murderous campaign, and the men he put in charge of those mass executions were these two. Voroshilov, known as the "Sword of Stalingrad," with two towns named after him, was chairman of the Presidium of the Supreme Soviet Commission of Defense from 1925 to 1940. In 1936, one year *after* the deal with Tesla, Voroshilov personally signed the execution orders for approximately two hundred military leaders whom Stalin saw as possible threats. And this method was continued in 1940, when Voroshilov oversaw the deaths of hundreds of Polish military leaders as well.

Not to be outdone, Kaganovich, Stalin's closest aide and deputy chairman of the Consul of Ministers, saw to his role in the Great Purge by organizing the arrest of thousands of political opponents, railroad administrators and other executives, suspected saboteurs and foreign spies, and he personally signed the execution order for hundreds of other supposed enemies.[8]

Tesla, along with most Americans, was unaware of these activities, but also it should be kept in mind that his deal was negotiated one or two years before this murderous purge took place. Having been aware of the Soviet Union's interest in the use of directed electromagnetic waves for destructive purposes since the 1920s and their inability to perfect such a weapon, after tantalizing them with the overview, Tesla waited for a response.

Stalin and his cronies discussed the possibility of inviting Tesla to Moscow so that he could oversee the construction of such a weapon. Of course, what they didn't tell him was that once he got there, he would never have been able to leave. Coincidentally, that very same year, 1934, another scientist, Peter Kapitza, fell into this very trap. Having emigrated to England to work with Ernest Rutherford at Cambridge University on the development of powerful magnetic fields, Kapitza had returned to Russia to visit his parents. Forced to stay in his home country for the next fifty years, Kapitza was given his own research facility and went on to win a Nobel Prize for his work in superconductivity.

And then there is the simple fact that three years after the deal was consummated, Stalin ordered the execution of at least three, and perhaps four presidents of Amtorg: Saul Bron, Pyotr Bogdanov, Ivan Vasilyevich Boev and possibly Isay Khurgin, the first three all executed in 1938 during the Great Purge and Khurgin dying in a suspicious boating accident.[9]

Had Tesla been younger, perhaps he might have been tempted to make the trip to oversee the production of the weapon. And had he done so, his fate would have paralleled that of Kapitza. However, the inventor was now an old man and therefore declined the offer. Instead, the aging but still sharp wizard was happy to direct the Soviet engineers through contact with Vartanian and through the mail if his plan was accepted.

In April of 1935, Tesla finalized the sale for the sum of $25,000, a deal structured with a stipulation to obtain additional funds should the device prove to be a success, and also the right to notify the United States and also sell the design to both the United States and his home country of Yugoslavia.

Voroshilov was concerned that Tesla might be scamming them. However, due to the inventor's patents and acclaim, the Politburo complied and sent Sinyavsky, "head of the Red Army Communication Department to America to allocate the capital from their Reconnaissance funds," whatever those might really be.

After conferring with his inner circle and his engineers, Stalin agreed that the risk to pay the inventor such a sum was worth it, but he also insisted on a deadline of four months to hand over the full plans.[10]

Dmitry Kruk, a Tesla researcher from Russia, states that this deal would have had to be okayed in Moscow by the Politburo. This supposition is supported by the fact that for the deal to be completed, Tesla negotiated not only with Vartanian of Amtorg but also with K. M. Narusky, "the Soviet consul in New York."[11]

DEATH RAY

While shooting scenes for *The Tesla Files* at the Tesla Museum in Belgrade in June of 2017, I saw the original contract and also the document Tesla gave the Soviets, and it turned out to be precisely his article "The New Art of Projecting Concentrated Non-dispersive Energy Through Natural Media," which I was familiar with because I had been present when it was first revealed. It was the summer of 1984 when Andrija Puharich stunned the crowd at the International Tesla Society Meeting held in Colorado Springs, where I also presented a paper, because he unveiled the long-sought but never-before-seen top secret document.

As Branimir Jovanovic, the museum curator, thumbed through the document, I saw the various diagrams of the weapon, which were the very same diagrams that Puharich had presented in 1984. This sale resulted in additional correspondence between Tesla and electrical engineers in the Soviet Union. For instance, on November 8, 1935, which was several months after the deal was consummated, Tesla wrote, "My dear Mr. Vartanian, I have finally managed to accumulate sufficient energy for an answer to the questions of your scientists and engineers in Moscow relative to the draft of my proposal dated May 16th of this year. The matter embraces many features and might be enlarged upon with profit, but my physical condition still compels me to confine myself to essentials. Your experts have erred in all their contentions simply because of their unfamiliarity with the subject. . . . Nevertheless, their adverse criticism of my statements is quite welcome, as it gives me an opportunity for further and clearer explanation.

"To be frank," Tesla went on, "I was not a little amused by the opinion expressed . . . that my electrostatic generator was a modification of an old and impractical device, which reminds me that some opponents of the alternating system claimed my induction motor to be nothing more than a 'slight modification' of the well known direct current motor. When your engineers become better acquainted with my machine, they will change their views. The belt generator has been experimented with for the past sixty years and was found of little value. Indeed, this type has already been abandoned and attempts were being made to replace the coil by a driving chain." [12]

What the Soviet engineers did not understand was precisely what MIT scientist John G. Trump also did not understand, namely that Tesla's driving chain was not a replication of what could be found in the old models but rather a new revolutionary design that replaced the circulating cardboard belt that Van de Graaff used to transfer the charge with "an ionized stream of air hermetically sealed in the vacuum chamber. Analogous to the way a shock can be created and transferred by rubbing one's shoes along a carpet on a dry day, the new fluid airstream belt achieved the same end but to a degree 'many times greater than a belt generator.' This charge, which apparently could be as much as 60 million volts, was in turn transferred to a bank of condensers in the cupola, the myriad small bulbs atop the tower, their round shape and internal structure constructed to augment the accumulation of energy." When the charged particles came into contact with this high-voltage stream, these microscopic projectiles were repelled out the barrel of the gun at a terrific speed and would cover a distance, according to Tesla, in excess of one hundred miles. Because this was a stream of minuscule pellets as opposed to a "ray," in theory there would be no dispersion, but rather the gun would operate much like a laser. [13]

What the Soviets did not initially also realize was that the actual cost of the weapon, which would be seventy-five feet in height, would amount to hundreds of thousands of rubles, and so it would take many years, well after Tesla's death, for the Soviets to develop a working model.

We know that the Soviets continued to work on perfecting this device because a schematic of it appeared in a 1977 issue of *Aviation Week and Space Technology* in an explosive article titled "Soviets Push for Beam Weapon."[14] The article reveals that "in recent public pronouncements," the newly retired head of Intelligence for the U.S. Air Force, Major General George H. Keegan, "has taken the CIA to task for having rejected Air Force intelligence information about Soviet beam weapon development. He also has spoken bitterly about a number of top U.S. physicists who refuse to accept even the possibility that the Soviets are involved in beam weapon development. . . . The U.S. attempted unsuccessfully to develop a charged-particle beam device under the project code name Seesaw. It was funded by the Defense Dept.'s Advanced Research Projects Agency but abandoned after several years."[15]

At that time, Tesla's paper was still classified, and that meant that no copies in America had been published. It wasn't until 1984, with Puharich's landmark presentation, that the particle beam paper became available to researchers and the similarity between the Soviet design and Tesla's design became apparent. Because Trump had downplayed the importance of the invention, I reasoned that may have explained why this highly classified paper was released to Kosanović in the early 1950s along with the rest of Tesla's estate. Since Yugoslavia was still behind the Iron Curtain, it seemed reasonable to conclude that the Soviets had simply obtained the paper from the Museum in Belgrade. However, now that I have personally seen the correspondence between Tesla and the Soviets, I have been able to establish that Tesla actually did sell the details to Russia in 1935 and that he probably did receive a large check, perhaps even the full $25,000, at that time for the sale of the details.

FLIGHT

Looking at this event as a kind of jigsaw puzzle, we can see that Tesla's highly classified particle beam weapon paper was studied by numerous key military and governmental officials in the United States, including colonels and brigadier generals from Wright-Patterson; top officials from Harvard, MIT, and Cal Tech; officials from the U.S. Army and Military Intelligence; scientists from Washington; the man in charge of the Manhattan Project; and probably the president of the United States. Simultaneously, this paper was also kept from public view for over fifty years and only came to light through clandestine means.[16] Yet, even though the War Department sat on

Tesla's estate for nearly a decade, eventually, during the height of the McCarthy period, the government also willy-nilly sent the paper to a communist regime! On some level, this makes little sense, except if the government also knew that the Soviet Union, and probably the Yugoslavian government, already had the details.

Kevin Burns, head producer of the TV show *The Tesla Files*, speculated that Trump's dismissal of the Tesla particle beam weapon paper might have been disinformation and that key details were in fact removed from Tesla's effects.

Support for this theory can be found in statements made by Sava Sremac, who, in 1951, was one of the first individuals to unpack the sixty-odd trunks of Tesla's estate that finally made their way to Belgrade. Reflecting back sixty-six years, Sremac stated that in the process of systematically organizing Tesla's correspondence and scientific treatises, he found that there were pages missing and sections from some of the notebooks had been ripped out. When pressed by interviewers, Sremac speculated that governmental officials from the United States may have removed these papers before Tesla's estate was shipped out. However, it seems just as likely that individuals in Belgrade, enamored of their hero, may have also removed occasional papers to save as trophies, or they simply made certain key documents inaccessible to protect particular lines of research.[17]

Solid evidence that some of Tesla's work has remained classified can be established by reviewing some recently declassified CIA documents which are labeled "Confidential/NOFORN," meaning that the information derived was so hush hush that it was not to be shared with foreign governments. Through an interview I did with an anonymous source in the Summer of 2020, I also discovered that in the early 1990s, DARPA hired several Tesla experts in a top secret experiment to attempt to replicate Tesla's particle beam weapon. Following Tesla's classified paper trail which Puharich had brought to the world, this DARPA crew made use of Tesla's invention of the valvular conduit "to maintain a vacuum." Through this means, air-flow was increased to "almost a sonic level" and copper vapor and then carbon vapor were used instead of tungsten to demonstrate the principle of the directed beam weapon with the goal of "depositing [this material] on a remote target." Similar experiments have been carried out by Lockheed Martin.[18]

Another clear example involves Tesla's "flivver plane," a helicopter-airplane he invented and patented in 1921 that evolved into the $75 million Osprey tilt-rotor aircraft. For a half-century, when one would research its etiology, one would be hard-pressed to find Tesla's name associated with its development. Had Tesla not patented the idea, it would have been next to impossible to establish his role in this aircraft's development.

Staying with aircraft, Tesla also created a flying wing that he called a "reactive jet dirigible," which was designed to be either lighter or heavier than air, circa 1910. It could be argued that this was a precursor to the stealth bomber or the as-yet-to-be-built space shuttle replacement, the X-37, and again Tesla's name cannot be found when exploring the history of flying wings or jet aircraft.

RADAR

Another example that was directly linked to the period in which Tesla died was in the field of radar. Starting in 1917, at the height of World War I, while working with sci-fi editor Hugo Gernsback, Tesla described precisely the invention of radar, based, to some extent, on his 1896 experiments with X-rays and also the work of Heinrich Hertz. Expanding on the theories of James Clerk Maxwell, Hertz speculated that electromagnetic waves could bounce off metallic objects. This idea was demonstrated by the German engineer Christian Hülsmeyer in a 1904 invention for "an obstacle detector and ship navigation device," which Hülsmeyer patented in several countries, but the idea was ignored by the kaiser's military.[19] A dozen years later, during the Great War, one of the biggest problems the Allies had was figuring out a way to defend against the German U-boats that were sinking ships at a rapid pace.

As we remember, Tesla had been working for Telefunken, helping the Germans improve their wireless communication system, until America entered the war in 1917. However, after the sinking of the *Lusitania* and Germany's overall treachery, including the use of poison gas and Germany's attack on Serbia, Tesla cut all ties. "Aghast at the pernicious ... regime," Tesla accused Germany of being an "unfeeling automaton, a diabolic contrivance for scientific, pitiless, wholesale destruction the like of which was not dreamed of before. ... Such is the formidable engine Germany has perfected for the protection of her Kultur and conquest of the globe."[20]

"Suppose," Tesla suggested, "we can shoot out [intermittently] a concentrated ray comprising a stream of minute electric charges at a tremendous frequency, say millions of cycles per second, and then intercept this ray, after it has been reflected by a submarine hull, for example, and cause [it] to illuminate a fluorescent screen. . . . Then our problem of locating the hidden submarine will have been solved." [21]

This article, which appeared in the *Journal Gazette* (Fort Wayne, IN), was derived from a cover story in Hugo Gernsback's highly popular *Electrical Experimenter* from August 1917, wherein Gernsback informed the reader "that several important electrical war schemes will shortly be laid before the War and Navy Departments by Dr. Tesla, the details of which we naturally cannot now publish."[22]

Tesla's dialog with the Navy cannot yet be confirmed, but based on additional evidence, it appears that Tesla also discussed with the military countermeasures to radar, such as a description of various components that would not reflect or absorb electromagnetic waves and other cloaking mechanisms. "The Teutons are clever, very very clever, but we shall beat them," said Tesla confidently.[23]

As difficult as it is to believe, this fundamental idea of how to employ radar was ignored by the Navy, not only in World War I but also through the first half of World War II. Vannevar Bush had devised a submarine detector method during World War I when he worked in the antisubmarine division in Connecticut with Frank Jewett, later the president of Bell Labs, and, like Tesla, his idea was also snubbed. Jumping ahead a quarter century, Bush was still facing incredible resistance during the Second World War from Admiral Ernst King, head of naval forces, as scores of Allied ships were continually being sunk off the East Coast throughout 1942 and the first few months of 1943. The obdurate admiral could not abide the idea of a civilian and, as he perceived, a line-cutter, dictating war strategy to his fleet. However, Bush, who had the backing of President Roosevelt, persisted, flatly telling King that he might understand military strategy, but he was not equipped to rule on technology. Bush, the premier scientist, quite simply had the upper hand. In a face-to-face confrontation, the admiral, by April of 1943, was begrudgingly forced to capitulate. Radar was implemented, and within two months, forty or more Nazi subs were sunk and the rest of the German fleet retreated.[24]

Time magazine covered the watershed event in April of 1944 within a cover story on Bush. "To U.S. scientists, the turning point of WWII was March 20, 1943. That was the day that U-boat sinkings began to sag like a beaten fighter." *Time* went on to say that "the struggle against U-boats was primarily a battle of technology. . . . When the chips were down, the U.S. scientists won."[25]

Curiously, even though Bush had his own submarine detector method going back to the First World War, radar took a very long time to develop.

Tesla's clear and full-bodied description of radar technology from 1917 coupled with the findings of Hertz and Hülsmeyer were sadly ignored, scores of ships were sunk, and hundreds, if not thousands, of needless deaths ensued because the Navy in this arena dragged its heels. Further, even though Tesla published his ideas in popular magazines and newspapers, one would be hard-pressed to find his name mentioned in traditional accounts of the history of radar.

However, this was only one type of radar system Tesla had invented. The technologist extraordinaire also had a worldwide system that dated to May of 1900, when he explained the idea in his patent #787,412, "The Art of Transmitting Electrical Energy Through the Natural Medium," granted in 1905. Within the patent, Tesla states explicitly his plan for a global radar scheme directly linked to the operation of his wireless tower. By varying the length and frequency of stationary waves, Tesla could not only pinpoint the transference of wireless energy to theoretically any location on the planet, he could also "easily determine without the use of a compass, the course or speed of a vessel. . . . [By installing stationary waves in fixed positions] in judiciously-selected localities, the entire globe could be subdivided into definite zones of electrical activity and . . . important data could thus be obtained by simple calculation or readings by suitably-graduated instruments."[26]

Well after World War II, from the 1950s through the 1980s, a similar mechanism using ELF waves was adopted by the Navy under a program known as Project Seafarer. From a centralized station, the Navy proclaimed, "virtually worldwide coverage" and the location of submarines could be achieved, even if the subs were several hundred feet under the water. Although they would have to surface to talk back, the centralized station could still track their whereabouts. Following Tesla's lead, perhaps such a system could have also been configured to detect enemy subs as well.[27]

Tesla's global tracking system was clearly spelled out in his fundamental wireless patent #787,412, which was crucial to the Navy for combating Marconi's wrong-headed assumption that his work was superior to and predated Tesla's. Because of this and related patents, Marconi was forced to capitulate, saving the Navy a considerable sum and, perhaps more importantly, saving their pride. But still, Tesla's work and contributions were kept underground. For many years Tesla became a nonperson in some quarters, a derisory footnote, and had it not been for his incredible photographs and John O'Neill's wonderful biography, *Prodigal Genius*, it is quite possible that Tesla's name would have all but disappeared, even though his work was central to so many key inventions that had military value.

16

Negotiations with the British Empire

Let me recall to you the air attacks on London during the war. Search-lights picked up the German raiders and illuminated them while guns fired, hitting some but more often missing them. But suppose instead of a searchlight you direct my ray? So soon as it touches the plane, this bursts into flame and crashes to Earth.[1]

—HARRY GRINDELL MATTHEWS

In 1914, the British War Office offered a reward of £25,000 for any-one who could come up with an invention that could search out and destroy incoming zeppelins. The inventor and jack-of-all-trades Harry Grindell Matthews successfully demonstrated a remote-controlled device that made use of a selenium cell to seek out and control a boat. This apparently satisfied the requirement, and he was thereby able to claim the prize. Having previously demonstrated a wireless system to converse with airplanes, Grindell Matthews upped his game considerably a decade later when he claimed to have invented a death ray that could take out incoming aircraft. He asked the British government for another £25,000, gained the backing of the press, and threatened to sell his "diabolical ray" to the French if the Brits did not come through.

Caught up in a death ray propaganda campaign, the Russians claimed "the potency of Grammachikoff's ray, so let Russia alone!" But this was topped by Marconi, who announced his own death ray for the Italians, corroborated by Mussolini's wife, Rachele, who claimed that Marconi stopped a variety of cars and trucks on a highway by sending out an electronic beam that disabled their instrumentation. Not to be outdone, the Germans proclaimed that they had not just one inventor but rather "three German scientists who perfected apparatus that can bring down airplanes, halt tanks and spread a curtain of death like gas clouds of the recent war!"[2] This pitch was

241

so successful that one German, Erich Graichen, received "invitations from Mexico and Chile" to demonstrate his weapon's power.[3]

The courts got involved and blocked Grindell Matthews from selling his device, which, unfortunately, was never successfully demonstrated. Wary of being scammed, it is possible that the British high command may have received some solace from the inventor in America, Nikola Tesla, who wrote, "It is impossible to develop such a ray. I worked on that idea for many years before my ignorance was dispelled and I became convinced that it could not be realized. This new beam of mine consists of minute bullets moving at a terrific speed, and any amount of power can be transmitted by them. The whole plant is just a gun, but one which is incomparably superior to the present."[4]

If we go back in time to the 1890s and to the etiology of Tesla's interest in this realm, two events stand out. Soon after the invention of the light bulb, Tesla experimented with something he called a button lamp; he placed carbon in the center of a globe and created a condition whereby the electrical rays rebounded off of the interior of the glass, thereby bombarding the filament to such an extent that it would actually disintegrate. Tesla also used a "ruby drop" as the filament. "Magnificent light effects were noted. . . . In this manner, an intensely phosphorescent, *sharply defined line* [emphasis added] corresponding to the outline of the drop [fused ruby] is produced." I made the case that he may have, in fact, invented the ruby laser at this moment, but never realized what he had achieved. Certainly all the key components to how a ruby laser works are inherent in this design. The second event was Roentgen's discovery of X-rays. Along with many other inventors, Tesla experimented with X-rays, and he was able to produce such images at distances of fifty feet or more from the source of the rays. Thus, we see early on Tesla's experiments with the destructive power of electric energy and also the ability to transmit rays over significant distances.[5]

When Tesla met King Peter II of Serbia during World War II, although weak, suffering fainting spells and near death, he predicted that "Hitler will fall, and soon Yugoslavia will be resurrected. . . . I hope that I shall live to see [that] most joyous moment . . . when you return to a great and happy Yugoslavia. Your Majesty, God bless you."[6] Unfortunately, Tesla's predictions would come only partially true. Peter would never return as king. Rather, he would be replaced by Josip Tito, a Croatian leader and revolutionary Communist, who, because he was married to a Serb, was able to cobble together a united Yugoslavia that lasted as such for fifty more years.

Due to Tesla's realization that Hitler was a real threat to the stability of Europe, five months after he sold the details to the Soviets, in September of 1935, Tesla began an impassioned correspondence with the British War Office and the very head of the Canadian armed forces, offering his services to them as well and insisting that his particle beam weapon was by far the best way for Great Britain to protect itself from a Hun invasion.

Having just turned 80, Tesla sent the following extremely detailed letter to the British War Office:

August 28, 1936
His Majesty's Principal Secretary of State for War Office
London, N.W. 1

Sir,

The following proposal to His Majesty's Government set forth in detail below, is respectfully admitted under the seal of secrecy.

EXPLANATORY REMARKS:

1. In the course of scientific . . . investigations carried on for many years, I have made important discoveries . . . enabling me to project particles of matter to great distances with high velocities which may even approximate that of light, thereby producing destructive and other effects of virtually any desired intensity.

   However unbelievable, it is a fact that in this manner, not only . . . the thickly settled districts in its vicinity, but all . . . visible surrounding country, can be rendered entirely immune from damage to life and property by an attack of aeroplanes no matter what their number, speed and altitude of flight. As to . . . Zeppelin[s] or other types, they could not approach the protected region without their crews being killed, the engines stopped and the bombs exploded. Protection may also be afforded in foggy weather and in the darkness of night.

2. The energy . . . emitted by my methods . . . is not subject to dispersion . . . but travels on and on in a state of unchanging concentration such that power amounting to thousands of kilo-watts may be transmitted through a channel of less than one hundred thousandths of a square centimeter . . . and, from the same machine, not only through the Earth's atmosphere but

also with equal facility through interplanetary space to unlimited distance.

3. This new art of transmitting . . . non-dispersive energy seems . . . destined to create a revolution in many fields and may, perhaps, be viewed by posterity as my best achievement. It is certain that the principle will be of profound influence on the future development of implements of war. Just to illustrate its immense possibilities . . . , consider a shell having a weight of 400,000 grams and a muzzle-velocity of 100,000 centimeters per second and imagine that the same weight were represented in minute particles which my machine can readily project with a speed of two hundred thousand times greater. In this case their aggregate kinetic energy would be forty billion times that of the shell. Evidently then, to equal the energy of the shell, a quantity of particles weighing only one hundred thousandths of a gram would be required.

4. The particles might be effective at thousands of kilometers but, of course, the useful trajectory is limited and will be almost perfectly rectilinear, the gravitational pull during the quasi-infinitesimal time of transit being negligible.

5. [As] the areas of impact . . . are very small, enormous instantaneous pressures and temperatures should be attained with the result of piercing not only thin aeroplane sheathing by also thick armor.

6. Obviously, in repelling attacks by armies . . . ten thousand men could be temporarily or permanently stopped at distances exceeding many times the range of an ordinary rifle with an expenditure of kinetic energy not greater than that imparted by one bullet. . . .

7. Most of the results contemplated are possible only by the use of electrical pressures much greater than any heretofore employed. This subject has received my concentrated attention for nearly forty years and in 1899 I had already mastered the art sufficiently to operate a large wireless transmitter at twenty million volts though not without difficulties and danger. It enabled me, among other things to flash a current around the globe, demonstrating terrestrial resonance and

produce a powerful "death ray." Numerous improvements in methods and apparatus which I have invented . . . make it easy to generate and control safely electric potentials from fifty to one hundred million volts by which I expect to realize practically wireless transmission of power. Noting with . . . pride the extensive use of my alternating system in His Majesty's domain I hope to achieve a similar success in this new field. . . .

8. It will also be capable of dispelling fog. . . .

9. A plant of one thousand kilowatts capacity, comprising a small power house and a tower twenty-five meters high, would be sufficient for the safety of London and the surrounding districts. To protect Great Britain a few of such plants would have to be used. . . .

PROPOSAL TO HIS MAJESTY'S GOVERNMENT:

This proposal is made with the sincere desire of producing a great service to the world, and . . . of advancing my work in the interest of science and humanity. I have an unbounded admiration for the people of Great Britain and believe that these . . . discoveries of mine would not be more beneficially employed than by protecting His Majesty's domain from a possible calamity which would be of disastrous consequences to all human progress. . . .

I am sure, I can render this great service to Great Britain, [and] feel . . . I should be aided in the attainment of my unselfish aims.

1. As the first [step], I propose to submit to His Majesty's Naval and Military experts, within three months from the date of acceptance, complete drawings and specifications of an operative plant referred to above, with all the details and necessary devices. This is . . . a special and very difficult task, to which I must devote my full attention, excluding other business. To serve all outlays . . . a sum of fifteen thousand pounds would have to be provided . . . half . . . paid in advance. . . .

2. My specifications [for cost of production]:

    (a) [An] immediate [expenditure] of five hundred thousand pounds [to build said plant].

(b) A guarantee by His Majesty's government that when the plant is built and tested and found to operate [so] it can protect London and the surrounding districts, . . . another payment will be made to me of ten million pounds. . . .

(c) His Majesty's government will further guarantee me . . . an annual royalty of one pound per kilowatt of the rated capacity of this as well as other plants . . . built according to my specifications.

(d) Coincident with the approval of my plans, I would offer my services for one year, [and] furnish additional drawings and specifications . . . for which I would receive ten thousand pounds.

Believe me, Sir,
Your obedient servant
Nikola Tesla[7]

After spending several weeks to consider Tesla's fantastical proposal, Mr. Coombs, the director of Artillery from the War Office in London, wrote back, stating that more time was needed to consider it.[8] Tesla acknowledged receipt of the communiqué and reiterated to the London War Office that he also had a more compact design of "a most powerful gun" that could be fitted onto ships and that he waited for their formal response. It came the following month.

H. L. Lewis, the director of Artillery from the London War Office, acknowledged Tesla's opening "explanatory remarks . . . made in connection with your discoveries and inventions for the projection of particles of matter to great distances of high velocity." The scheme, he said, was "carefully considered by an Inter-Departmental Scientific Advisory Committee," but the committee "was unable to form an opinion as to its scientific possibilities [due to] lack of necessary technical details." Lewis requested "a fuller description of the means you propose to employ in order to produce the destructive effects you claim. . . . Further details are necessary before the expenditure of public funds can be considered."[9]

Tesla seemed to stall for time, and nearly a year went by as, meanwhile, his comprehensive proposal got bumped up to higher circles. Major General A. E. Davidson, director of mechanization, who had kept up correspondence with Tesla during the interim, met with Lieutenant General Sir Hugh Elles of the British War Office as well as other military officials and, according to Tesla, possibly Neville Chamberlain to discuss the best way to proceed with this complicated proposal.

Wounded in the Great War in the Second Battle of Ypres, Lieutenant General Sir Hugh Elles had been a machine gunner and commander-in-chief of a tank brigade who personally led a squadron of 350 tanks against the Germans at the Battle of Cambrai, continuing in this position until the Germans were defeated in 1918. Two decades later, Elles tried to retire but was called back because of yet another German threat, and he was placed in charge of civil defense. Having studied the correspondence, he decided to call in General Andrew McNaughton, an electrical engineer, army officer, head of secret weaponry development via the National Research Council of Canada (essentially Vannevar Bush's Canadian counterpart), and "a favorite of Winston Churchill."

Like Elles, McNaughton had also been wounded in the Great War, and his expertise went beyond tanks. Having made "advances in the science of artillery," McNaughton invented the cathode-ray compass for airplanes and "introduced the idea of using an oscilloscope as a forerunner of radar, selling the invention to the Canadian government for ten dollars."[10] He was the perfect choice to continue the dialogue with the Serbian inventor.

McNaughton, who would go on just a few years later to command the Canadian Army during World War II and make the short list with Lord Mountbatten and General Eisenhower to be supreme commander of Allied forces in all of Europe, was described as "having unforgettable eyes that stab and brood" and as a man with "a perpetual cock in his left eyebrow that sets off a face that has a deep touch of thought."[11] During the war, due to some unpleasantries and a clash of egos, the general ceded his command, but after the war, McNaughton would advance to become minister of defense, a member of the United Nations Atomic Energy Commission, and temporary president of the U.N. Security Council. An exceptional leader who, during the war, would soon be meeting with Winston Churchill, Charles de Gaulle, and, in Washington, Vannevar Bush and President Roosevelt to discuss secret weapons and to help coordinate efforts against the Nazis,[12] McNaughton took Tesla's invention seriously.

Writing directly to the inventor in July of 1937 in a "Secret and Personal" communiqué, McNaughton let Tesla know that he had met with Sir Elles and had read through the correspondence the inventor had sent to the secretary of war. "I understand from Sir Elles that I am to send two of our scientists to visit you in New York for the purpose of discussing this invention with you, and that they are to prepare a confidential report on its technical features for transmission to the War Office so that the whole matter may be given further consideration.

"I have entrusted this task to Dr. D. C. Rose and Mr. B. G. Ballard of our Division of Physics and Electrical Engineering and I would inquire as to when and where it would be convenient for you to receive them.

"I can assure you," McNaughton concluded, "that I welcome this opportunity of making direct contact with you for, as a young engineer, I studied your great contributions to the art of electrical transmission, and particularly your work leading to the development of the rotating magnetic field and the induction motor."[13]

Tesla was elated to receive this communiqué from one of the leaders of the free world, a man he very much respected, and he was impressed that the general had assigned such esteemed scientists as Donald Charles Rose and Bristow Guy Ballard to the proposal. Where Ballard was a physicist, electrical engineer, and after the war vice president of the Nuclear Regulatory Commission, Rose was chief superintendent of Canadian Army Research and science advisor to the General Staff during World War II. Much like his counterpart, Robert Millikan in the United States, Rose was head of cosmic ray research for the Nuclear Regulatory Commission.[14] Clearly, unlike Bush's people in the country where Tesla resided, McNaughton had gone all out to understand precisely how Tesla's revolutionary invention worked.

From Tesla's point of view, apparently trying to rewrite history, we can see from his initial letter to the British that almost a half century after his rash decision to rip up his royalty clause with George Westinghouse during a heated moment in the so-called Battle of the Currents, the mysterious Serbian genius designed his proposal so as to not make that same mistake twice. Unlike the portrait often painted of the inventor as someone completely altruistic and not concerned about monetary gain, although entering his eighties, he was very much interested in money and placed a hefty price on this new *Star Wars*–like defensive shield.

From Tesla's point of view, the deal had been structured in a reasonable fashion, asking for, in today's dollars, an advance of about $600,000, informing the British War Office that the cost of production would be about $40 million in today's dollars (£500,000 in 1935) and then asking for a huge bonus, approximately $85 million in today's dollars, and a continuing royalty should his plans be successful. One way or another, the British had to come up with the advance if they wanted to get the ball rolling.

Unfortunately, during these critical negotiations, Tesla was clobbered by a taxicab. Knocked to the ground yet rolling back to his feet, the inventor cracked a few ribs and suffered internal injuries. However, rather than go see a doctor, he medicated himself.

July 18, 1937
A. G. L. McNaughton, Esq.
President, National Research Council
Ottawa, Canada

Sir:

Your letter of the 6th . . . found me in a trying situation. Five weeks before I had met with an accident and sustained serious injuries. My birthday was approaching and two ambassadors were coming from Washington to deliver to me . . . certain high distinctions [Order of the White Lion from Czechoslovakia]. . . . Although in great pain and unable to move, I managed . . . to recover sufficiently and went through the ordeal on July 10, 1937 with a smile on my face. But when I retired the trouble returned and I have been confined to my room ever since. Just as soon as I am well enough I shall communicate with you and place myself at the disposal of your experts.

You are a man who has gone through the war doing more than his duty and you will appreciate that all my previous achievements are insignificant compared to what I am giving the world now. I am anxious to help the cause of peace and hope that active work can be started very soon.

With expressions of admiration, I remain, Sir,
Yours most respectfully,
N. Tesla[15]

The following day, Marconi died of heart failure, and in honor of his passing, BBC radio and its affiliates created a radio silence that lasted two minutes. Meanwhile, correspondence with the British Empire continued, with Tesla showing due respect to McNaughton, who was diligently trying to get the elusive wizard to meet with his two scientists, Drs. Rose and Ballard. Well aware of their expertise, Tesla, one would think, would have made every effort to accommodate them, but instead, he kept finding ways to postpone the meeting. The wizard's delaying tactics could have been caused by either his wish to cement a financial deal before revealing the details or because he was unsure that he could adequately explain this advanced invention without actually constructing and demonstrating it. There is also the possibility that Tesla was bluffing, but the complete discussion herein is a counterbalance to that facile position.

In September of 1937, Tesla stated that he was "ready to furnish all information which can be reasonably expected from me at this juncture [but] I feel that any agreement between us contemplating secrecy would become automatically a scrap of paper in case of war. You would be on the side of England and reveal everything as you should, mindful of your military duty."[16]

Concerned that an invasion of England seemed imminent and therefore to promote the deal, it is possible that Tesla may have forwarded an article from *Popular Mechanics* from July of 1934 on a related idea of generating "Invisible Dust which can halt airplane engines in midair and can be hung as a vast curtain or protective wall about a nation's borders to guard against sky invaders. . . . Ten thousand planes flying into one of these curtains would all be destroyed, the inventor averred, and [he envisioned] force projecting plants set up every 200 miles along a nation's borders, each shooting rays 100 miles on either side."[17]

Unable to comprehend why the Brits continued to balk, Tesla boldly wrote, "I am at a loss to understand why His Majesty's Government has not acted more promptly in a matter of such vital moment. Perhaps my price is considered too high." Then, like a salesman, Tesla moved in for his close. Having already stated that his price would involve many millions of pounds for construction and a pretty penny for himself, he said he was "emphatically of the opposite opinion [that the price is too high]. My friend Carnegie received 350 millions of dollars for his plants and my friend J. J. Hill 165 millions for ore lands discovered by me and purchased by him for an insignificant sum on my advice. My proposal is of importance transcending any industrial one. If I were making it now I would certainly ask much more, for I am offering something that no other living man can give."[18]

Tesla's throwaway line about railroad magnate James J. Hill was certainly intriguing. Founder of the transcontinental Great Northern Railroad, by the late 1890s, Hill had expanded his base by pairing the Great Northern with J. P. Morgan's Northern Pacific Railroad. From about 1897 through 1899, at a time when Tesla and the railroad baron were corresponding, Hill had completed the purchase of a number of iron ore mines in the state of Minnesota that did indeed yield him a fortune. Perhaps Tesla played a role in confirming one or more mining purchases by using a combination of terrestrial frequencies, such as ELF waves, to monitor this locale in his experiments involved in global radar and/or telegeodynamics, which was a way for Tesla to locate ore by discerning underground mineral deposits using oscillators "bolted to rocky protuberances" to resonate the Earth for underground seismic exploration.[19]

Tesla ended the letter "assuring [McNaughton] that as soon as circumstances permit, I will communicate with you in regard to our proposed meeting." [20]

Giving Tesla the benefit of the doubt, the general followed up two days later with yet another "Secret and Personal" communiqué, "looking forward to seeing you later at any time of your convenience," [21] and then patiently or impatiently waited five more months before trying again. In the interim, Tesla received yet another communiqué from the War Office declining to make any payments until Tesla met with representatives of the British government so that he could explain and also prove his brazen claims. Displaying a genuine respect for the aging inventor, they were also showing unusual restraint.

Meanwhile, on the home front, the British offered a reward of £100 to anyone who could kill a sheep at a hundred yards by directing an electrical charge. If the Brits could come up with their own death ray, they would no longer need Tesla's expertise. It was for this reason that they called on the Scottish scientist Robert Watson-Watt to look into the matter. Unable to design such a device, Watson-Watt instead told his English overseers that their money would be better spent developing radar.

Not completely convinced by Watson-Watt's failure and having received a letter from Tesla on December 7, 1937 completely dismissing the idea that a "death ray" constructed from electrical waves would ever work, key individuals in the British War Department continued to press Tesla for details before they would agree to write a sizeable check. From the Hotel New Yorker came this rather heated, distasteful, and yet prophetic response:

February 8, 1938
The Director of Mechanization
The War Office
London, S.W., 1
84/T/3458 (M.G.O. 4 b)

Sir:

The uncertain allusion to terms in your letter of January 7, 1938 has greatly astonished me. With all deference due to an official of exalted position I cannot refrain from remarking that I am no Jew or Gypsy but a man of principle unused to bartering. I have named my price, Sir, and if His Majesty's Government offered me one half-penny less I would not give it my discoveries and inventions for electric protection and my co-operation in their application.

The best way to prevent another international cataclysm, possibly greater than the World War of twenty years ago, [is] to make England immune from all kinds of attack. . . . But in view of your statement, I can no longer hold in abeyance attractive proposals from other countries. . . . Although my system of electric power, light and heat distribution . . . is used on a huge scale and entirely indispens[ible] to the comfort and safety of the people [of England], this means nothing to His Majesty's Government which is amazingly short-sighted and penurious. Such policy exposes the country to the gravest of dangers.

You are, undoubtedly, versed in history and know what happens when an infuriated people rise thirsty for blood and vengeance. This would be likely to happen in England if they should suddenly realize that the country is at the mercy of others owing to the Government's failure to accept repeated offers for protection. The inevitable sequel would be a revolution which might engulf the Empire. As ever before, those who are in command would be made the scapegoats and I can predict, with almost mathematical certitude, that your own distinguished career would be quickly and tragically terminated.

I am unmoved by any personal interest and perfectly sincere in my warning . . . not to throw away lightly a unique chance. It would be well for you to read a German book on "Electrical Warfare" recently published in Vienna. . . . [The author] question[s] whether a perfect electric arm for military use can be produced, and if there [is] anywhere on this globe a man [who] can do it, . . . [it] is Nikola Tesla. What I have practically developed surpasses the author's wildest dreams. . . .

You have written me repeatedly suggesting a meeting with Major-General McNaughton's representatives. But in such a conference without preparation I would be unable to . . . prove essential principles to the satisfaction of his experts who have their reputations at stake.

Their report under such circumstances could not materially change the situation. If we are to achieve success, I must meet them prepared to give an elaborate reply to every question and leave no doubt whatever . . . as to rationability and operativeness of every part of my . . . plan. Then, as in all similar cases in my professional career, they will make an enthusiastic report which will awake the War Office from its sleep. To do this I have to prepare a condensed specification and drawings . . . good enough to serve the purpose. This work would take about six weeks but can be done . . . for a relatively small sum of money.

I shall write more fully . . . to Major-General McNaughton, and you will hear from him.

In the meantime I have the honor to remain,
Sir, your obedient servant,
N. Tesla[22]

Simultaneously, as he promised, Tesla penned to McNaughton a more restrained rebuttal. A key complaint Tesla had was that he expected to be paid for his consultation. Simultaneously, to the general, he dropped his advance from £7,500 to £1,200:

February 10, 1938

Major-General McNaughton
President, National Research Council
Ottawa, Canada

Sir:

Since our correspondence of July and September, 1937, I have received two letters from the War Office in which stress [is placed] on my proposed conference with your representative[s]. Knowing that a meeting between us was in conformity with your wishes, I would have put myself gladly at their disposal had I not found, on careful reflection, that the outcome would be unsatisfactory . . . and could only hurt my cause. Permit me to explain.

My system I have perfected . . . is the result of many years of theoretical and experimental research and happy inspiration. Its practical realization was made possible through revolutionary discoveries I was fortunate to make . . . of new methods and means for the generation, control and transmission of non-dispersive energy under tension exceeding twenty million volts I attained in 1899 and of such intensity as to destroy attacking aeroplanes at great distances and explode the enemy's shells while in flight. The system embraces an immense variety of subjects in the domain of mechanics, electricity, physics and other branches of industry and science so that many volumes might be written in describing it.

To your experts this is all terra incognita. In a brief conference, I could not describe [the] many novel devices, much less demonstrate, to their full satisfaction, the principles embodied in my system. . . . Under such circumstances, it is obvious that our proposed meeting would fail in its purpose.

If something is to be done I must meet your representatives well prepared to answer every question and to prove everything [and] furnish, virtually, all information to be contained in my full specifications which the disgusting stinginess of the British Government has prevented me from producing—a view that may cause the fall of the Empire. To this end, it would be necessary to prepare condensed specifications, drawings and diagrams.

You are a man of action who does not sleep over a proposition and for this reason I have devoted each thought to determine the time and cost of such a preparation, finding that the papers might be completed in six weeks for twelve hundred pounds, not including a reasonable compensation for myself.

In view of pending . . . proposals I am anxious to dispose of this matter speedily and I shall be obliged if you will communicate with London and signify your pleasure at your earliest convenience.

With expressions of highest regard, I remain,
Very truly yours,
Nikola Tesla[23]

This statement dating the onset of his particle beam weapon to his time at Colorado Springs squares with an interview Tesla gave the local *Gazette*

Telegraph shortly after he arrived on June 2, 1899, wherein he stated that he had an "instrument" at his lab that had the capability of slaying "30,000 men in an instant."

McNaughton, ever the gentleman, replied two days later, assuring the aged wizard in another "Secret and Personal" communiqué that "you can be assured our technical officers are looking forward to their visit with you" and that the general would "communicate at once with the war office authorities," as Tesla suggested, presumably to pave the way for Tesla to be paid for his consultation.[24]

Two months later, after receiving another communiqué from the British War Office, Tesla further explained his position. The Brits assumed that he had a working model of his weapon, but Tesla explicitly stated (1) that his mention of its point-and-shoot feature was not based on his actual weapon, but it was analogous to how an ordinary rifle operated, and (2) that although he had "demonstrated experimentally and in other ways every feature of his plan . . . since the destruction of my plant, I have not been in a position to repeat my demonstrations."[25]

What Tesla is apparently saying is that he never constructed a fully working model of his particle beam weapon. It is therefore reasonable to conclude that his "other ways" were theoretical. Further, his plant was destroyed in 1917, but according to my research, it became incapacitated long before that. In 1904, the abilities of his wireless tower were severely crippled when the Westinghouse Corporation removed Tesla's main generator for lack of payment of $25,000.

However, even considering Tesla's own timeline and the fact that Tesla's lab at Wardenclyffe, for all intents and purposes, was still somewhat viable up until about 1914, he still suggests not only that he never built a fully functioning particle beam weapon but also that whatever experimental work he did accomplish was undertaken easily a quarter of a century earlier, although he may have worked on aspects of the device in the 1920s. We know, for instance, that as far back as 1896, Tesla was taking X-rays at distances exceeding fifty feet from the source of the rays, and Tesla experimented with various other kinds X-ray tubes, "producing rays of much greater intensity than it was possible to obtain with the usual appliances."[26] Through the years, he no doubt built numerous aspects of the gun. For instance, in May of 1899, while lecturing before the Commercial Club of Chicago, Tesla "said he had succeeded in converting a column of gas into a solid column with the rigidity of steel."[27] Although that was not a particle beam weapon, this statement certainly suggests that Tesla was involved in curiously analogous experiments for many years and had created, as far back as 1898, the means to seal off an open-ended vacuum tube with a rapidly flowing column of air.

According to electrical engineer and Tesla researcher Leland Anderson, Tesla had constructed open-ended vacuum tubes in the 1920s probably at Czito's machine shop. As we see above, Tesla certainly had the means of sealing the open end, using, in his own words, "a gaseous jet of high velocity" which would still allow the ejected particles to pierce through.[28] That said, it is still highly unlikely that he ever constructed a device capable of taking down an airplane. Most likely, he experimented with small prototypes.

According to Nancy Czito, Julius Czito's daughter-in-law, in about 1918 and probably into the early 1920s, Tesla would go out with Julius with a device "to bounce electronic beams off the moon."[29] This, of course, is not a particle beam weapon, but it does suggest that Tesla was still experimenting and constructing apparatuses along those lines well after his laboratory and tower were destroyed. Another point to keep in mind is that the only source I have for this work is the interview I did with Mrs. Czito when I met her at a showing of the movie *The Secret of Nikola Tesla* that premiered in Washington, DC, in the early 1980s, which was shown in association with the unveiling of block-of-four postage stamps honoring Nikola Tesla and three other inventors, Philo T. Farnsworth, Charles Steinmetz, and Edwin Armstrong.

Tesla's actual experiments in this realm were top secret. He never published findings associated with his experiments with particle beam weapons or with bouncing beams off of the moon.

Add to that the inventor's serious decline in health since he was hit by the taxicab in June of 1937, and it is fair to conclude that Tesla was being either wholly unrealistic or incredibly optimistic to think he had the energy and wherewithal to truly construct a working model of his grand scheme, even if the Brits did come through. Simply put, sadly, age had caught up to him. Nikola Tesla was just too old and frail to be of much physical help.

Nevertheless, based on his track record and the cogency of these reports, it is also reasonable to conclude that had the Brits invested serious funds at this time, there is every reason to believe that Tesla would have taken them a long way towards accomplishing his goal, namely of constructing a working model that may very well have had the capability to shoot down enemy craft at great distances.

In March of 1938, the Anschluss occurred, with Nazi Germany annexing Austria, followed quickly, in September, with Hitler taking over the Sudetenland, snatching it from the Czechs with the pathetic acquiescence of British prime minister Neville Chamberlain. With the Nazis arresting the Jews or driving them out of German-occupied lands at a rapid pace, the horror was magnified a quantumfold with "the shot heard round the world," when on November 9, 1938, Herschel Grynszpan, a Jew, shot and killed his Nazi lover, Ernst vom Rath, at the German consulate in Paris. Having reneged on a promise to help Grynszpan save his parents' fate of internment in Poland by denying them a visa to France, vom Rath paid the ultimate price.

This assassination triggered an anti-Semitic rage known as Kristallnacht, or "the Night of Broken Glass," during which 1,000 synagogues and as many as 7,500 Jewish businesses were destroyed, 30,000 Jews were arrested and nearly 100 German Jews were shot or beaten to death. Germany was a tinderbox, and so Tesla's warning about the vulnerability of the British empire became that much more dire.

Add to this the fact that at Christmastime 1938, two radio chemists from Berlin, Otto Hahn and Fritz Strassmann, changed the course of history when they discovered nuclear fission.[30] Those in the know, such as General McNaughton, who was Vannevar Bush's counterpart in Canada, and numerous theoretical physicists could now easily envision the nightmare scenario that the Germans might develop an atomic weapon before the West did.

In July of 1938, writing from Canada, McNaughton sent Rose to New York to confer with the inventor. Having "recently returned to us from the Cavendish laboratory in England [en route to Bell Labs], it occurred to me that you might be interested in having a talk with him. He is charged to carry to you an expression of my highest regards and the hope that sometime we may have the honor of a visit from you at the National Research Council [here in Canada]."[31]

Several months later, the undersecretary of state of the British War Office thanked Tesla for his previous communiqués and requested more detailed information. Assuming that any information provided "would not be . . . in service of corporations" but rather used for military purposes, Tesla complied. Having reviewed his thinking on how to deal with the problem that an electronic ray would disperse over distance, Tesla noted that "particles very much larger than molecules have to be projected through an open tube [to destroy] any obstacle in their way. . . . My first tests were conducted with a tube one end of which was closed by an airtight metal cap carrying a small pipe for supplying the particles . . . [with an electrostatic charge of something like] 20,000 volts per centimeter." Using Maxwell's equations, Tesla went on to calculate "precisely the surface density of such minute metal spheres of suitable roundness." And then he discussed the details of the electrostatic force required to repel these tiny pellets out the barrel of the gun to "produce destructive actions at great distances. . . . I have employed the device for years in my investigations and consider it a revolutionary invention which besides affording an ideal means for the attainment of the [particle beam weapon] will also yield other scientific and practical results of incalculable consequence."

The weapon, as mentioned earlier, would be embedded in a Van de Graaff–like tower, or "charged elevated spherical terminal of radius R," approximately seventy-five feet in height and able to generate as much as 7,500,000 volts. And then the letter ended on this highly technical note:

> In order to obtain a higher potential, it was [indispensable] to increase the radius and in my earliest designs conformable to the general scientific view, I planned a terminal of 1000 centimeters radius for a working pressure of 10,000,000 volts. But subsequently, I discovered new means and methods enabling me to charge a terminal of 250 centimeters radius for a pressure of 60,000,000 volts. . . . [Such a terminal would generate] the highest [voltages] produced in any known way. They have been thoroughly demonstrated on long continued practical tests and are second in importance only to the open tube. By taking advantage of these two far-reaching inventions, it is perfectly practical to impart to a particle over 200 million times the energy and propel it 40,000 times faster than by the means heretofore available.[32]

Tesla went on to say that even though he had achieved this great success, "I was still very far from my desired goal." Two problems he outlined were the need to insulate the apparatus from the extreme temperatures and the need to locate the perfect compound to create his microscopic bullets.

> These [problems] consumed much of my time and effort until I was fortunate developing devices of elemental simplicity supplying perfect spheres of heavy metal which may be of any use from [measurable] to molecular dimensions. Their flow can be

controlled and thus [their] charge, energy, speed and rate of performance [can be] instantly varied merely by manipulation of the open tube serving as projector.

All these and other particulars will be accurately described in my specifications and should convince your Scientific Advisory Committee that I approached you not with presentable claims but with accomplishments. . . .

It would be easy to provide the moderate sum necessary to cover the cost of the drawings and specifications. You surely realize that no matter how big your army and fleet, you would not prevent some enemy planes from breaking through the barrage. The difficulty in protecting a city or a battleship from aerial attack is largely due to the fact that the means for defense cannot be effectively employed until brutal conflict with the enemy is established.

My inventions make it possible to stop the attacks at great distance all the more readily as it is highly probable that the new agent of destruction will produce a very demoralizing effect on them.

Trusting that you will honor me with a reply at an early date,
I remain, sir, your obedient servant,
Nikola Tesla[33]

In reading through and analyzing this extraordinary correspondence, it becomes apparent that Tesla, at the age of eighty-two, was still brilliant and keen of mind. He had not only conceived of a weapon that, similar to a laser, would direct a pencil-thin stream of microscopic particles charged to a particular rate traveling at a terrific speed, he had also calculated their size using advanced mathematics, (e.g., Surface = $4 \times 3.1416 \times 4/10^{26} =$ square centimeters) and also calculated their lethality, the tiny stream of charged particles being able to bring down incoming zeppelins or airplanes by punching holes in their encasements. With this kind of precision, the wizard had thus accomplished his goal, namely of laying out a well-thought-out plan of how to protect England from a possible invasion.

It would seem, however, with the Second World War on the horizon, coupled with the frailty of Tesla's health, the problem of bureaucracies, and the general reluctance of the British War Office to truly commit the kind of resources necessary to undertake this enormous task, that Tesla was, as he had been so many times before, simply too far ahead of the curve for these plans to come to fruition. Clearly, the British War Office took this scheme seriously, but they needed a commitment of the brightest of minds to fully comprehend the wizard's ideas as well as the time, energy, finances, and patience to undertake such a complicated task that certainly would have involved numerous stumbling blocks along the way. Building a particle beam weapon along the lines Tesla envisioned would probably have taken a commitment comparable to something like the Manhattan Project or sending a man to the Moon.

THE YUGOSLAV CONNECTION

Although Tesla's dialogues with General McNaughton and the British War Office were established in writing as "secret and personal," this did not deter the aging wizard from sharing the full cache of his 1937 British negotiations with the Yugoslavian government via its American ambassador, Konstantine Fotic.

On the one hand, Tesla had indeed breached his promise of secrecy; however, giving him the benefit of the doubt, Tesla had his eye on the larger picture, protecting not only England, but also his sacred homeland from the growing possibility of an enemy invasion.

When looking at the history of World War II, its start is usually placed as September of 1939, when Germany and Russia invaded Poland. But the seeds of this second world war were planted years earlier, with the assassination of King Alexander of Yugoslavia by a Bulgarian fanatic during the king's state visit to France in 1934, and by Italy's invasion of Ethiopia and Mussolini's corresponding treaty of cooperation with the Third Reich and Adolf Hitler in 1936.

A further issue of concern was Josip Tito, who was gaining power by heading up the Yugoslav Communist party which had close ties to Joseph Stalin and the Soviet Union. These were dangerous times for Yugoslavia, and Tesla was fully convinced that if his homeland began construction of his particle beam weapon, it would be protected from incursions whether they came from the west or the east.

Like all Serbs including Tesla, Ambassador Fotic was a staunch supporter of the new sovereign ruler of Yugoslavia, King Peter II, even after Tito took over the country, a move that began in 1943 and culminated in 1945. In November of 1937, Fotic sent Tesla a check for $400, and again, in 1941, for $2,000, this time via D. Stanojevic, Royal Yugoslav Consulate General.[34]

Fotic read carefully the British War Office communiqués and those with General McNaughton and noticed how harsh Tesla was in some of the dialog. At the same time, with the advantage of hindsight, Tesla aptly predicted for the British precisely what did occur. To the British War Office, on December 7, 1937, he wrote, "I think that the British people worry too much about aeroplane attacks with poisonous gas. The nation that would attempt this would commit suicide. But a cunning enemy would, in all probability, resort to a different means of attack by which you would be absolutely defenseless." Such an enemy would not use "aeroplanes, guns, battleships, submarines or armies." Instead, this type of nemesis "would strike from great distance and risk nothing himself. His sole object will be to raze all cities and places of importance while avoiding loss of life." This of course predicts the advent of the ultimate drone strikes, the Nazi V-2 rockets which did indeed destroy a good portion of England's urban centers.

"As to my compensation," Tesla revealed to Fotic that he sought ten thousand pounds for the drawings and specifications from the British, and an additional "ten million pounds immediately upon satisfactory demonstration. . . . I feel," Tesla went on, "that the terms submitted are very reasonable especially when considering the vastness of the British Empire which covers over thirteen million square miles, with almost one-quarter of the whole world's population." [35]

In response to the ambassador's concern over Tesla's haughty tone in a number of Tesla's letters to the British covered earlier in this chapter, the inventor doubled down. "You are correct that the English were not happy to read my notes and I am sorry that I did not add: 'and how the descendants of the fearless old Britons run like scared rabbits at the slightest gesture of an upstart in power.' . . . They will certainly inform General McNaughton . . . and I would be surprised if I received immediate response." [36] But at the same time, Tesla himself feared for his own life because the Nazis had made several overtures to obtain the secrets to his weapon, his apartment had been ransacked, and simultaneously, he was under scrutiny by the FBI.

"Opportunities for the use of my discoveries . . . are limitless," Tesla wrote Fotic, but "the danger to my life is great. I hope that you will soon receive news from Yugoslavia." [37] Unfortunately, just as the Brits and the U.S. military continued to pass on this invention, the Yugoslavs also followed suit.

Four years later, Hitler made good on the threat by invading Belgrade, killing, according to Tesla's calculations, "twenty thousand innocent lives . . . and the best part of the beautiful city laid in ruins." Begrudgingly describing the Führer as an *Übermensch*, having on the one hand taken his defeated nation to a world power, Tesla also saw the German leader as woefully misguided. "This fiendish act [of leveling Belgrade] proved Hitler to be an enemy of mankind who should be destroyed."[38]

Fotic would come to denounce Tito and resign his ambassadorship. Authoring *The War We Lost: Yugoslavia's Tragedy and the Failure of the West*, Fotic echoed the consternation that most Serbs felt, referring to their disappointment in the West's support for Tito and their corresponding failure to come to the aid of Serbia's rightful ruler, King Peter II, who was deposed in the process.

In analyzing this period with regards to Tesla's invention of a particle beam weapon, particularly from the American side, what these documents suggest is a real split in the philosophy and assessment of the viability of Tesla's particle beam weapon. One group, headed by John G. Trump and Frank Jewett, completely dismissed this aspect of Tesla's work, while another group, headed by Bloyce Fitzgerald, Colonel T. B. Holliday, and Brigadier General Laurence Cardee Craigie, fully embraced what was now known in inner circles as "Project Nick." Clearly, the very highest echelons of government were involved in considering this device. In Russia, there is every indication that the final decision to purchase the details ended up on Joseph Stalin's desk. In Great Britain, Tesla tells his biographer, John O'Neill, the decisions reached as high as Neville Chamberlain, who was prime minister at the time. We know for sure that key generals were also involved, including Lieutenant General Sir Hugh Elles, head of civil defense for Great Britain, and General Andrew McNaughton, the highest ranking military leader in Canada, head of secret weaponry development for a branch of the British Empire, a friend of Winston Churchill, and close associate of Elles.

A question remains as to how Vannevar Bush read the situation and whether he sided with Trump, Jewett, and Homer Jones in dismissing Tesla's invention or if he supported in any way the engineers at Wright-Patterson or the professors at MIT. Bush probably accepted the Trump conclusion that Tesla's death ray was bogus and turned his attention to the Manhattan Project, but it is also reasonable to conclude that he probably was aware that his counterparts in Russia, Great Britain, and Canada, and most likely in Germany as well, took Tesla's plans seriously. According to Bush biographer Pascal Zachary, at the height of World War II, "in this dark time, Bush began pushing to give his researchers a bigger role in forming strategies that took full advantage of new weapons."[39] Tesla's particle beam device certainly fit that bill.

Zachary suggests that Bush was being influenced by Waldemar Kaempffert, science editor for the *New York Times* and a long-time editor at *Scientific American*, who, coincidentally, had written a number of articles on Tesla spanning four decades. Having once called Tesla "an intellectual boa constrictor," Kaempffert pointed out that "outsiders spurred change; it was these interlopers who triggered the collapse of systems and the rise of new ones." In war, he insisted, "all the revolutionary means of killing on a wholesale scale came from *outsiders*, that is from technologists who were not professional soldiers."[40]

As mentioned earlier, even before the war, Bush was meeting regularly with McNaughton, as the two countries were working closely together on numerous projects. Since many of these meetings took place while Tesla was alive and gaining a lot of press for his so-called death ray and since McNaughton had been in close contact with Tesla for a number of years, circa 1936 to 1939, trying to figure out the particulars of the invention, it is quite possible that the topic of Tesla and his particle beam weapon came up. We know, for instance, that a few months after McNaughton met with Roosevelt in Washington, the president wrote a memo suggesting a meeting with Tesla, which unfortunately was proposed just a week before the inventor died.[41]

Support for the supposition that Tesla's particle beam weapon was never abandoned can be found not only in the talks I've had with a former contractor for DARPA who stated overtly that this invention was clandestinely worked on in the 1990s, but also in today's futuristic rail gun, and also in some conceptual way in Israel's so-called Iron Dome, an anti-defense missile system able to intercept short range rockets and artillery shells. The development of advanced weaponry systems can take decades to perfect, and in this case, it took more than fifty years.

RAIL GUN

According to Robert Afzal, senior fellow at Lockheed Martin, his company "has a long history of developing laser weapon systems for defensive purposes. The modern technology," he states, "is enabling us to build high-powered beams in a compact size for the battlefield." Since this technology is now a reality, this enabled Lockheed Martin to develop a prototype laser weapon system that they call Athena. "Although you can't see the actual beam itself because it is infra-red, you can see the effect of the beam," says Afzal. "From over a mile away . . . the tests that we've run are showing its utility against trucks, rockets and unmanned vehicles." With regards to the contributions made by Nikola Tesla, Afzal stated that "Tesla envisioned the use of electromagnetic energy for defensive applications, and although it's not exactly the same as he originally envisioned, that's essentially what we are doing today with a laser."[42]

Along with the schematic drawings and mathematical analysis, Tesla's particle beam weapon employed a number of novel and unusual features. The key one was its mechanism for creating and repelling a nondispersive beam of particles out the barrel of the gun.

"I perfected means for increasing enormously the intensity of the effects," Tesla wrote, "but was baffled in all my efforts to materially reduce dispersion and became fully convinced that this handicap could only be overcome by conveying the power through the medium of small particles projected, at prodigious velocity, from the transmitter. Electrostatic repulsion was the only means to this end. . . . Since the cross section of the carriers might be reduced to almost microscopic dimensions, an immense concentration of energy, irrespective of distance, could be attained."[43]

This design can be traced back to Tesla's childhood and his use of a popgun. This common toy, which Tesla used to shoot crows, worked by pumping in forced air to repel a cork out of the barrel, and this same principle can be found in today's rail gun. The main difference, according to physicist Travis Taylor, is that Tesla's gun used electrostatic energy for the repelling force, whereas the rail gun uses electromagnetic energy.[44] Tesla certainly knew the difference between the two, and he lectured on key features of these differences in the 1890s when he spoke at numerous conferences in London, New York, St. Louis, Philadelphia, and Chicago.

> Some of the Navy's futuristic weapons sound like something out of "Star Wars" with lasers designed to shoot down aerial drones and electric guns that fire projectiles at hypersonic speeds. . . . Rail guns, which have been tested on land in Virginia, fire a projectile at six or seven times the speed of sound— enough velocity to cause severe damage. The Navy sees them as replacing or supplementing old-school guns, firing lethal projectiles from long distances. . . . The rail gun requires vast amount of electricity to launch the projectile, said Loren Thompson, defense analyst at Lexington Institute. . . . Both weapon systems are prized because they serve to "get ahead of the cost curve." . . . In other words, they're cheap.[45]

Captain Mike Ziv, "program manager for directed energy and electric weapon systems for the Naval Sea Systems Command, [said], 'It fundamentally changes the way we fight.'" Thompson noted that "rail guns require vast amount of electricity to launch the projectile," which was precisely what Tesla stated in his numerous and detailed communiqués to the British War Office. The big problem, Thompson noted, was how to produce enough energy for the rail gun to work. According to the journalist David Sharp, Navy destroyers such as "the Zumwalt, under construction at Bath Iron Works in Maine [have enough power to run a rail gun.] The stealthy ship's gas turbine-powered generators can produce up to 78 megawatts of power. That's enough electricity for a medium size city—and more than enough for a rail gun."[46]

In Tesla's case, in order to produce the enormous voltages needed, he had redesigned his Wardenclyffe-like towers for just such a purpose. Atop Tesla's domed citadel, which was planned to be seventy-five feet in height, was the particle beam weapon. Nestled in a turret as a supergun, the entire apparatus apparently was also constructed for nonmilitary purposes such as for transmitting streams of electrical energy to distant places, much like microwave wireless telephone trunk lines do today. It takes a tremendous effort to develop new technologies, and one can only speculate whether or not Tesla's plan for a particle beam weapon was ultimately workable given that realization. The simple fact of the matter was that time caught up to him and he was simply too old to really be of much on-site help, if indeed any of the countries he was negotiating with would have proceeded in the effort.

However, as we can see from today's Navy weaponry, Tesla's ideas did eventually filter up to the top. As he told his secretary George Scherff on a number of occasions, "Do not worry about me, I'm a hundred years ahead of the other fellows."

PART V

God Particle

I prefer to be remembered as the inventor who succeeded in abolishing war. That would be my highest pride.

—Nikola Tesla,
"Tesla's Latest Wonder:
Magician of Electricity Says He Will Abolish War,"
Reading Times,
November 10, 1898, p. 2

17

The Birth of the New Physics

It followed from the special theory of relativity that mass and energy are both but different manifestations of the same thing—a somewhat unfamiliar conception for the average mind. Furthermore, the equation E is equal to MC squared, in which energy is put equal to mass, multiplied by the square of the velocity of light, showed that very small amounts of mass may be converted into a very large amount of energy and vice versa. . . . This was demonstrated by Cockcroft and Walton in 1932, experimentally.

—ALBERT EINSTEIN[1]

Einstein's original theory of relativity stemmed from the nineteenth-century discussion of the so-called ether, the medium through which light traveled, and the inability of scientists to prove its existence. Isaac Newton (1642–1727) made use of the ether as a medium to explain gravity's action at a distance, but "whether this agent be material or immaterial, I have left to the consideration of my readers."[2]

Unlike Newton, Michael Faraday (1791–1867) stated that the ether was a material force that resulted in crisscrossing rays of starlight emanating in all directions to infinity. It was the medium that light and other forms of electromagnetic energy traveled through.

According to Oliver Nichelson, by the 1800s, ether was perceived as "a thick gas . . . pictured traditionally as a non-material substance that could be found everywhere. . . . In 1659, Robert Fludd described the ether . . . as [a] far subtler condition than is the vehicle of visible light."[3] Nicholsen notes that Fludd tracks theories of the ether back to the time of Plotinus and the third century A.D., "where the ether is described as being so fine that it 'doth penetrate . . . nourish and preserve all bodies.'"[4]

The definitive experiment for measuring the ether was performed in the

1880s by Albert A. Michelson and Edward W. Morley. A ray of light was partly reflected onto two mirrors situated twenty-two miles away set up at right angles to each other. "This means that one half of the ray moved with the normal speed of light and the other with the speed of light plus the rotation of the Earth. [After bouncing back, the return rays would meet] at the union of the divided ray, [where it was hoped] there would [be a] difference in speed [thereby] showing the relative movement between the Earth and the aether, [thereby], proving the existence of the aether."[5]

Since no substantive differences were detected, the conclusion drawn was that either the ether did not exist or that by its nature it could not be detected. "This seemed conclusive, but it had the embarrassing consequence of depriving electromagnetism of a most successful theory and leaving nothing in its place."[6]

James Clerk Maxwell (1831–1879), whose textbook on electromagnetic theory Tesla kept through his entire life, formulated his concepts about twenty years before the Michelson-Morley experiment, and these required an ether. According to Maxwell, "There can be no doubt that the interplanetary and inter-stellar spaces are not empty, but are occupied by a material substance or body, which is . . . the most uniform body of which we have any knowledge."[7]

Maxwell's ideas stemmed from the work of Faraday and his predecessor Hans Christian Oersted (1777–1851), each of whom looked at the structure of the ether in slightly different ways. Oersted had realized that when electricity is sent down a wire, a magnetic field is created around the wire. This combination of electric and magnetic forces created "a circulation action" that had polarity. These two forces, which interact in conflicting ways or in harmony, could thereby spread out to cover all of space.[8]

Oersted hypothesized that the nature of this interaction of two fundamental forces could account for five separate phenomena: electricity, magnetism, light, chemical action, and heat. All space is now filled with a patchwork of forces "which manifest themselves according to the various conditions that exist locally in any state."[9] Faraday expanded on this idea. The ether, for Faraday, was really "a three-dimensional web of lines of force crisscrossing to infinity."[10] Light would thereby be carried as a vibration along these energy vectors, and spin would play a role in their construction.

A big problem that physicists were having at the turn of the twentieth century was reconciling these views with the constancy of the speed of light and its link to the Earth traveling through this hypothetical medium.

Einstein states that the velocity of light does not depend on the rate of movement of the light source. It's correct. But this principle can exist only when the light source is in [a] certain physical environment (ether), which cuts down velocity of light due to its properties. Ether's substance cuts down the velocity of light in the same way as air substance cuts down acoustic speed. If the ether did not exist, then velocity of light would strongly depend on the rate of movement of the light source.

—Quote attributed to Tesla
but actually written by Mikhail Shapkin[11]

Tesla discussed the very problem concerning the transmission of electromagnetic energy through space in all three of his lectures that he gave on high frequency phenomena in the early 1890s. As an inventor and experimentalist, Tesla considered whether the ether was motionless or in motion and suggested that this primary realm could be "capable of transmitting such vibrations when they range into hundreds of million millions per second."[12]

Tesla assumed the presence of an ether, and because of this, he was able, in the early 1890s, to transmit electrical energy through the air and also light up fluorescent and neon vacuum tubes by means of high-frequency AC currents and wireless power transmission. When the vacuum tube resonated at a particular frequency, it emanated light. What Tesla had done was do away with the need for the Edison filament; tubes could be lit whether or not they contained rarified gases, for Tesla could simply remove the filament from Edison's vacuum bulbs and illuminate these as well. The vacuum, Tesla said to the dismay of his rival Edison, was more important than the filament.

It is certainly more in accordance with many phenomena observed with high-frequency currents, to hold that all space is pervaded with free atoms, rather than to assume that it is devoid of these. . . . Is then energy transmitted by independent carriers or by the vibration of a continuous medium? . . . Especially light effects, incandescence or phosphorescence, involve the presence of free atoms and would be impossible without these.[13]

Tesla's talk was given a few years before one of his cohorts, J. J. Thomson, discovered the electron. His idea that energy may be transmitted by vibration of independent carriers resurrected Newton's treatise on optics in which this dual nature of light was first introduced, and these ideas were presented to physicists like Elmer Sperry, inventor of the gyroscope, Alexander Graham Bell and Nobel Prize winner Robert Millikan at a lecture at Columbia University in 1891 and in 1892 to members of the Royal Academy of Science, including Lord Kelvin, Sir William Crookes, Ambrose Fleming, James Dewar, Sir Oliver Lodge, Sir William Preece, and the future Nobel Prize winners J. J. Thomson and Lord Rayleigh.

Coming from a world that accepted the all-pervasive ether as a given and also realizing that photons could act as waves or particles, Einstein emphasized the latter concept, and that resulted in a different view. If light traveled like little bullets from, say, the Sun to the Earth, no ether was necessary, and over time, this view came to prevail. For many scientists, the ether no longer existed, replaced by an empty vacuum. However, part of the problem is really one of semantics, for what Einstein really did was replace the nineteenth-century 3-D Euclidian ether with a twentieth-century 4-D non-Euclidean space-time continuum that had its own medium. This conceptualization, however, has its own shortcomings, because it ignores the self-evident fact that space cannot be empty. It certainly contains the light emanating from all the stars and the crisscrossing forces that Faraday and Oersted describe. And Einstein's theory about the ultimate structure of the universe is also missing such other attributes as *tachyonic* (faster than light speed) dimensions and holographic aspects whereby each point in space, containing the intersecting light from every star, codes for every other point in space, and the conscious component whereby the mind, by its nature, transcends time.[14]

IMAGINARY NUMBERS, RELATIVITY, AND QUANTUM THEORY

> Relativity theory has shown that space is not three-dimensional and time is not a separate entity. Both are intimately and inseparably connected and form a four-dimensional continuum which is called 'space-time.'"[15]
>
> —FRITJOF CAPRA

Ronald Clark's biography of Einstein makes it clear that Einstein owed much of his success to his college math teacher Hermann Minkowski (1864–1909). "Whether Einstein would ever have done it without the genius of Minkowski we cannot tell," said Ebenezer Cunningham, author of *The Principle of Relativity*. "Minkowski gave a mathematical formulism to what had been a purely physical conception of special relativity." His mathematical theories allowed, in Einstein's words, "the time coordinate to play exactly the same role as the three space coordinates." Thus, three-dimensional space would not exist without it being anchored to time. "The two were indivisible."[16]

What Minkowski did was to introduce the mathematical unit $\sqrt{-1}\ ct$ as the time coordinate, and this in turn could relate symmetrically to the three space coordinates. The important point to keep in mind is that $\sqrt{-1}$ (square root of negative one) is an *imaginary* unit, that is, it has no physical counterpart. Yet, the use of this imaginary number explained quite efficiently the mathematical formulism for relativity whereby "a happening in 3-d space physics becomes, as it were, an existence in the 4-dimensional world."[17]

In 1903, Minkowski addressed a scientific convention in Cologne:

> Gentlemen! The ideas on space and time which I wish to develop before you grew from the soil of experimental physics. Therein lies their strength. Their tendency is radical. From now on, space by itself and time by itself must sink into the shadows, while only a union of the two preserves independence."[18]

Unfortunately, a year later Minkowski died of a burst appendix. He was only forty-four. At the same time, the European scientists, heavily influenced by Minkowski's mathematics, invited Einstein, along with Marie Curie, to Geneva to be presented with honorary doctorates.[19]

The square root of negative one, (i), an imaginary unit, is also called a hypernumber. Adopted by Carl Friedrich Gauss in the early 1800s for use in non-Euclidean or non-flat geometry, $\sqrt{-1} = i$ is one higher dimension above Euclidean geometry. Abstract mathematician Charles Musès states that numbers, as a language and description of reality, "are then seen to be powers of transformation." The introduction of the hypernumbers, he adds, "more than doubled the entire mathematical power of all previous centuries. . . . They became intimately related to the new physical discoveries in electronics, atomic theory and twentieth century chemistry."[20]

COMBINING RELATIVITY WITH QUANTUM MECHANICS

It was Paul Adrien Dirac (1902–1984) who introduced the use of the hypernumber to quantum physics. Shortly after the discovery of the electron by J. J. Thomson, Niels Bohr published his theory on the structure of the atom. Bohr, much like Tesla before him, likened the electron's paths to the motion of the planets around the Sun, with circular orbits around the nucleus. However, this initial theory did not completely correlate with the observed frequencies in the line spectra of the hydrogen atom.

German physicist Arnold Sommerfeld "extended Bohr's ideas to the case of elliptical orbits,"[21] and this addition to the theory of the structure of atoms better explained the motion of the electron. Wolfgang Pauli continued the study by formulating the Pauli principle, which distinguished three quantum numbers to correspond to (1) the diameter of the electron's orbit, (2) the eccentricities of the orbit (elliptical and azimuthal shape of orbit), and (3) the space orientation of the orbit (orientation of angular momentum). However, the data from the line spectra was still not completely satisfied.

"Studies of the Zeeman Effect (the splitting of spectral lines by strong magnetic fields) revealed that there are more components than the three integers could account for, and to explain their existence, a fourth quantum number was introduced. . . . In 1925 two Dutch physicists, Samuel Goudsmit and George Uhlenbeck, made a bold proposal. This surplus line-splitting, they suggested, is not due to any additional quantum number describing the electron's orbit but to the electron itself."[22] Goudsmit and Uhlenbeck suggested that the electron was spinning on its own axis, and George Gamow, in his book *Thirty Years That Shook Physics,* informs us that their measurements exceeded the speed of light! Thus, it violated relativity. This aspect shall be covered in the following chapter. What is important here is that this orthorotational component became the fourth quantum number. Now an electron's orbit could be described from clear parameters. (Interestingly, Samuel Goudsmit would later immigrate to the United States and gain work at MIT. During WWII, under the auspices of Army and Navy Intelligence and Vannevar Bush's OSRD, Goudsmit would become head of the Alsos Mission which brought him back to Europe during the war to seek out German scientists involved in chemical and biological weapons.)

Faced with a problem similar to Minkowski's, but now on a microscopic scale, Dirac was trying to create an equivalence between the time and space coordinates for the spinning electron and at the same time combine quantum physics with relativity. Minkowski's use of $\sqrt{-1}$ allowed an equivalence between the three space coordinates and time on the macroscopic scale, thereby solving the problem of creating symmetry in which 3-D space (x, y, and z, or height, width, and depth) became equivalent to the one time coordinate (t), and this same essential procedure, whereby x, y, and z (or x^2, y^2, and z^2) stood for the location of the electron in its orbit and t (or t^2) stood for its spin, was used by Dirac to solve the problem of integrating relativity with quantum physics for subatomic particles. By substituting $\sqrt{-1}$ for the tachyonic spin of the electron, the problem of integrating relativity with quantum physics was neatly sidestepped and Dirac received a Nobel Prize for the effort. In the process, Dirac also hypothesized the existence of the antielectron that contained a positive charge that appears in the "holes" created when an electron is knocked out of that orbit.[23] Simultaneously, the finding of Goudsmit and Uhlenbeck that electrons orthorotate at speeds in excess of the speed of light went the way of the passenger pigeon.

TODAY'S ATOM

> Electrostatic force is that which governs the motion of atoms.
> . . . It is the force which causes them to collide and develop
> the life-sustaining energy of heat and light, and which causes
> them to aggregate to an infinite variety of ways according to
> nature's fanciful designs, and form all these wondrous struc-
> tures we see around us. It is, in fact, if our present views be
> true, the most important force for us to consider in nature.
>
> —NIKOLA TESLA, August 22, 1893[24]

The original design for the structure of the atom consists of four elementary particles: the electron, proton, neutron, and photon. Electrons, which are negatively charged (–), orbit the nucleus. These are fundamental and appear not to consist of anything smaller. Protons, however, are made up of six quarks. Inside the nucleus are protons (+), and neutrons (+–). The latter are neutral because they, in turn, are made up of a proton-electron pair. Photons, which have no mass and no charge, are tiny wave packets of light that are used to bind one atom to another. In that sense, photons can be seen as the glue that holds molecules together. They are bundles of energy determined by the wavelength of light. When photons collide with electrons, they act like solid bodies and can knock the electron out of orbit. Visible and ultraviolet (UV) wavelengths of light are too large to collide with electrons. Since neutrons are made from protons and electrons, one could say that all of matter consists of just two major building blocks, a positive and negative charge, and a third block, the photon, which is the glue that carries no charge.

This overarching model was further refined into the subatomic realm after numerous studies were conducted with particle accelerators used to smash these four entities. The present model of the atom is constructed from (1) *fermions*, which are particles, and (II) *bosons*, which are binding forces.

1. There are two kinds of fermions: quarks (with six subtypes) that form the protons, and leptons, which are electrons (and other particles, e.g., muons, tau, and two kinds of neutrinos). Protons can be broken down further, whereas electrons are still seen as fundamental.

2. Bosons are binding forces, and there are three that are known and one that is hypothesized. These four bosons account for the four known forces in the universe: electromagnetism, strong nuclear force, weak nuclear force, and gravity. The photon, as stated before, holds atoms together; gluons holds the nucleus together; the Z and W bosons hold neutrons together; and the graviton, also called the Higgs boson, which is a theoretical particle associated with gravity, gives fermions their inertia or mass.

According to recent data, the Higgs boson has been observed to decay into "photons, tau-leptons and W and Z bosons," and there is strong speculation that the Higgs should also decay into "a pair of b-quarks," thereby supporting the hypothesis "that the Higgs mechanism is responsible for the masses of quarks . . . [and] hint of a new physics beyond our current theories."[25]

Now that we have described the particles and their binding forces, let's consider their sizes. The proton is 1,836 times the size of the electron, yet 100,000 times smaller than the size of an atom. The first atom from the periodic table of elements, the hydrogen atom, consists of an electron orbiting a proton/nucleus. If this atom were the size of a room, the nucleus would be so small it would be invisible. If the atom were the size of St. Peter's Cathedral in Rome, the nucleus/proton would be the size of a grain of sand, and if the atom were the size of the planet, the nucleus/proton would be the size of a three-quarter-acre house lot. As we can see, an atom, and therefore all of matter, is mostly made up of space.

DARK ENERGY

> Each force . . . excluding gravity . . . is transferred by an additional particle: i.e., the photon carries electromagnetism.
>
> —GUY WILKINSON,
> University of Oxford,
> former spokesperson for CERN[26]

Astrophysicists tell us that a large percentage of the energy or matter in the universe cannot be accounted for. They don't know where it is. If we go back to the beginning of time, one way or another, when the universe was born, billions and then trillions of galaxies were formed and light from every star within each galaxy began radiating. If photons act like particles and can knock electrons out of their orbits, perhaps they do indeed have a tiny bit of mass. Einstein states that mass is equivalent to energy, and since photons have energy, then by definition, they also have mass. And if that were the case, this tiny addition of matter inside the trillions and zillions of photons that have been emanating from the universe since the beginning of time might indeed account for the missing mass. If that were the case, then so-called dark energy would actually be "light" energy.[27]

THE HIGGS FIELD

Numerous articles echo the simple fact that physicists have yet to integrate gravity into their modern-day space/time model. Einstein spent the latter half of his life looking for a unified field theory to achieve this end, but he never succeeded. In prestigious forums such as *Science News* and *University of Chicago Magazine*, it seems clear that the nineteenth-century ether theory has reappeared in the guise of the theories of Peter W. Higgs, a physics professor from Edinburgh University.

> To patch these flaws in the standard model, theorists proposed the existence of some sort of influence that permeates all of space, weighing down particles passing through it. This cosmic molasses is called the Higgs field . . . a pervasive field in the universe that . . . could bestow mass on all fundamental particles that have mass.[28]
>
> —PETER WEISS

The quote by from Peter Weiss is from an article he wrote for *Science News*. He also wrote, "Higgs bosons suffuse the field and are drawn to the particles; the more energetic particles attract more bosons, the less energetic ones attract fewer ones. This clustering gives the particles the solidity we associate with matter." Using sophisticated particle accelerators and spending billions of dollars, physicists around the world continued to look for this Higgs boson, which some have actually dubbed the "God particle." Through literally trillions of collisions, the great binding particle proved to be incredibly elusive. However, according to reports stemming from Geneva, the Higgs boson was apparently found and measured by two separate groups at CERN during the years from 2011 to 2013.

> "My God!" Fabiola Gianotti exclaimed, jumping up in her chair after she was brought the readouts proving that the Higgs had been found. Maybe it was just an exclamation, but the empiricist nonetheless took care to correct herself at the press conference later. "Thanks, Nature!" she called out, but it was too late; the cat was out of the bag.[29]

This particle, which they tell us exists for 1.56×10^{-22} seconds, decays into several quarks, two W and two Z bosons and two photons. Apparently, the odds of finding a Higgs boson is one in 10 billion particle collisions. And since the Higgs exists for so brief a time, that increases the difficulty of capturing it.

In 2014, the Tesla expert and German documentarian Michael Krause traveled to CERN to interview many of the particle physicists who captured and apparently photographed the Higgs boson. Using a 100 megapixel digital camera that weighs fourteen thousand pounds and takes 40 million pictures a second, the researchers at CERN can photograph the collisions they create with this gigantic particle accelerator. Considering that the Large Hadron Collider can create 600 million collisions per second, one can begin to consider how truly miraculous this equipment and these studies really are.[30]

> The standard model of physics describes with remarkable precision all the known particles in the universe and the interactions that occur between them. Only gravity has yet to be integrated into the model. . . . For all its success, however, the theory omits a rather pivotal trait of particles—their mass.[31]
>
> —PETER WEISS

MASS = E/C²

The more we learn about the Higgs boson, the more we will learn about "the mass-giving mechanism" and thus help pave the way for helping us understand not just how the universe works but also "why the universe works the way it does." According to Jeffrey Kluger, science writer for *Time*, "the Higgs boson paves the way for the very future of physics."

According to this theory or model, there would be at least two densities for the Higgs field, one that would be the "source of mass shared by all particles" and a "second . . . thicker molasses that only affects quarks—the constituents of protons and neutrons—and give them much more mass than the Higgs field does."[32]

My question is, If a boson is a "binding force" and the Higgs boson is the particle that gives matter its mass, what is the Higgs derived from, to achieve this magical end? The answer they give, of course, is the Higgs field. Such theoretical physicists as Matt Strassler assures us that this particle is not a "pseudo particle." That's good to know, but it seems to me that what the modern physicists have really done is simply rename this mysterious, all pervasive ether the Higgs field, and once they did that, they avoided all the controversy about the existence of a background medium supplying particles their integrity that simply has to exist. It is the nature of this medium, which may very well exist in a tachyonic dimension, that would be the next step in the full understanding of the process that creates physical reality. According to Einstein's equation, if matter can be converted into energy (e.g., the atomic bomb) and the reverse is also true, the speed of light squared remains an inherent component (see opening quote of this chapter). Thus we see that in order for this physical universe to exist, a tremendous amount of energy involving mathematically the speed of light squared, and thus a tachyonic component, is involved.

AS ABOVE, SO BELOW

One of the boldest attempts to refute our current notion of gravity was proposed by P. D. Ouspensky, a unique genius, mathematician, and mystical researcher. Ouspensky pointed out that Newton's law of gravitation was descriptive of a relationship between two bodies (planets) but that this relationship was not necessarily a mysterious force of attraction.

Both [Newton] and ... Leibniz, definitely gave warning against attempts to see in Newton's law the solution of the problem of action through empty space, and regarded this law as a formula of calculation. Nevertheless the tremendous achievements of physics and astronomy attained through the application of Newton's law caused scientists to forget this warning, and the opinion was gradually established that Newton had discovered the force of attraction.

—P. D. OUSPENSKY[33]

Ouspensky's explanation for the motion of the planets stemmed from a radically different view of the structure of the universe. Conceiving of the solar system in its dynamic form, as a gestalt pattern hurtling through space, Ouspensky wrote:

If we wish to represent graphically the paths of this motion, we shall represent the path of the Sun as a line, the path of the Earth as a spiral winding round this line, and the path of the Moon as a spiral winding around the spiral of the Earth.

The Sun, the Moon, the stars, which we see, are cross-sections of spirals which we do not see. These cross-sections do not fall out of the spirals because of the same principle by reason of which the cross-section of an apple can not fall out of the apple.[34]

Spirality may be a fundamental property of the universe, which manifests itself geometrically on subatomic levels (e.g., electron spin) and on macroscopic levels as well (e.g., the orthorotation of the Earth and the structure of the solar system). Naturally, the galaxy must have a direct effect on the fate of the Earth. The Coriolis force, or the analogous force of the galaxy, would generate spin in solar systems and components thereof. Assuming the galaxy to be one entity spinning in unison, one can see that the innermost sector would spin at a rate slower than the outermost sector. One way or another, this overarching spinning force will influence all of its components. On the macroscopic scale, these would include eye-of-the-storm hurricanes, tornadoes, waterspouts, and whirlpools, and on the subatomic scale, the very structure of atoms.

Ouspensky's idea would replace the gravitational law of attraction with an altogether different concept related to a principle of symmetry, that is, the "movement from the centre along radii."[35] For Ouspensky, the Earth is no more attracted to the Sun than the left eye of an animal is attracted to the right. This is a unique insight derived from a more Pythagorean view of a dynamic universe that takes into account the momentum caused by the enormous speed at which the solar system is traversing the galaxy, that is, over five hundred thousand miles per hour!

18

Tesla's Dynamic Theory of Gravity

Grand Unification, the God Particle, Tesla, and Einstein

All attempts to explain the workings of the universe without recogniz-ing the existence of the ether and the indispens[ible] function it plays in the phenomena . . . [are] futile and destined to oblivion.

—NIKOLA TESLA,
July 10, 1937, comments said to the press
prior to being interviewed

One of the most important findings associated with Tesla was his belief that ELF waves would follow the curvature of the Earth. During the late 1890s, it was assumed by Marconi and others that radio waves would fly off into space because they moved in straight lines and the Earth was curved. From looking at this situation, several theories emerged:

1. Impulses would bounce around the Earth between the ground and the ionosphere.

2. If an impulse could be beamed into the ionosphere, then this rarefied atmosphere, much like an evacuated tube, would act as an excellent conduit for carrying impulses around the globe.

3. Ground waves would simply adhere to and follow the ground. Since the Earth was curved, electromagnetic (EM) waves would follow the curvature.

4. EM waves in a resonant frequency with Earth currents would follow pathways already embedded within the electrical makeup of the planet.

It is the third possibility that is linked to Tesla's highly elusive and, as we will see, highly controversial dynamic theory of gravity. The story begins in May of 1891, when Tesla gave an amazing lecture at Columbia College (later University) on high-frequency phenomena and wireless communication. Those who attended, including Alfred S. Brown, Tesla's first partner after he quit working for Edison; Cornell University Professor William Anthony, the first scientist to test Tesla's induction motor and give it the seal of approval; Elmer Sperry, who would go on to morph Tesla's rotating magnetic field into the gyroscope; telephone king Alexander Graham Bell; Francis Upton and William Stanley, Tesla's colleagues from the Edison and Westinghouse plants; Tesla's archcompetitors Elihu Thomson and Michael Pupin, who according to Tesla, were disrespectful during his talk; and Robert Millikan, the future Nobel Prize winner for his work in cosmic rays, would all remember this watershed moment for the rest of their days.

> July 10, 1931
> California Institute of Technology
> Pasadena, California
>
> Dear Dr. Tesla,
>
> When I was a young man of 25 as a student in Columbia University, I attended a downtown public lecture in New York at which you made one of the first demonstrations of your Tesla coil and its capabilities. Since then I have done no small fraction of my research work with the aid of principles I learned that night. So that it is not merely my congratulations that I am sending to you but with them also my gratitude and my respect in overflowing measure.
>
> Cordially yours
> Robert Millikan[1]

Aside from demonstrating his wireless fluorescent tubes that gave off light when different frequencies were produced, Tesla also discussed far-reaching theoretical considerations.

There is no subject more captivating, more worthy of study, than nature. To understand this great mechanism, to discover the forces which are active, and the laws which govern them, is the highest aim of the intellect of man. . . . The assumption of a medium pervading all space and connecting all gross matter, has freed the minds of thinkers of an ever present doubt, and, by opening a new horizon—new and unforeseen possibilities—has given fresh interest to phenomena with which we are familiar of old. . . . It has been for the enlightened student of physics what the understanding of the mechanism of the firearm or of the steam engine is for the barbarian. Phenomena upon which we used to look as wonders baffling explanation, we now see in a different light.[2]

Tesla goes on in great depth to discuss how fascinated he is with electricity and magnetism, "their seemingly dual character, unique among the forces in nature. . . . We are now confident that electric and magnetic phenomena are attributable to ether." Electricity, for Tesla, could be considered "bound ether," and he felt that "the so-called static charge of the molecule is ether associated in some way with the molecule. Looking at it in that light, we would be justified in saying that electricity is concerned in all molecular actions."[3] Tesla then goes on to create a "mental picture" of what he sees as the most likely structure for matter. The following quote, given in 1891, was published a half a decade before J. J. Thomson identified the electron:

An infinitesimal world, with the molecules and their atoms spinning and moving in orbits, in much the same manner as celestial bodies, carrying with them static charges, seems to my mind the most probable view, and one which, in a plausible manner, accounts for most of the phenomena observed. The spinning of the molecules and their ether sets up the ether tensions or electrostatic strains; the equalization of ether tensions sets up ether motions or electric currents, and the orbital movements produce the effects of electro and permanent magnetism.[4]

As far as I know, no standard text on the history of physics mentions Tesla, even though these ideas would lead to Nobel Prizes when they were further developed by Rutherford and Bohr (with their solar-system description of the atom with electrons orbiting the nucleus) and as Einstein's discovery of the photoelectric effect, which was equivalent to Tesla's wave and particle-like description of light, which stemmed directly from the ideas of Newton and Kelvin.

> I, of course, am well aware of the great contributions Nikola Tesla has made to electrical engineering in many directions, and in particular I was greatly impressed in my younger days by his experiments in high frequency currents. I have often made use of the Tesla transformers as a method of producing high voltages in my research.
>
> —ERNEST RUTHERFORD,
> Cavendish Laboratory, March 19, 1936[5]

For reasons difficult to fully understand, these ideas that Tesla discussed in the presence of so many elite scientists of the day, such as Sir William Crookes, Sir William Preece, Sir Oliver Lodge, J. J. Thomson, James Dewar, Ambrose Fleming, Andre Blondel, Lord Kelvin, Robert Millikan, and Lord Rayleigh, were abandoned by modern physicists, including, in particular, the link between an all-pervasive ether and the structure of matter. This led to a number of key differences between Tesla's worldview as compared with that of Einstein (1879–1955). Tesla disagreed with the findings of Einstein's theory of relativity in a number of ways. As far back as the turn of the century, Tesla thought that he had intercepted cosmic rays emanating from the Sun that attained velocities "vastly exceeding that of light." In the last decade of his life, he also claimed that these cosmic rays could be harnessed to generate electrical power. Tesla also saw radioactivity as evidence of a material body (such as uranium) *absorbing* energy as much as it was giving it up. Thus, from Tesla's point of view, the process of radioactivity was due to its acting as a conduit, thereby resulting in the leaking of this primary energy into the environment.

The following quote, which is directly linked to Tesla's dynamic theory of gravity, that is, the particle (or process) that gives matter its mass, can also be seen as evidence of Tesla's messiah complex. For he lays out clearly here how the human could attain Godlike qualities:

> According to an adopted theory, every ponderable atom is differentiated from a tenuous fluid, filling all space merely by spinning motion, as water whirls in a calm lake. Once set in movement this fluid, the ether, becomes gross matter. Its

movement arrested, the primary substance reverts to its normal state. It appears, then, possible for man through harnessed energy of the medium and suitable agencies for starting and stopping ether whirls to cause matter to form and disappear. At his command, almost without effort on his part, old worlds would vanish and new ones would spring into being. He could alter the size of this planet, control its seasons, adjust its distance from the Sun, guide it on its eternal journey along any path he might choose, through the depths of the universe. He could make planets collide and produce his suns and stars, his heat and light; he could originate life in all its infinite forms. To cause at will the birth and death of matter would be man's grandest deed, which would give him the mastery of physical creation, make him fulfill his ultimate destiny.[6]

Concerning the tachyonic issue, about two decades later, the inventor would come to state that the impulses transmitted from his turn-of-the-century Wardenclyffe wireless tower traveled at velocities in excess of the speed of light. He likened the effect to the Moon's shadow spreading over the Earth.

It is very difficult to explicate Tesla's speculations concerning tachyonic cosmic rays and his view of radioactivity. However, with regard to the third claim, this suggestion that he transmitted energy at speeds in excess of the speed of light, this assertion can be discussed from a variety of viewpoints. As the Earth has a circumference of roughly 25,000 miles and light travels at about 186,000 miles per second, one can see that it would take light approximately one-seventh of a second to circle the Earth. But does the Earth itself exist in its own realm that by the nature of its size transcends the speed of light? For example, does the North Pole interact/exist with the South Pole instantaneously? If so, in a sense, the theory of relativity is violated because nothing, according to this theory, can "travel" faster than the speed of light, yet the Earth's very electromagnetic unity belies that theory. The same, of course, would hold for the larger planets and the Sun.

Regarding the ponderous decision throughout the twentieth century to abandon the ether, there simply has to be a medium between the Sun and the Earth for light to travel through, and as we see below, Einstein agreed with this premise. Keep in mind that Hendrik Lorentz, George Fitzgerald, Michael Faraday, Lord Kelvin, and many others also assumed an ether.

COSMOGENESIS

The fact that Tesla has a spiral galaxy as a letterhead is significant, given that he is the discoverer of the rotating magnetic field. The big bang does not account for the prevalence of *spin* in the creation of the universe. An alternative hypothesis for creation would be two powerful alternating forces, yin and yang, crisscrossing in a syncopated or out-of-phase rhythm, correlating with Tesla's rotating magnetic field.

Light traveling at 186,000 miles per second would take approximately one hundred thousand years to cross any one of these galaxies. Allowing for the space between the galaxies, when traveling at light speed, it would take light approximately a half billion years to travel from one end of a galaxy to the other end, and yet we can view the entire panorama in one simple shot.

There are many things to consider about this thought problem. First of all, the size of any galaxy is really incomprehensible, and yet we can look at a cluster of galaxies with our telescopes. To add to this thought problem, when we look out at the night sky, we are seeing various stars as they were thousands and tens of thousands of years ago, because that's how long it took the light from those stars to reach us.

We are therefore looking at an entity that is tens of thousands of light years long, yet we can see the entire galaxy and view it as a single entity. It therefore stands to reason that there must be some mechanism that easily usurps the speed of light because a galaxy is one thing. Its angular momentum, and/or its gestalt properties instantaneously connect one side of itself to the other side. Based on simple calculations, c^3, or the speed of light cubed, would, for all intents and purposes, be a "speed" that would connect one side of a galaxy instantaneously to its other side. Clearly, Einstein's theory of relativity, which states that nothing can travel faster than the speed of light, must be an incomplete theory. This is self-evidently true by simply looking upon a galaxy and considering it as one thing.

Thus if Tesla states that he has measured cosmic rays that travel fifty times the speed of light, we can certainly see that this idea is not completely radical because there simply must be some mechanism that enables energy to traverse vast stretches of the cosmos at "speeds" in excess of the speed of light. The biggest problem for me is that there is no record that I am aware of that explains how and when Tesla captured these cosmic rays and no record of how he could have possibly measured their speed.

JOSEPH ALSOP

The Waldorf-Astoria
Fifth Avenue and 33rd and 34th St.
and Astor Court, New York

Monday, March 27, 1899

My dear Mrs. Robinson,

Would you give me the great pleasure of your and Mr. Robinson Junior's company on Wednesday evening next?

I have a party of friends to dinner at the Waldorf 7:30 in the Astor dining room and we are to go to my humble laboratory afterwards where I shall endeavor to entertain you with some experiments. These will be a great success if you come, if not it will be the worst fiasco any experimenter has ever had.

Kindly let me know if possible by telephone during the day tomorrow. My telephone address is 299 Spring.

With kind regards believe me, sincerely yours,
N Tesla[7]

In my quest to understand Tesla's various theories, I came across a very unusual statement that Tesla made to Joseph Alsop, a young reporter for the *New York Herald Tribune*. The year was 1934. Just three years earlier, Tesla had made the cover of *Time* magazine; it was his seventy-fifth birthday. And because of this, the rather obscure wizard became a headliner again. Taking full advantage, Tesla decided to reveal one secret after the next, each successive year on his birthday. In 1934, Tesla decided to describe his particle beam weapon, and he did so to reporters for the *New York Sun*, the *New York Times*, the *New York World Telegram*, the *New York Herald Tribune*, and *Every Week* magazine. The meeting was organized as a birthday celebration where hors d'oeuvres were served in one of the banquet halls of the Hotel New Yorker. Having just moved in, Tesla was ecstatic because he had successfully cajoled the Westinghouse Corporation into covering his rent and had also received a check for $1,000 from the city of Philadelphia for winning the John Scott Award.

Tesla had threatened to sue Westinghouse for using his wireless patents for their ubiquitous radio broadcasting stations without compensating him, and Hugo Gernsback was able to help negotiate this favorable settlement.

Unfortunately, Tesla was not able to get them to cover the back rent he owed the Hotel Governor Clinton, but he had solved that problem by giving the Clinton an electronic device as collateral, which he claimed was the top secret and highly dangerous particle beam weapon. He put a value on it at $10,000 and signed a note to that effect, which the hotel kept in a safe file.

In the case of the John Scott Award, Tesla had fielded a call from an old acquaintance, the metaphysician and artist Walter Russell, who was on the board of directors of the city of Philadelphia. Unable to comprehend why his "spiritual mainstay . . . was not proclaimed the greatest man in science of his day," and realizing that "[Tesla's] soul was low from doubtings and attacks by lesser minds,"[8] Russell successfully campaigned to get the city to confer on Tesla the award. In a friendly mood, Tesla responded by telling Russell an amusing story about Mark Twain, whom the inventor had cured of constipation by having Twain step aboard a vibrating platform that almost instantly stimulated peristaltic action. "The aftermath of your delightful conversation is still with me," Russell wrote, "with the thrill of your Mark Twain reminiscences still repeating themselves in my memory like the aftermath of a symphony." What Russell was laughing about was the image of Twain racing to the lavatory so that he wouldn't relieve himself in his pants![9]

Three months later, in October of 1934, Tesla was invited to a dinner and award ceremony in Philadelphia honoring Dr. Nikola Tesla for his inventions of the induction motor and rotating magnetic field; Professor Alfred Newton Richards for research on the kidney; and Dr. Robert Mehl, from Carnegie Institute of Technology, "for his development of gamma ray photography," and there is little doubt that Tesla sat with Russell at that time. After receiving the John Scott medal and accompanying check, Tesla spoke briefly, stating simply that "I have new things which before long, will revolutionize industry." When pressed for further details, the aging wizard declined "to divulge the nature of his experiments."[10]

Russell had written a highly original treatise concerning his belief that the periodic table of elements was set up much like a musical scale with nine octaves, and the idea that each element was created by compressing different rings of light. Whether this treatise played a role in Tesla's thinking is not known. However, what is certain is that Tesla knew Russell and that they did indeed spend time together.

Tesla, however, was focused on promoting his particle beam weapon. He was in negotiations with several countries, and he therefore sought to give his invention more visibility. Titles for these articles ranged from "Invents

Peace Ray" and "Tesla on Power Development and Future Marvels" to "Death Ray Machine Described." And Tesla did indeed reveal in detail the particulars to his Buck Rogers–like futuristic weapon, a mechanism that, he told his biographer, John O'Neill, would be a weapon that would end all war. Concerning the details, the *New York Sun* reported "four machines combined in the production and use of this destructive beam. . . . First, apparatus for producing manifestations of energy in free air instead of a high vacuum. . . . Second, a mechanism for generating a tremendous electrical force. . . . Third, a method of intensifying and amplifying the force. . . . Fourth, a new method for producing a tremendous repelling force. This would be the projector, or gun of the invention."

The reporter ended this article with a curious paragraph:

> Another addition to the anniversary message of the famous inventor was a positive declaration that he expected soon to construct apparatus that would disprove the theories of modern astronomers that the Sun, gradually was cooling off and eventually the Earth would be unable to sustain life as it would grow too cold.[11]

The *New York Times*'s headline read "Tesla at 78, Bares New 'Death Beam,'" and the subhead stated, "Powerful Enough to Destroy 10,000 planes at 250 Miles Away, He Asserts Defensive Weapon Only. . . . Says Will Kill Without a Trace."[12]

The longest and most detailed article was published in the *New York Herald Tribune* as written by Joseph Alsop. A young reporter just a year out of college, Alsop described "the aging inventor, as a tall, thin, almost spiritual figure in the sort of brown cutaway suit that older men wore before the World War. . . . Before he would speak of his present work, he reviewed his past achievements which entitle him more than Edison, Steinmetz or any other, to be called the father of the power age."

Alsop then portrayed the inventor's declaration as a "Jules Vernean announcement. Dr. Tesla disclosed that he has lately perfected instruments which flatly disprove the present theory of the high physicists that the Sun is destined to burn itself out until it is a cold cinder floating in space. . . . He had, he said, detected 'certain motions in the medium that fills space, and measured the effects of these motions.' The results of the experience had led him 'inescapably' to the conclusion that such bodies as the Sun are taking on mass more rapidly than they are dissipating it by the dissipation of energy in heat and light."[13]

This passage appears in *Wizard*, which was published in 1996, but I did not include Tesla's term "dynamic theory of gravity" in relationship to this statement, and I'm not quite sure why I didn't. Said in another way, I don't really know when I put everything together to unravel the theory that is being developed in this chapter.[14]

There are several points to make here. First of all, Tesla was the subject of my doctoral dissertation, a treatise that was 725 pages long and took about six years to produce. And after that, I just kept going to produce *Wizard: The Life and Times of Nikola Tesla, Biography of a Genius*, which was the result of easily another ten years of daily effort. There are over one thousand endnotes, including about four hundred letters quoted. I traveled to Tesla archives in New York; Washington, DC; Belgrade and elsewhere, used interlibrary loans to obtain scores of other articles, used the Freedom of Information Act to get hundreds of other pages of documents, interviewed numerous Tesla experts, and spoke at a dozen or more Tesla conferences, and in that capacity, I obtained additional highly obscure Tesla papers. In all that time, culling though all that data, this statement by Tesla that the Sun was absorbing more energy than it was radiating can only be found in this one article. There is one more passage that I uncovered in the Kenneth M. Swezey Papers at the Smithsonian Institute in Washington, DC, where Tesla does indeed refer to his dynamic theory of gravity, but this was an unpublished and highly obscure document, although it does also appear at the tail end of John O'Neill's Tesla biography:

> After . . . a brief visit to my home in Yugoslavia I returned to this country in 1892 eager to devote myself to the . . . study of the universe. During the succeeding two years of intense concentration, I was fortunate enough to make two far-reaching discoveries. The first was a dynamic theory of gravity, which I have worked on in all details and hope to give to the world very soon. It explains the causes of this force and the motions of heavenly bodies under its influences so satisfactorily that it will put an end to idle speculation and false conceptions, as that of curved space. . . . But even if it existed, it would not explain the motions of the bodies as observed. Only the existence of a field of force can account for them and its assumption dispenses with space curvature. All literature on this subject is

futile and destined to oblivion. So are also all attempts to explain the workings of the universe without recognizing the existence of the ether and the indispens[ible] function it plays in the phenomena.

My second discovery was a physical truth of the greatest importance. As I have searched the scientific records in more than a half dozen languages for a long time without finding the least anticipation, I consider myself the original discoverer of this truth, which can be expressed by the statement: There is no energy in matter other than that received from the environment.[15]

There are two aspects to this passage as it relates to Tesla's Dynamic theory of gravity. The first has to do with what Tesla apparently conceived of, namely, a way to explain "the force and motions of the heavenly bodies," which I think refers to the way the planets circle the Sun and perhaps the way the Sun circles the galaxy. As far as I know, there are no publications of Tesla's to explain his theory on the cause of these celestial events. However, Tesla did indeed describe the motion of the planets through the heavens in a talk and demonstration he gave at the 1893 Chicago World's Fair. One exhibit of his in particular was similar to his Egg of Columbus, which illustrated his rotating magnetic field by causing a brass egg to spin in the field as it produced electromagnetic rotation. Tesla extended this idea to suggest that the same process was responsible for celestial and planetary motion. "In this experiment one large, and several small brass balls were usually employed. When the field was energized all the balls would be set spinning, the large one remaining in the center while the small ones revolved around it, like moons about a planet, gradually receding until they reached the outer guard and raced along the same.

"But the demonstration which most [impressed] the audiences was the simultaneous operation of numerous balls, pivoted discs and other devices placed in all sorts of positions and *at considerable distances from the rotating field*. When the currents were turned on [in the auditorium] and the whole animated with motion, it presented an unforgettable spectacle. Mr. Tesla had many vacuum bulbs in which small light metal discs were pivotally arranged on jewels and these would spin anywhere in the hall when the iron ring was energized."[16]

Concerning Tesla's "second discovery," that "there is no energy in matter other than that received from the environment," Mach's principle, which is often stated as "local physical laws are determined by the large-scale structure of the universe," is an analogous theory that does indeed pre-date Tesla.[17] In a sense, Mach argued for a kind of super gravitational force to explain that local inertial forces were somehow connected to all the masses in the universe.[18] So it has always surprised me that Tesla never refers to Mach, given that Mach taught at both the University of Graz and the University of Prague when Tesla was attending both colleges. Mach's principle was a well-known axiom that even played a key role in Einstein's various theories.

Be that as it may, taking Tesla at his word, one way or another, he came to the conclusion that this discovery, stated in the way he did, was unique and, further, that it was also for him a closely guarded secret. As far as I can see, except for Joseph Alsop, he told no one about it and never published any treatise on its details or ramifications.

So, the question arose as to why Tesla told this twenty-three-year-old fellow this highly protected confidential revelation. This led me to study just who Joseph Alsop was. I knew he became a celebrated news reporter, but I didn't know that he was the grandson of Mrs. Douglas Robinson, one of Tesla's dearest friends!

The Waldorf-Astoria

Monday, March 6, 1899

Dear Mrs. Robinson,

In response to your kind invitation just received, I wish to say that I shall anticipate the pleasure of seeing you again on Thursday the 23rd.

On this occasion I comply with the agreeable duty at thanking you for the enjoyment you have given me yesterday. It was a great privilege to meet your brother and to listen to his enlightening conversation.

Very sincerely yours,
N Tesla[19]

There are a number of letters between Tesla and Mrs. Robinson, easily spanning ten years. They apparently met frequently, often for dinner, and as we have seen earlier, she was invited back to the wizard's lab, and he, in turn, was invited to the wedding of her daughter, also named Corinne, to Alsop's father.[20] Corinne was a shaker and doer, able, as far back as the early 1870s, to talk J. Pierpont Morgan into becoming one of the first sponsors of the Metropolitan Museum of Art. This letter mentions her brother, who turns out to be Teddy Roosevelt. It should be noted that this meeting with the governor of New York, soon to be vice president and then president of the United States, took place just a few months *after* Teddy returned from his fighting as a Rough Rider during the Spanish-American War. A true warrior, Roosevelt had led a charge up San Juan Hill in the fight in Cuba where many of his soldiers were either killed or wounded.

Corinne's first cousin was Eleanor Roosevelt, who, of course, was a distant relative to Eleanor's husband, Franklin Roosevelt, who was the U.S. president at the very time of the Alsop 1934 interview.

Thus, it seems, the aging wizard had a warm spot for Joseph, a youngster whom he may very well have known as a child, who was now a rising journalist related to two U.S. presidents. So he told the budding journalist a secret, one so closely kept that it simply doesn't appear anyplace else in all of his writings. When Tesla told Teddy Roosevelt's grandnephew that the Sun was absorbing more energy than it was radiating, what he was actually describing was a key aspect to his coveted dynamic theory of gravity.

Due to the prejudice against the concept of the ether, which was really brought about by Einstein's idea that photons could travel as particles through empty space, Higgs renamed the ether the Higgs field, and now it was cool again to talk about this primal field made up of Higgs bosons that give matter their mass.

Tesla's dynamic theory of gravity *is simply the absorption of ether by large bodies,* like the Sun, stars, and the Earth. However, *all* matter absorbs ether all the time. That really is what gravity is. We are conduits, standing waveforms, transformers of etheric energy into physical matter. And this process, it seems to me, is also responsible for particle spin. Tesla alludes to this in an unpublished article from 1936 simply titled "Nikola Tesla: New York."

> There is no energy in matter other than that received from the
> environment. . . . It applies rigorously to molecules and atoms
> as well as to the largest heavenly bodies, and to all matter in
> the universe in any phase of its existence from its very forma-
> tion to its ultimate disintegration.[21]

Clearly, Einstein's theory can in no way account for what we are talking about. If light travels as a particle instead of a wave, no ether is necessary. But as we all know, light can act like both a wave and a particle. Because there simply must be a medium between all the stars and galaxies, a medium to transfer light, it becomes obvious that Einstein never really abandoned the ether, even though that is often how his theory is portrayed. In point of fact, Walter Isaacson, in his recent Einstein biography, includes the letter from Einstein to Hendrik Lorentz wherein *Einstein overtly states*, "I agree with you that general relativity admits of an ether hypothesis." [22]

What Einstein really said was that by its nature, the ether cannot be detected. That is a very different thing than saying it doesn't exist. On this point, Einstein realized that if the ether can be detected, then relativity is missing a piece. "If Michelson-Morley is wrong," Einstein reflected, "than relativity is wrong." [23]

Tesla, of course, insisted that the ether existed, and he also greatly criticized Einstein's theory of relativity. For instance, Tesla thought it was just nutty to think that space could be curved. What Tesla realized was that light bent around stars and planets because these bodies were *absorbing ether* and light was being sucked into the influx. That was why Tesla realized that the ground connection in wireless communication was so crucial. Wireless waves would follow the curvature of the Earth because of this process, and Tesla made use of that knowledge, although he was highly secretive about his true underlying understanding of this phenomenon.

> The puzzling behavior of the ether as a solid to waves of light and heat, and as a fluid to the motion of bodies through it, is certainly explained in the most natural and satisfactory manner by assuming it to be in motion, as Sir William Thomson [Lord Kelvin] has suggested; but regardless of this, there is nothing which would enable us to conclude with certainty that, while a fluid is not capable of transmitting vibrations of a few hundred or thousand per second, it might not be capable of transmitting such vibrations when they range into the hundreds of million millions per second.
>
> —NIKOLA TESLA[24]

If, indeed, the ether is oscillating in a range that Tesla suggests here, changing from positive to negative in the billions or even trillions per second, this extremely rapid oscillation would be undetectable to our present-day instrumentation, in a way similar to how we perceive a continuity of consciousness and see or detect no flicker, even though our brains are reversing their polarity from positive to negative (i.e., oscillating) at from 7 to 60 or more cycles per second while we are awake. Yet at the same time, the ether may also account for the creation of matter by transforming its energy from this hyper or primary realm into the standing wave forms that we know as the elementary particles, which were discovered first by J.J. Thomson in 1896. The Nobel Prize winner had attended Tesla's lectures just four years earlier in 1892.

The discovery of particle spin for J. J. Thomson's electrons occurred in spectacular fashion about three decades later, as explained by George Gamow. Born in 1904 and schooled at the University of Leningrad and then at Göttingen, Germany, Gamow was a popular science writer. One of the founding fathers of quantum physics, his closest associates included Niels Bohr, Ernest Rutherford, Werner Heisenberg, Wolfgang Pauli, Enrico Fermi, Albert Einstein, Edward Teller, and many others.

What Gamow tells us in his masterwork, *Thirty Years That Shook Physics*, was that the fellows who measured electron spin, circa 1925, Samuel Goudsmit and George Uhlenbeck, threw a monkey wrench into Einstein's theory. "It turns out," Gamow wrote, "that in order to produce the necessary electromagnetic field, the electron would have to rotate so fast that the points on its equator would move at much higher velocities than the speed of light!" [25] This was not a problem for quantum physics, Gamow explained, it was a problem for relativity, which required that nothing could travel faster than the speed of light. However, since Einstein's theory was fast becoming sacrosanct, the physicists did their best to reconcile the problem, and Paul Adrien Dirac found the solution. [26]

Following in the footsteps of Hermann Minkowski, Einstein's math teacher, who used $\sqrt{-1}$ as the time component to equal the three space coordinates for the model of space-time, Dirac used the same imaginary number, $\sqrt{-1}$, to stand for particle spin as equivalent to the three space coordinates for the electron, and in that way he was able to combine quantum physics with relativity and get a Nobel Prize for the effort!

Now that the theory of relativity was elegantly combined with that of quantum physics, Einstein set out to achieve the holy grail: *grand unification,* which would be the way to combine the first three forces, *strong and weak nuclear forces and electromagnetism,* all three of which were being united with the stickler, *gravity*, a task he was unable to achieve, even after forty years of thought on the matter.[27]

Simply stated, the strong nuclear force holds the nucleus of the atom together, the weak nuclear force holds the neutrons together, and electromagnetism holds molecules together. So what the heck is gravity? Gravity is often defined as "the force that attracts a body towards the center of the Earth or any other physical body having mass."

This doesn't really tell us what gravity is, but rather what it does. And the idea gets a bit tricky when we consider two large masses such as the Earth and the Moon. Are they attracted to each other? And if so, why don't they collide? Newton, of course, came up with the law of gravitational attraction, which is M_1M_2/D^2. For instance, in this case, the mass of the Earth times the mass of the Moon divided by the distance between them, 240,000 miles, squared.

Tesla's dynamic theory of gravity is quite different. Stated simply, the reason we fall back to the Earth in this view is not because we are attracted to the center of the Earth, but rather because we are in the way of the influx of energy that the Earth is absorbing. This is the idea of the God particle, the particle (or rather process) that gives matter its mass. *All matter is constantly absorbing ether all the time,* so the Moon is absorbing ether and the Earth is absorbing ether, and it would be that competition between these two large bodies to constantly draw in ether that keeps them near each other. In a sense, they are both competing for the same primal substance, namely ether.

In the case of particle spin, Gamow has told us that Goudsmit and Uhlenbeck calculated that particles are indeed spinning at rates exceeding the speed of light. No physicist talks about this anymore. What this means is that the entire evolution of twentieth-century and nascent twenty-first-century physics ignored and is ignoring this key Goudsmit and Uhlenbeck finding. However, if we take Goudsmit and Uhlenbeck's finding at face value, the ramifications suggest that elementary particles, by their nature, interface dimensions, the so-called physical subatomic world, with the tachyonic realm stemming from the ether.

ETHER AND GRAVITY

On a body as large as the Sun, it would be impossible to project a disturbance of this kind [e.g., radio broadcasts] to any considerable distance except along the surface. It might be

inferred that I am alluding to the curvature of space supposed
to exist according to the teachings of relativity, but nothing
could be further from my mind. I hold that space cannot be
curved, for the simple reason that it can have no properties. It
might as well be said that God has properties. He has not, but
only attributes and these are of our own making. . . . To say
that in the presence of large bodies space becomes curved, is
equivalent to stating that something can act upon nothing. I
for one, refuse to subscribe to such a view. . . .

I may state that even waves only one or two millimeters
long, which I produced thirty-three years ago [in Colorado
Springs in 1899] . . . can be transmitted around the globe.
This . . . is not so much due to refraction and reflection [off
the ionosphere] as to the properties of a gaseous medium and
certain peculiar actions *which I shall explain some time in the
future*. . . . (emphasis added) [However,] this bending of the
beam projected . . . does not affect in the least its behavior in
other respects. . . . It acts just as though it were straight . . .
[but] the downward deflection always occurs, irrespective of
the wave length . . . all the more pronounced, the bigger the
planet. On a body as large as the Sun, it would be impossible
to project a disturbance of this kind to any considerable distance except along the surface.

—NIKOLA TESLA, 1932[28]

These ideas concerning ground waves that follow the curvature of the
Earth were related to Tesla's original theories on gravity, which do not
seem to have ever been published but can be ascertained by decoding
related articles by or about Tesla from the 1930s and 1940s. They also
coincide with some of the most recent theories on physics, gravity, and
magnetism that challenge Einstein's claim that nothing can travel faster
than the speed of light. E. Lerner, in the article "Magnetic Whirlwinds" in
Science Digest in 1985, stated that "magnetism is as fundamental as gravity." Citing the research and theories of plasma physicist A. Peratt of Los
Alamos National Laboratory, Lerner noted, "Astronomers using [a] . . .
radio telescope [have] . . . observed filaments of gas arcing far above the
galactic plane . . . held together by a magnetic field . . . stretching across
500 light years. . . . Such magnetic vortices [may] play a major role in the
universe . . . as important . . . as gravitation."[29] The fact that these gestalt

forces interact in realms that are hundreds of light years in length suggests the need for a new paradigm that takes into account this tachyonic aspect.

In a quote that seems to resemble Mach's principle, Tesla writes, "There is no thing endowed with life—from man, who is enslaving the elements, to the nimblest creature—in all this world that does not sway in turn. Whenever action is born from force, though it be infinitesimal, the cosmic balance is upset and universal motion results."[30]

It seems to me that this interconnectedness is directly linked to the concept of the ether. Similarly, this view also aligned itself with that of the Theosophists:

> Long ago [I] recognized that all perceptible matter comes from a primary substance, of a tenuity beyond conception and filling all space—the Akasa or luminiferous ether—which is acted upon by the life-giving Prana or creative force, calling into existence, in never ending cycles, all things and phenomena. The primary substance, thrown into infinitesimal whirls of prodigious velocity, becomes gross matter; the force subsiding, the motion ceases and matter disappears, reverting to the primary substance.[31]

Removing the spiritual component from "Akasa," Tesla postulated that everything in the universe derived its energy from external sources. This corresponded to his model of the automata or remote-controlled robot, which received commands from the electrician, and also the model of himself, that is, of the human condition as well. Each hierarchical entity in his system was not endowed with a soul, per se, but rather a self-directed electrical component moved by attraction or repulsion "entirely under the control of external forces." As a nonpsychologist, Tesla negated, by necessity, the unconscious and the instincts or Freudian id as primary motivators. So, for instance, a dream would always ultimately derive from some extrinsic factor, never from an inner source. Tesla addressed this factor with his construction of the first prototype of a thinking machine—his remote-controlled robotic boat displayed before key investors at Madison Square Garden in 1898.[32] In essence, for Tesla, the mind was, at its neuronal basis, a binary electrical system of attractions and repulsions, self-acting yet stimulated from an outside source and wholly compatible with Pavlov's stimulus-response reflex model for cognitive processes.

PREMATERIAL ETHER

> According to an adopted theory, every ponderable atom is
> differentiated from a tenuous fluid, filling all space merely by
> spinning motion, as water whirls in a calm lake. Once set in
> movement this fluid, the ether, becomes gross matter.
>
> —Nikola Tesla[33]

Returning to a more traditional nineteenth-century view and taking into
account the findings of Oersted and Maxwell, who equate ether with complex electromagnetic fields, the ether itself may be a matrix of tiny vortices
analogous to the web-like illustrations that so often appear in textbooks
drawn to describe the bending, or sinking, of space around large bodies
(e.g., stars, planets). Maybe this grid of hyperspatial spinning vortices is the
warp and woof that gives the ether its stability as well as its complex structure and undetectable nature.

My guess is that the ether is in actuality a primary AC current, oscillating at a tachyonic rate. Clearly, the speed of light separates dimensions;
the vast realm that our physical bodies inhabit and some primary domain
from which the physical world is made manifest. This text suggests, as
many people intuit, that tachyonic realms must exist, and Gamow inadvertently gives us one proof, the orthorotational speed of electrons, which,
according to Goudsmit and Uhlenbeck, significantly exceeds the speed of
light. This whirling feature, intimately linked to a pre-matter state, bears
some relationship to the origins of light and gravity and what some people
call hyperspace and perhaps consciousness. Concerning the elementary
particles:

> Let us put it bluntly: Every charge in the universe already
> freely and continuously pours out EM energy in 3-space in
> all directions, without any observable EM energy input. That
> is the well-concealed *source charge problem*, known but ignored
> by the leaders of the scientific community for a century. *All*
> EM fields and potentials and their energy come from those
> source charges, according to electrodynamics itself. Either
> we must give up the conservation of energy law entirely, or
> else we must accept the fact that unobservable virtual EM
> mass and energy are continuously absorbed from the vac-
> uum by the source charge, transduced into real observably

EM energy, and then re-radiated in 3-space in all directions as observable EM energy creating the associated fields and potentials reaching out across the universe.

—TOM BEARDEN[34]

Why is it that magnets never lose their charge? What I think Tom Bearden in suggesting is that the conservation of energy laws can't explain this.

SMASHING ATOMS

According to the physical truth I have discovered, there is no available energy in atomic structures, and even if there were any, the input will always greatly exceed the output, precluding profitable, practical use of the liberated energy.

—NIKOLA TESLA, 1936,
completely missing the boat
when it came to foreseeing the atom bomb[35]

Tesla also differed with Einstein and the quantum physicists in his view of the structure of the elementary particles and the possible consequences caused by the smashing of atoms. "I have disintegrated atoms in my experiments with a high potential vacuum tube . . . operat[ing] it with pressures ranging from 4,000,000 to 18,000,000 million volts. . . . But as to atomic energy, my experimental observations have shown that the process of disintegration is not accompanied by a liberation of such energy as might be expected from present theories."[36]

To Tesla, the theory of relativity was just "a mass of error and deceptive ideas violently opposed to the teachings of great men of science of the past and even to common sense. The theory wraps all these errors and fallacies and clothes them in magnificent mathematical garb which fascinates, dazzles and makes people blind to the underlying error. The theory is like a beggar clothed in purple whom ignorant people take for a king. Its exponents are very brilliant men, but they are metaphysicists rather than scientists." Writing a decade before the explosion of the atom bomb, Tesla brazenly concluded, "Not a single one of the relativity propositions has been proved."[37]

It would be shortsighted to simply judge Tesla wrong and Einstein and the quantum physicists right, for at least two reasons. First, both relativity and quantum theory have been established as incomplete and, in some sense, incompatible theories on the structure of the universe. Second, Tesla was discussing these phenomena from a different perspective that was not completely analogous to the one espoused by the theoretical physicists. In Colorado Springs, for instance, Tesla was generating over four million volts, whereas only about one million volts are required for separating electrons from the nucleus of an atom. Thus, Tesla *was* indeed disintegrating atoms, but in an entirely different way than was postulated by Einstein or the quantum physicists, for Tesla did not destroy the nucleus. No atomic explosion could ever occur with his type of apparatus. Tesla completely misunderstood the ramifications of Einstein's equation $E = mc^2$, and the corresponding suppositions of the equivalence of mass and energy and the vast difference between the process of separating electrons from the nucleus—which is what he did—and creating an atomic bomb, which released millions of times more energy. Unfortunately, he would never live to see the proof that tremendous amounts of power were locked inside the tiny space occupied by the nuclei of atoms.

GRAVITY

For your information, The Library of Congress reports that it has a list of works, writings, and research studies by Dr. Tesla concerning inventions and research on gravity and related fields but nothing on the cosmic ray as such.

—LEWIS E. RUBIN,
associate to Chief Henry G. Hilken,
Intercustodial and Property Branch,
U.S. Government, November 6, 1951[38]

A key aspect of this ether theory that derives from Tesla and numerous other modern writers such as Lew Price and B. Herbert Gibson, Ron and Ed Hatch, Vencislav Bujić, Warren York, and David Wilcox, is that matter is constantly absorbing ether all the time.[39] Michael Hodges, a scientist from New Zealand who studies "aether theory" and gravity, said it this way: "An atom's electrons' spin would generate a vacuum force that collectively produce gravitational attraction."[40] This idea of an electron spin "vacuum force" linked to tapping primary space is a concept worth considering.

Regarding electron spin, Gamow tells us, as already noted, that according to the findings of Goudsmit and Uhlenbeck, *"In order to produce the necessary EM field, the electron would have to rotate so fast that the points on its equator would move at much higher velocities than the speed of light!"* What Gamow is telling us here is that traditional physicists understood that tremendous energy is involved in creating or producing and maintaining electron spin and further that this very process of orthorotation generates electromagnetic energy, but when the measurements and calculations were completed, in order for this to be achieved the particles needed to spin at tachyonic rates. What is important to realize here is that the entire premise of this chapter, namely that ether absorption occurring at a tachyonic rate (that is, what we are calling gravity) generates electromagnetic energy, is inherent in this very statement made by George Gamow, one of the founders of modern quantum physics. Further, the proof that tremendous amounts of energy are involved in the construction and maintenance of atomic structures is the invention of the atom bomb, and that discovery/invention was a direct outcropping of Einstein's famous equation: $E = mc^2$. The fact that the speed of light squared is part of the mix also suggests that tachyonic energies are involved.

GRAND UNIFICATION

According to the etheric view as espoused by Price and Gibson et al. ether is easily detected. If you are driving in a car and accelerate greatly, you will feel a g-force. This is an increased absorption of ether. That's what a g-force is. *Ether flowing into matter is gravity; matter flowing rapidly through ether is acceleration* (experienced as a g-force).

Concerning this issue, Einstein would write the young mathematician Karl Schwartzschild on January 9, 1916, "Inertia is simply an interaction between masses, not an effect in which *space* of itself is involved, separate from the observed mass." Schwartzschild, as Isaacson points out, disagreed. Then, four years later, in 1920, after reconsidering the necessity of the ether, for instance, as a means to propagate light, "Einstein changed his mind." He abandoned Mach's principle and in a second letter to Schwartzschild said that he now saw that a rotating body did not obtain its inertia from and in relation to all the rest of the matter in the universe (Mach's principle), but of its own accord due simply to "its state of rotation [because] space is endowed with physical qualities."[41]

If we look at the development of Einstein's thinking, the reason he changed his view in his two letters to Schwartzschild was because of the power of Louis de Broglie's emphasis on the wave aspect of wave-particle theory, which was coming into vogue circa 1916. Einstein shifted gears to be current. Once again ahead of the curve, he even lectured on the ether at Leiden University. Einstein never came to view gravity as the absorption of ether by elementary particles and electromagnetism as a product of this process, because to do so would be to abandon relativity. However, as we can see from this second letter to Schwartzschild, Einstein came very close. If the "rotating body" maintained its state of rotation because "space is endowed with physical qualities," that is precisely the key component to Tesla's dynamic theory of gravity. Einstein was never able to integrate gravity into his grand unification scheme because he did not explore as deeply as he might have these "endowed qualities" of space that he alludes to. As Isaacson points out, the solution to grand unification eluded the great physicist the entire second half of his life.

Once it is realized that electrons spin at speeds in excess of the speed of light, a *new paradigm* emerges. The idea simply is that the elementary particles, by their nature, spinning at tachyonic rates, are absorbing ether all the time. This influx is what *gravity* is. As ether is absorbed, two things happen: (1) the process enables the elementary particles to maintain their spin, and (2) simultaneously, this tachyonic etheric energy is transformed into electromagnetic energy through the ongoing, perpetual process of ether absorption and particle spin.

> **GRAND UNIFICATION:** *Through the process of ether absorption (e.g., gravity) resulting in particle spin, elementary particles transform primal etheric energy into electromagnetism.*

This simple and exquisite theory, which is very much in accord with traditional physics as explained by George Gamow regarding tachyonic particle spin, better explains the idea of the so-called God particle, the particle that gives matter its mass. A modification of what has been called the Higgs boson, the Teslaic view elegantly explicates precisely how matter gets and maintains its mass. This is *grand unification*, Einstein's dream of how to combine gravity with electromagnetism.

Tesla understood ether theory a lot better than Einstein did, but obviously, Tesla also did not truly understand the ultimate nature of atomic structure, that is, in particular, the structure of the nucleus and thus the ramifications of Einstein's famous equation: $E = mc^2$. He dismissed it as mathematical poppycock. Had he lived a few more years to see the explosion of the atom bomb, Tesla would have been forced to reevaluate what he had discarded, and had Einstein reevaluated the full ramifications of Tesla's ether theory, he may have been able to achieve his grand dream of unifying gravity with electromagnetism, a process explainable by a full understanding of ether theory coupled with Tesla's dynamic theory of gravity.

A large number of thinking physicists believe that an ether of sorts exists. This, in fact, was the prevailing paradigm in the 1800s. Further, many scientists today also accept the premise that tachyonic or (essentially) instantaneous action is possible.

Quantum Entanglement

Scientists have hailed the advance as a significant step towards the goal of creating an unhackable quantum internet.

"Space-scale teleportation can be realised and is expected to play a key role in the future distributed quantum internet," the authors, led by Professor Chao-Yang Lu from the University of Science and Technology of China, wrote.

The feat sets a new record for quantum teleportation, an eerie phenomenon in which the complete properties of one particle are instantaneously transferred to another—in effect teleporting it to a distant location.

—HANNAH DEVLIN[42]

Quantum entanglement ironically evolved from a criticism Einstein had leveled against quantum physics. Paraphrasing, what Einstein essentially said was that there had to be some flaw in quantum mechanics because if the theory was completely correct, then this would imply the possibility of *instantaneous information transfer*, or what Einstein called "spooky action at a distance," which, of course, he said was impossible. Working with Boris Podolsky and Nathan Rosen, Einstein wrote a paper explaining the problem, and this became known as nonlocality or the EPR effect.

Ironically, where Einstein saw this as a flaw in quantum physics theory, researchers have been able to verify the EPR effect! In the case of the Chinese, they recently beamed photons from a ground station in Tibet to a satellite three hundred miles above the Earth. "The research hinged on a bizarre effect known as quantum entanglement, in which pairs of particles are generated simultaneously meaning they inhabit a single shared quantum state. Counterintuitively, this twinned existence continues even when the particles are separated by vast distances: Any change in one will affect the other." In this manner, information was, according to the article, "teleported" instantaneously from the ground base to the satellite.[43] According

to physicist and rocket scientist Travis Taylor, his American group working with NASA have duplicated routinely this experiment of instantaneous
information transfer from ground to satellite numerous times.[44] The end
result is that scientists have established that the tachyonic realm can now
be employed.

> As a man who has devoted his whole life to the most clear-
> headed science to the study of matter, I can tell you as a result
> of my research about atoms this much: There is no matter as
> such. All matter originates and exists only by virtue of a force
> which brings the particle of an atom to vibration and holds
> this most minute solar system of the atom together. We must
> assume behind this force the existence of a conscious and intel-
> ligent spirit. This spirit is the matrix of all matter.
>
> —MAX PLANCK[45]

Once one begins to study and unravel Tesla's dynamic theory of gravity,
profound new insights concerning particle spin, zero point energy, the fundamental structure of matter and space, the constancy of light speed, and
the link between gravity and electromagnetism begin to emerge. According
to this theory, the ether exists. It is a primary substance, most likely AC,
oscillating at a tachyonic rate, and it is a substance that is constantly flowing
into what we call the physical universe, a universe made up of atoms, which
in turn are made from elementary particles. For these particles, (e.g., electrons) to generate electromagnetic energy, they are, according to Goudsmit
and Uhlenbeck as reported by Gamow, spinning faster than the speed of
light. Based on their size, Gamow tells us, in order for elementary particles
to generate energy, they simply have to spin at tachyonic rates, and in this
process of ether absorption, they generate EM energy. That is how to combine gravity with electromagnetism and that is what grand unification is.
Elementary particles interface the etheric dimension with physical reality.[46]

   In Tesla's words, "There is no energy in matter other than that received
from the environment."

19

Final Thoughts

December 22, 1895

A number of millionaires have been after me but I have resisted the temptation. . . . Xmas I want to be at –327 Lexington Ave.–with my friends–my dear friends the Johnsons. If you will prepare a dinner for half a dozen and invite nobody it will suit me–the dinner. We shall talk of blessed peace and be merry until then.

Yours sincerely
N Tesla[1]

In trying to obtain a full-bodied depiction of Tesla's personality, one encounters many challenges and contradictions. If we keep in mind that Tesla was first and foremost a Serb, born in a hostile environment from a kingdom decimated a half a millennium earlier by battle, bloodshed, and carnage, we can understand why Tesla was so drawn to war and to figuring out ways to end it. Shortly after Mussolini invaded Ethiopia in October of 1935, Tesla wrote a detailed piece on how to combat the invasion. After suggesting that Addis Ababa should be abandoned so the inhabitants could hide out in the hills to launch a guerilla counterattack, Tesla wrote that if the Italians brought in tanks, "If I were the Lion of Judah, under my guidance . . . [I would] quickly perfect an efficient defense, possibly rifles of large bore, adapted to fire charges of mercuric fulminate. This is an extraordinarily powerful explosive. The extreme suddenness of its detonation is such that, even if unconfined, it will punch a hole in a thick steel plate."[2] Clearly, Tesla had spent considerable time thinking about various types of weapons, such as his particle beam artillery for defense and large steel-piercing rifles for counterattack.

On a more personal level, Tesla claimed he was celibate, a confidence the seventy-one-year-old inventor shared in an interview he did with Dragoslav

Lj. Petkovic, a young Serb, at a restaurant in New York in April 1927. For dinner, "Mr. Tesla . . . gives an order every day what should be cooked for him, and this time he ordered fish, in addition to the soup and celery with other vegetables, milk, apples and California prunes in honey."

Petkovic asked Tesla about Marconi. "With disturbed expression on his face and in his voice he said, 'Mr. Marconi is a donkey. . . . When I sent electrical waves from my laboratory in Colorado, around the world from the height of [six thousand feet], Mr. Marconi was experimenting with my apparatus unsuccessfully at sea. Afterword Mr. Marconi came to America to lecture on the subject, stating that it was he who sent those signals around the globe. I went to hear him, and when he learned that I was present he became sick, postponed the lecture, and up to the present time has not delivered it.'"

In one revealing moment during a discussion of hypnosis and Tesla's ability to use "magical power" to remain unhypnotizable, Petkovic wrote, "On his face was an expression of great pain and finally almost in a whispering tone he said, 'I have never touched a woman. As a student, and while vacationing at my parents' home in Lika, I fell in love with one girl. She was tall, beautiful, and had extraordinary understandable eyes.'" [3]

Why does the old wizard tell this young man this bit of information? One possibility would be that he was making a pass at the man, another possibility was that he was letting this youngster know that in an alchemical sense, Tesla worked to transform his sexual energy into other creative pursuits, or maybe he simply let his guard down and told a compatriot something that was indeed well-known, namely that he was a celibate.

It can be argued that this attitude caused a displacement of his libido whereby he tended to shower his affection on the city pigeons that he fed religiously in Bryant Park by the New York Public Library, and that he reduced his diet to such an extent that he became anorexic in later life. The true details of Tesla's sexual life, however, must remain a mystery.

Tesla was certainly a well-rounded individual who attended the theater, lectured, dined with the glitterati, and invited them back to his laboratory. Here was a scientist and inventor interested in literature, friends with numerous married individuals, and concerned about the ramifications and changing dynamics of an evolving society.

Was Tesla homosexual? If he had that attitude, it is doubtful that he acted on those urges. He was indeed attracted to muscular men such as boxers. He was a boxing fan. Tesla also sought physical contact when he went to his favorite barber, a man whom he had gone to for years, who, according to John O'Neill, used to give Tesla vigorous head massages two or three times a week to stimulate the inventor's brain cells; the barber most likely used a mobilized hand-glove (see the 1945 movie *Weekend at the Waldorf*). When Tesla's eight-year-old grandnephew William Terbo met the inventor, circa 1938, he was greeted with a warm kiss, according to the grown Terbo. But Tesla was also attracted to actresses such as Sarah Bernhardt, Elsie Ferguson, and Greta Garbo, whom he apparently met. And he certainly had a flirtatious streak, promising Marguerite Merington to purchase her a country home in shouting distance of downtown as soon as his ship came in, and maintaining a lifelong friendship with J. P. Morgan's daughter Anne. So to pigeonhole him one way or the other is clearly a mistake. What we can say for sure is that he lived a life of self-denial and, in that sense, self-transformation. Tesla was committed to his work. When it came to his sexual nature, he showered his affection on the city pigeons, he had intimate friendships with people of both sexes, and he devoted his life to changing the planet for the betterment of mankind through the powers of his cerebrations. When asked about his relationship to money, Tesla said that he shoveled it out the window as soon as it came through the door.[4]

We know that Tesla stopped paying his rent at most of the hotels that he lived in, particularly the Waldorf-Astoria, the Hotel Pennsylvania, the Hotel Governor Clinton, and also the Hotel St. Regis. There were several reasons why he didn't cover these debts. One was he no longer had the funds that he used to have when he was a younger man. The First World War, also, was an enormous blow, as Tesla had highly lucrative contracts with two separate German wireless enterprises: the Sayville plant out on Long Island owned by Atlantic Communications, a division of Telefunken, where he was earning $1,500 a month, and Homag, which ran the Tuckerton station, where Tesla had negotiated 5 percent royalties on all gross income from that New Jersey plant. Had the war not intervened, Tesla's revenues would have been substantial and long-term. Since the French superior courts had ruled Tesla's priority over Marconi, both France and Germany were recognizing his priority. As he had legal suits against Marconi in America, had the war not intervened, he may very well have obtained additional revenues from the colossal Marconi concern as well. Another reason that Tesla may have used to not pay his hotel bills was, perhaps, the aging wizard's way of getting back at society. Since he had invented an electrical power system that ran the world, as a prima donna, in a sense, he felt society owed him. And maybe in some metaphysical way they did. But in the real world, one has to pay one's rent.

Tesla's penchant for walking out on debts can be traced back to his youth. While attending Graz University in 1876 and 1877, Tesla was given a stipend by the government of 420 guilders per month in exchange for a promise, after graduation, to spend eight years in the military. Never graduating, instead, Tesla took off for Maribor, where he worked as an electrician. Living an easy life, playing pool and gambling, he somehow got himself arrested. Shipped back home to Gospic, somewhat disgraced, with his father's help, he used the time to seek another scholarship so that he could continue his education at the University of Prague.[5] There, he undertook courses in foreign languages, physics, electrical engineering, and higher mathematics. But as with Graz, he also never graduated from Prague. Due to a severe illness, which may have been cholera, that he suffered between the time he returned home and his stay in Prague, his father may have helped him use that excuse to get out of the debt, an obligation that would have involved a long stay in the military.

Later in life, perhaps Tesla's celebrity status enabled him to avoid covering many of the debts he never repaid. He certainly remained bitter with regards to his decision to rip up his royalty contract with George Westinghouse during the tensions that arose during the War of the Currents. And it is very possible that displaced aggression played a role in his continuing decision to skip out on what he owed to the various hotels.

University of Chicago
July 8, 1931

My dear Mr. Tesla:

As one of the millions who have benefitted from the product of your inventive genius, may I add my word of congratulation to those of your other friends on the occasion of your seventy-fifth birthday. To men like yourself, who have learned first hand the secrets of nature and who have shown us how her laws may be applied in solving our everyday problems, we of the younger generation owe a debt that cannot be paid.

Sincerely,
Arthur H. Compton[6]

Certainly, Tesla had an extraordinary mind that was appreciated by such Nobel Prize winners as Arthur Compton, who would go on to become a prominent participant in the Manhattan Project. Tesla was highly literate, memorized entire books by heart, spoke more than a half-dozen languages, and was an eternal optimist, expecting to live to age 140 as he worked around the clock. He claimed to sleep only two hours a night! He was highly original, deriving many of his ideas from studying nature and the works of his predecessors. There is almost no comparable figure in history; Leonardo da Vinci comes to mind. A true renaissance man. If you look at Tesla's achievements, unlike many other inventors, his creations are in diverse fields, from machinery construction to wireless communication, artificial intelligence, particle beam weapons, and original airplane designs. It is hard to square his rather limiting view of man as a self-acting automaton with his actual accomplishments. His greatest failure, of course, was Wardenclyffe, that catastrophe directly related his obdurate decision to either do the entire project or not do it at all. And so the stubborn wizard watched inferior wireless designs take over the planet as he shifted gears to work on patents for rain-making devices and the supposed ability to illuminate the high seas for shipping lanes instead of inaugurating his own cell phone–type technology on at least a modest scale.

> Tesla, in his 80s, was still manifesting the superman complex, and on even more elaborate a scale than when in his 20s. In his earlier dreams his visions were terrestrial, but in later life they were extended to embrace the entire universe.
>
> —JOHN O'NEILL[7]

His brother Dane's untimely death when Tesla was five could explain, in a psychoanalytic way, Tesla's obsession with feeding the city pigeons, as it would be a way for him to reconnect with his life on the farm when his brother was still alive. Tesla was also adamant about contacting extraterrestrials. Based on the work of physicists Jim and Ken Corum, the best guess for what occurred in Colorado Springs in 1899 was that Tesla received pulsed frequencies stemming from one of the moons of Jupiter, not the planet Mars. Another possibility, as I suggested in *Wizard*, was that Tesla intercepted Marconi's experiments, which were occurring at the same time six thousand miles away off the coast of England. Since Tesla claimed that he received three beated pulses and Marconi was sending Morse code for the letter *S*, namely *dot-dot-dot*, this is also a reasonable possibility. Either way, it is highly unlikely that these beated frequencies came from extraterrestrials.

When it came to Tesla's great emphasis on creating weapons of war, namely his remote-controlled robotic boat, which was the precursor to drone warfare, and his particle beam weapon, in both instances, he claimed these devices were so dangerous that if every nation had them, war would become obsolete. A notion that he shared with none other than Mark Twain back in 1898 when Twain requested permission to sell the former device to foreign nations for the very same reason, this concept could be seen as a forerunner of the MAD doctrine (mutually assured destruction) regarding nuclear weapons.

Thus, we are left with a seeming contradiction. Tesla constructed weapons of war because his goal was to promote peace. Quite aware of the true horror of war, having lived it himself, paradoxically, the aging wizard understood that one of his greatest contributions would be to use his vast knowledge of high technology to create a weapon so deadly that if every country had such a device, it would simply be folly to engage in war. Combined with his invention of a world telegraphy system, Tesla's hope and partially realized goal was to unite the entire world so that boundaries between nations would be transcended, and people from every continent could communicate with each other freely no matter the distance, as if they were all sitting in the same room. This Teslaic vision is spectacularly realized today via the Internet with software like FaceTime, Skype, and Zoom. To take the particle beam weapon out of that larger context is to miss entirely the essence of who Tesla was.

> The scientific man does not aim at an immediate result. He does not expect that his advanced ideas will be readily taken up. His work is like that of the planter—for the future. His duty is to lay the foundation for those who are to come, and point the way. He lives and labors and hopes.
>
> —NIKOLA TESLA

In general, my view of Tesla is that he was a bon vivant; O'Neill said he was the best-dressed man on Fifth Avenue in his heyday. We can see his penchant for wit, his artistic nature, and his sense of camp by way of his fantastic double-exposure photographs of himself surrounded by lightning and his ability to complete Wardenclyffe, at least on paper, by working with sci-fi aficionados such as editor Hugo Gernsback and premier artist Frank

R. Paul. A star of the Gilded Age with a multifaceted personality, living at the Waldorf-Astoria, the most posh hotel in all the world at its height, the Serbian inventor enjoyed close friendships with numerous individuals of both sexes and of many ages, many of them world-reknowned movers and shakers themselves. He had a caring nature and sense of loyalty, and he was driven by his wish to help the human race progress in a way that did not sap the Earth of its natural resources, was nonpolluting, made use of renewable energy, and worked within its rhythms. Although he constructed novel weapons for war, his goal was always to create a situation that would demand peace. Seeking the utopian ideal, in one communiqué to J. Pierpont Morgan, in trying to appeal to the financier's philanthropic side and the inventor's wish to complete his great world telegraphy tower at Wardenclyffe, he pointed out that such a goal of peaceful communication with all peoples of the planet was actually achieved in the case of the postal system. Such was the vision and understanding of this unusual individual.

Perhaps the best way to understand Tesla's attempt to reveal himself and "illuminate the world" is to let him speak for himself in a way he once characterized poetically as a "Fragment of Olympian Gossip."[8]

"Man's Greatest Achievement"
Nikola Tesla
Milwaukee Sentinel, Sunday, July 13, 1930

When a child is born its sense-organs are brought in contact with the outer world . . . and in this act a marvelous little engine, of inconceivable delicacy and complexity of construction, unlike any on Earth, is hitched to the wheel-work of the Universe.

The little engine labors and grows. . . . Inspired to this task he searches, discovers and invents, designs and constructs, and covers with monuments of beauty, grandeur and awe, the star of his birth.

He descends into the bowels of the globe to bring forth its hidden treasures and to unlock its immense imprisoned energies for his use. . . . He subdues and puts to his service the fierce, devastating spark of Prometheus, the titanic forces of the waterfall, the wind and the tide. He tames the thundering bolt of Jove and annihilates time and space. He makes the great

Sun . . . his obedient toiling slave. Such is his power and might
that the heavens reverberate and the whole Earth trembles by
the mere sound of his voice.

What has the future in store for this strange being, born of
a breath, of perishable tissue, yet immortal, with . . . powers
fearful and divine? What magic will be wrought by him . . . ,
what is to be his greatest deed, his crowning achievement?

Long ago he recognized that all perceptible matter comes
from a primary substance, of a tenuity beyond conception . . .
filling all space—the Akasa or luminiferous ether—which is
acted upon by the life-giving Prana or creative force, calling
into existence, in never ending cycles, all things and phenomena. The primary substance, thrown into infinitesimal whirls
of prodigious velocity, becomes gross matter; the force subsid-
ing, the motion ceases and matter disappears, reverting to the
primary substance.

Can Man control this grandest, most awe-inspiring of all
processes in nature? Can he harness her inexhaustible energies
to perform all their functions at his bidding, more still—can he
so refine his means of control as to put them in operation simply
by the force of his will? If he could do this he would have powers almost unlimited and supernatural. At his command . . . old
worlds would disappear, and new ones of his planning would
spring into being. He could fix, solidify and preserve the ethe-
real shapes of his imagining, the fleeting visions of his dreams.
He could express all the creations of his mind, on any scale,
in forms concrete and imperishable. He could alter the size of
this planet, control its seasons, guide it along any path he might
choose through the depths of the Universe. He could make planets collide and produce his suns and stars, his heat and light. He
could originate and develop life in all its infinite forms.

To create and to annihilate material substance, cause it
to aggregate in forms according to his desire, would be the
supreme manifestation of the power of Man's mind, his most
complete triumph over the physical world, his crowning
achievement which would place him beside his Creator and
fulfill his ultimate destiny.[9]

LIST OF ABBREVIATIONS

BDA	Barbara Daddino Archives
BLCU	Butler Library, Columbia University
CERN	European Organization for Nuclear Research (Organisation Européenne pour la Recherche Nucléaire)
CIA	Central Intelligence Agency
CSLI	From Colorado Springs to Long Island, NT 2008
cps	Cycles per second
DARPA	Defense Advanced Research Projects Agency
ELF	Extremely low frequency
EM	Electromagnetic
ER	Electrical Review
EW	Electrical World
FBI	Federal Bureau of Investigation
FOIA	Freedom of Information Act
GW	George Westinghouse
IRE	Institute of Radio Engineers
ITS	International Tesla Society
JHH Jr	John Hays Hammond Jr.
JJA	John Jacob Astor
JPM	J. Pierpont Morgan
JPM Jr	J. P. Morgan Jr.
KJ	Katharine Johnson
KSP	Kenneth M. Swezey Papers at Smithsonian Institute
LA	Leland Anderson Papers
LOC	Library of Congress
MIT	Massachusetts Institute of Technology

MJS	Marc J. Seifer Archives
NA	National Archives
NDRC	National Defense Research Committee
NT	Nikola Tesla
NTM	Nikola Tesla Museum
NYS	New York Sun
NYT	New York Times
OAP	Office of Alien Property Custodian
OSRD	Office of Scientific Research & Development
OSS	Office of Strategic Services
RUJ	Robert Underwood Johnson
SWP	Stanford White Papers, Avery Library, Columbia U.
TAE	Thomas A. Edison Archives, Menlo Park, NJ
VLF	Very low frequency

Frequently Cited Sources
by or about Nikola Tesla

NT. *Electric Power Transmission Patents, Tesla Polyphase System.* Pittsburgh, PA: Compliments of Westinghouse Electric Manufacturing Co., 1893.

NT. *The Inventions, Researches, and Writings of Nikola Tesla.* Edited by T. C. Martin. New York: Electrical Engineer, 1894.

NT. "The Problem of Increasing Human Energy." *The Century,* June 1900, 175–211.

NT. *Nikola Tesla: On His Work with Alternating Currents and Their Application to Wireless Telegraphy, Telephony, and Transmission of Power.* Edited by L. Anderson. Denver CO: Sun Publishing, 1992. First printing, 1916.

NT. *My Inventions: The Autobiography of Nikola Tesla.* Edited by Ben Johnston. Williston, VT: Hart Bros, 1981. First printing, 1919.

NT. "The New Art of Projecting Concentrated Non-dispersive Energy Through Natural Media." in *Proceedings of the Tesla Centennial Symposium*, edited by Elizabeth Raucher and Toby Grotz, 144–50. Colorado Springs, CO: ITS Publishing, 1984. First printing of Tesla article, 1937.

NT. *Nikola Tesla: Lectures, Patents, Articles.* Belgrade: Nikola Tesla Museum, 1956.

NT. *Tribute to Nikola Tesla: Letters, Articles, Documents.* Belgrade: Nikola Tesla Museum, 1961.

NT. *Colorado Springs Notes and Commentary.* Edited by Alexander Marincic. Belgrade: Nikola Tesla Museum, 1979.

NT. *Solutions to Tesla's Secrets.* Edited by John T. Ratzlaff. Milbrae, CA: Tesla Book Co., 1981.

NT. *Tesla Said.* Edited by John T. Ratzlaff. Milbrae, CA: Tesla Book Co., 1984.

NT. *Nikola Tesla: From Colorado Springs to Long Island.* Edited by Vladimir Jelenkovič. Belgrade: Nikola Tesla Museum, 2008

NT. *The Unresolved Patents of Nikola Tesla.* Edited by Snezana Sarboh. Belgrade: Nikola Tesla Museum, 2013.

John T. Ratzlaff and Leland I. Anderson. *Dr. Nikola Tesla Bibliography.* Palo Alto, CA: Ragusan Press, 1979. Contains writings by and about Tesla, 1884–1978.

Joe Kinney, ed. *Tesla Archives.*

Derek Worthington, ed. *Tesla Articles – Forty Years.* Aether Force, 2016.

Marc J. Seifer. *Wizard: The Life and Times of Nikola Tesla: Biography of a Genius,* New York: Citadel Press, 1998.

Bibliography

Abramovič, Velimir. "Nikola Tesla." YouTube.

Adams, Edward Dean. *Niagara Power: 1886–1918*. Niagara Falls, NY: Niagara Falls Power Co., 1927.

Aitken, Hugh. *The Continuous Wave Technology*. Princeton, NJ: Princeton University Press, 1985.

Anderson, Leland, ed. "John Stone Stone on Nikola Tesla's Priority in Radio and Continuous-Wave Radiofrequency Apparatus," *Antique Wireless Association Review* 1 (1986): 19–42.

———. *Nikola Tesla: Guided Weapons and Computer Technology*. Breckenridge, CO: 21st Century Books, 1998.

———. *Nikola Tesla: Lecture Before the New York Academy of Sciences, 4/6/1897*. Breckenridge, CO: 21st Century Books, 1994.

Barnett, Lincoln, *The Universe and Dr. Einstein*, New York: Time Inc. Book Division, 1948/62.

Bearden, Tom. *Energy from the Vacuum*. Santa Barbara CA: Cheniere Press, 2002.

———. "Solutions to Tesla's Secrets and the Soviet Tesla Weapon." In NT 1981, 1–45.

Bernstein, Jeremy. *Three Degrees Above Zero: Bell Laboratories in the Information Age*. New York: Scribner, 1984.

Blum, Howard. *Dark Invasion: 1915: Germany's Secret War and the Hunt for the First Terrorist Cell in America*. New York: HarperCollins, 2014.

Bush, Vannevar. *Endless Frontier*. Washington, DC: Public Affairs Press, 1946.

Carlson, W. Bernard. *Tesla: Inventor of the Electrical Age*. Princeton, NJ: Princeton University Press, 2013.

Cheney, Margaret. *Tesla: Man Out of Time*. Englewood Cliffs, NJ: Prentice Hall, 1981.

Chernow, Ron. *The House of Morgan*. New York: Grove Press, 2010.

Childress, David Hatcher, *The Tesla Papers*, Kempton, IL: Adventures Unlimited, 2000.

Christgau, John. *Enemies: WWII Internment*. Lincoln: Bison Books / University of Nebraska Press, 2009.

Clark, Ronald. *Einstein: The Life and Times*. New York: World Publishing Co., 1971.

318

Collins, Rodney. *Theory of Celestial Influence*. London: Vincent Stuart Publishers, 1970.

Conant, Janet. *Man of the Hour: James Conant, Warrior Scientist*. New York: Simon and Schuster, 2017.

Crockett, Albert Stevens. *Peacocks on Parade*. New York: Sears, 1931.

Culbert, David, ed. *Information Control and Propaganda: Records of the Office of War Information: 1942–1945*. Annapolis, MD: Naval Institute Press, 2004.

Douglas, Alan. *Radio Manufacturers of the 1920's*. Vol. III. Phoenix, AZ: Sonoran Publishing, 2006.

Dutka, Alan. *Misfortune on Cleveland's Millionaire Row*. Mount Pleasant, SC: Arcadia Publishing, 1915.

Einstein, A.; B. Podolsky; and N. Rosen. "Can Quantum Mechanical Description of Physical Reality Be Considered Complete?" *Physical Review* 47 (1935).

Fenn, Charles. *At the Dragon's Gate: With the OSS in the Far East*. Annapolis, MD: Naval Institute Press, 2004.

Fritjof, Capra. *The Tao of Physics*, Berkeley, CA: Shambhala Press, 1975.

George, Willis. *Surreptitious Entry*. Boulder CO: Paladin Press, 1990.

Hammond, John H., Jr. "The Future in Wireless," *National Press Reporter* 15, no. 110 (May 1912).

Hunt, Inez, and Wanetta Draper. *Lightning In His Hands: The Life Story of Tesla*. Hawthorne, CA: Omni, 1964–1977.

Isaacson, Walter. *Einstein: His Life and Universe*. New York: Random House, 2007.

Johnson, Robert Underwood. *Remembered Yesterdays*. Boston: Little Brown, 1924.

———. *Songs of Liberty*. New York: *The Century*, 1897.

Jolly, W. *Marconi*. New York: Stein and Day, 1972.

Josephson, Matthew. *Thomas Alva Edison*. New York: McGraw-Hill, 1959.

Jones, John Price, and Paul Merrick Hollister. *The German Secret Service in America*. Boston: Small, Maynard, and Co., 1918.

Koestler, Arthur. *The Sleepwalkers*. New York: Random House, 1959.

Kosanović, Nicholas, trans. and ed. *Nikola Tesla, Correspondence with Relatives*. Belgrade: Nikola Tesla Museum, 1995.

Krause, Michael. *CERN: How We Found the Higgs Boson*. Berlin: Rich and Famous Publishing, 2014.

Kruk, Dmitriy. *Nikola Tesla: The Force Awakens*. Amazon Books, 2017.

Lessing, Lawrence. *Man of High Fidelity: Edwin Armstrong*. New York: Lippincott, 1956.

Marconi, Degna. *My Father Marconi.* New York: McGraw Hill, 1962.

Martin, T. C., ed. *The Researches, Writings, and Inventions of Nikola Tesla.* New York: Electrical Engineer, 1894.

Miessner, Benjamin Franklin. *Radiodynamics: The Wireless Control of Torpedoes and Other Mechanisms.* New York: Van Nostrand, 1916.

Mrkich, Dan. *Nikola Tesla: The European Years.* Ottawa, Canada: Commoner's Publishing, 2003.

Munson, Richard. Tesla: *Inventor of the Modern,* New York: W.W. Norton, 2018.

Musès, Charles, and Arthur Young. *Consciousness and Reality.* Avon Books, 1972.

O'Neill, John. *Prodigal Genius.* New York: Ives Washburn, 1944. Reprinted by David McKay, 1972.

Ouspensky, P. D. *New Model of the Universe.* New York: Random House, 1971.

Puharich, Andrija. *Beyond Telepathy.* Garden City, NY: Doubleday, 1962.

———. *Tesla's Magnifying Transmitter.* Ossining, NY: 1980.

Rauscher, Elizabeth, and Toby Grotz. *Proceedings of the Tesla Centennial Symposium.* Colorado Springs, CO: ITS Publishing, 1984.

Rhodes, Richard. *The Making of the Atom Bomb.* New York: Simon and Schuster, 1986.

Satterlee, Herbert. *J. Pierpont Morgan: An Intimate Biography.* New York: Macmillan, 1939.

Seifer, Marc. *Crystal Night.* Kingston, RI: Doorway Press, 2018.

———. "Secret History of the Wireless." In *Nikola Tesla's Electricity Unplugged,* edited by Tom Velone, 63–92. Kempton, IL: Adventures Unlimited Press, 2016.

———. "Tesla's Dynamic Theory of Gravity: Grand Unification, the God Particle, Tesla, and Einstein." The Wardenclyffe Conference Speech, October 24, 2011. *Tesla Magazine,* July 2014, 18–23.

———. "Tesla vs. Einstein: The Ether and the Birth of the New Physics." *New Dawn* (Australia), March–April 2009, 47–52; also *Infinite Energy,* January, 2010; translated into Danish, *Danish Institute of Ecological Techniques,* May 2010, 3–11.

———. *Transcending the Speed of Light.* Rochester, VT: Inner Traditions, 2008.

———. *Wizard: The Life and Times of Nikola Tesla, Biography of a Genius.* New York: Citadel Press, 1996.

Seifer, Marc, and Howard Smukler. "The Puharich Interview." *Gnostica* 47 (September 1978): 21–25.

———. "The Tesla/Matthews Connection: Part 1." *Pyramid Guide,* May 1978, 5.

———. "The Tesla/Matthews Connection: Part 2." *Pyramid Guide,* July 1978, 5.

Smith, Harris. *The OSS: The Secret History of America's First Central Intelligence Agency.* Los Angeles, CA: UCLA Press, 1972.

Storm, Margaret. *Return of the Dove.* Baltimore, MD: Margaret Storm Publication, 1956.

Strouse, Jean. *Morgan: American Financier.* New York: Random House, 1999.

Talbot, David. *The Devil's Chessboard: Allen Dulles, the CIA, and the Rise of America's Secret Government.* New York: Harper, 2015.

Thistle, Mel, ed. *The Mackenzie-McNaughton Wartime Letters.* Toronto, Canada: University of Toronto Press, 1975.

Thompson, Silvanus. *Polyphase Electric Currents.* New York: American Technical Book Co., 1897.

Valone, Tom, ed. *Nikola Tesla's Electricity Unplugged.* Kempton, IL: Adventures Unlimited Press, 2016.

Waller, Douglas. *Wild Bill Donovan: The Spymaster Who Created the OSS.* New York: Free Press, 2011.

Wasik, John F., *Lightning Strikes: Timeless Lessons in Creativity from the Life and Work of Nikola Tesla,* New York: Sterling Publishing, 2016.

Wheeler, G. *Pierpont Morgan and Friends: Anatomy of a Myth.* Englewood Cliffs, NJ: Prentice-Hall, 1973.

Williams, L. *Origin of Field Theory.* New York: Random House, 1966.

Zachary, G. Pascal. *Endless Frontier: Vannevar Bush: Engineer of the American Century.* New York: The Free Press, 1997.

Notes

Preamble

1. Velimir Abramovič, transcribed with his permission from YouTube.

Introduction: Wizard at War

1. Irving Jurow, letter to Marc J. Seifer, July 5, 1993.
2. D. Ladd, FBI memorandum re: L. C. Smith aligned with OAP taking charge of Tesla papers, January 11, 1943, FOIA.
3. NT, "Talking with the Planets," *Colliers*, February 9, 1901, 405–6; *Current Literature*, March 1901, 429–31.
4. NT, letter to the Red Cross, Christmas 1900.
5. NT, "'Only a Matter of Patience Now,' Says Tesla." *New York American*, March 11, 1906.
6. Seifer and Smukler, "Puharich Interview," *Gnostica*, Sept, 1978, v. 47, pp. 21–25.
7. Puharich, *Tesla's Magnifying Transmitter*, xvii.
8. Nicholas Kosanović, trans., *Nikola Tesla, Correspondence with Relatives*, Tesla Museum, 1995; NT to Nicholas Kasanovic, 03/01/1941.
9. "Nikola Tesla, Pioneer Radio Engineer Gives Views on Power. Tesla Says Wireless Waves Are Not Electromagnetic, but Sound Waves in Nature. Holds Space Not Curved. Predicts Power Transmission to Other Planets." *New York Herald Tribune*, September 11, 1932.
10. Trump Report, February 13, 1943, OAP Archives, FOIA.
11. NT to Sava Kosanović, 3/4/1941. In Nicholas Kosanović, ed., *Nikola Tesla: Correspondence with Relatives*, Lackawanna, NY: Tesla Memorial Society, 1995. This Kosanović book states "60 billion volts," but that was an error. It should read "60 million volts." This issue was confirmed by viewing the original telegram sent by Milica Kesler from the Tesla Museum, Belgrade, and translated by Nemanja Jevremovic, as stated in emails to M. Seifer, 12/2/2019.
12. Trump Report, February 13, 1943, letterhead, OAP Archives, FOIA.
13. "Nazi Saboteurs in the Amagansett Sands," NYT, June 13, 1942.

14. Homer Jones, letter re: Tesla papers, 2/4/1943, FBI Archives, FOIA.
15. John O'Neill, *Prodigal Genius*, New York: Ives Washburn, 1944. Reprinted by David McKay, 1972; BLCU.
16. Elizabeth Rauscher and Toby Grotz, *Proceedings of the Tesla Centennial Symposium*. According to Grotz, the top secret particle beam weapon paper that Puharich obtained was in microfiche form, which suggested that Bergstresser took the paper from Tesla to be photographed, which is pretty close to what he told me. He said that Tesla gave him the paper to make copies. Grotz further said, according to Bergstresser, whom he had met, that Tesla had contacted the U.S. War Department, and since he was an electrical engineer, that is a key reason why the assignment to meet with Tesla was given to him; personal interview with Toby Grotz, Jamestown, RI, 9/30/2018.
17. Andrija Puharich, *Tesla's Magnifying Transmitter*, xvii.
18. "Chronicle: Plenty of Demand for Poe, Kennedy, and Tesla," NYT, January 27, 1992.
19. W. Bernard Carlson, *Tesla: Inventor of the Electrical Age*, Princeton, NJ: Princeton University Press, 2013, 126.
20. Ibid., 115–16, 360, 412.
21. NT. "Our Future Motive Power," *Everyday Science and Mechanics*, December 1931.
22. "What's Going on at the Tesla Museum?" The Oatmeal.com, 2012.
23. NT, "The Transmission of Electric Energy Without Wires," *Electrical World and Engineer*, March 5, 1904, 429–31 (condensed).
24. Michael Krause, *CERN: How We Found the Higgs Boson*, 45.

1. Interview with Nikola Tesla

1. NT to Robert Underwood Johnson, 4/5/1900, BLCU.
2. Zmai Jovanovic, "Luka Filipov," in *Poems*, by Robert Underwood Johnson, New York: *The Century Collection*, 1902, 146–49.
3. NT, in *Poems*, by Robert Underwood Johnson, New York: *The Century Collection*, 1902, 135–36; interview with Professor Mike Markovitch, Long Island University, 1984.
4. NT to Katharine Johnson, 4/8/1896 and 4/9/1896, BLCU.
5. NT to Katharine Johnson, 4/1/1901, BLCU; "Gunga Din," poem by Rudyard Kipling, 1890.

6. Branimir Jovanovic, *Wireless: The Life, Work, and Doctrine of Nikola Tesla*, Belgrade: Vulcan Press, 2016, p. 16.

7. John O'Neill, *Prodigal Genius*, New York, Ives Washburn, 1944, introduction.

8. NT, "A Story of Youth As Told by [Old] Age, 1939, Hotel New Yorker," in NT 1984, 283–84.

9. M. K. Wisehart, "Making Your Imagination Work for You, Interview with Nikola Tesla," *The American Magazine*, April, 1921. Ibid.

10. NT, "A Story of Youth As Told by [Old] Age, 1939, Hotel New Yorker," in NT 1984, 283–84.

11. Op. cit., NT in Wisehart, 1921.

12. Branimir Jovanovic, *Wireless*, p. 9.

13. Op. cit., NT in Wisehart, 1921.

14. Edward Dean Adams, *Niagara Power*, Niagara Falls, NY: Niagara Falls Power Company Publ., 1928.

15. George Westinghouse, *Transmission of Power, Polyphase System, Tesla Patents*, Pittsburgh, PA: Westinghouse Electric and Manufacturing Company, 1893, iv.

16. Op. cit., M. K., in Wisehart, 1921.

2. The Wizard's Lab

1. "Nikola Tesla," *Courier-Journal* (Louisville, KY), Sunday, August 22, 1897.

2. NT, letter to Edward Dean Adams, 7/8/1895(date uncertain), NTM.

3. "Tesla Talks of His Work," *Boston Sunday Globe*, June 7, 1896, p. 21, col 5.

4. NT, "Tuned Lightning," *English Mechanic and World of Science*, March 8, 1907, 107–8; also in NT 1984, 94–95.

5. Robert Underwood Johnson, *Remembered Yesterdays*, Boston: Little Brown, 1924, p. 401.

6. Edward Dean Adams to NT, 6/28/1897, NTM.

7. Henry Fairfield Osborn, dean, Columbia College, to Seth Low, re: Tesla, January 1894, in NT 1916, 71.

8. Email from Tim Eaton to Marc Seifer, 5/28/2017, re: Watson analyzing Tesla as an artist.

9. Branimir Jovanovic, *Wireless: The Life, Work, and Doctrine of Nikola Tesla*, Belgrade: Vulcan Press, 2016, 132.

10. Ibid., p.136; NT to American Pipe Construction Co., 4/4/1910, NTM.

11. Marion Crawford to his wife, Elizabeth Berdan, 4/27/1894, from Bratislav Stojiljkovic, curator, Tesla Museum; parts of the letter translated with the help of Marc J. Seifer; original at Houghton Library, Harvard University.
12. NT to RUJ, 05/02/1894, BLCU.
13. "Can See Through A Man: Nikola Tesla's Remarkable Achievements with X Rays and a Luminous Screen," *New York Herald*, April 10, 1896.
14. Richmond P. Hobson to Grizelda Hull, 12/22/1903, Hobson papers in Carlson, *Tesla: Inventor of the Electrical Age,* 354.
15. "The Merrimac Destroyed," NYT, June 4, 1898, 1:4.
16. Hobson to NT, 5/1/1905, KSP.
17. Kelvin letter to NT, 05/20/1902, BLCU.
18. KJ to NT, 6/6/1898, NTM.
19. NT to KJ, 6/11/1900, BLCU.
20. NT, "Of Interest to Women," *Butte Daily Post,* January 29, 1902, 10 (condensed), BDA.
21. Fritz Lowenstein to NT, 4/18/1912, KSP.

3. Gay Nineties

1. Stanford White to NT, 3/2/1895, SWP.
2. "Mr. Tesla's Great Loss, Firemen Unable to Save His Laboratory," NYT, March 14, 1895.
3. NT to Clara Reed Anthony, March 25, 1895 from The Gerlach Hotel, Swann Galleries, 10/28/2021.
4. Phoebe Hearst to NT, 5/12/1896, NTM.
5. "Tesla in Jersey," *Rochester (NY) Express*, April 5, 1895, TAE.
6. Theodore Dreiser, *Newspaper Days*, New York: Horace Liveright Publ., 1931, 182.
7. Theodore Dreiser, *Sister Carrie*, New York: Random House Modern Library, 1932.
8. Theodore Dreiser, letter to NT, 9/29/1929, NTM.
9. The phrase "the tragedy of desire" refers to *An American Tragedy* and *A Place in the Sun*, Imogen Sara Smith, 2018, loa.org.
10. Evelyn Light to NT, re: Theodore Dreiser, 10/24/1932, NTM.
11. George E. Hale to NT, 3/17/1896; NT to George Hale, 3/20/1896, NTM.
12. NT to George E. Hale, 6/4/1908, NTM.

13. "Artificial Sunlight: The Stupendous Problem that Tesla Is Trying to Solve," *Pioche (NV) Record,* July 31, 1903, 3, BDA.

14. "Sending Messages to the Planets Predicted by Dr. Tesla on His Birthday, NYT, July 11, 1937, 13:2; cited in Seifer, *Wizard,* 422.

4. Electric Bath

1. "Tesla's New Invention to Preserve the Beauty of Youth Through Life," *The World,* Sunday, October 31, 1897.

2. Robert Connolly, "The Vital Light of Nikola Tesla: Healing Power of UF Rays," *Vitality,* November 25, 2019.

3. Robert Rowen and Howard Robins, "A Plausible 'Penny' Costing Effective Treatment for Corona Virus—Ozone Therapy," *Journal of Infectious Diseases and Epidemiology* 6, no. 2 (2020).

4. "Science Turns to Electricity in Cancer War," *New York Herald Tribune,* September 7, 1932, 36.

5. Daisy Maud Gordon to NT, 11/2/1897, NTM.

6. Alan F. Dutka, *Misfortune on Cleveland's Millionaire Row*, Charleston, SC: History Press, 2015, pp. 34–35.

7. Ibid.

8. Ibid., 34.

9. Daisy Maud Gordon to NT, 11/18/1898. Correspondence continued mostly involving visits to the lab, introducing friends to Tesla and dinner engagements, including letters written on 12/23/1898 and1/3/1899 and several telegrams NTM.

10. Ibid., Wednesday and Monday, circa 1898, NTM.

11. Ibid., undated Thursday handwritten epistle, circa 1898.

12. Ibid., two telegrams dated 12/8/1898; *Wikipedia.*

13. Ibid., 12/15/1898, handwritten letter signed Daisy Maud.

14. Ibid., 12/23/1898.

15. Dutka, op. cit., 36.

16. Mrs. Elizabeth G. Pelton (a.k.a. Daisy Maud Gordon), obituary, *The New York Sun,* July 7, 1919.

17. NT to Louis C. Tiffany, 5/16/1914, NTM.

18. "Earth Electricity to Kill Monopoly: A Way to Harness Free Electric Currents Discovered by Nikola Tesla," *World Sunday Magazine,* March 8, 1896, NTM; Tesla and the Autoharp, *The Music Trade Review,* 1897, p. 21. This article with the accompanying illustration which resembles cell phone technology states that Tesla sent "electric signals

through the earth a distance of twenty miles." Although not overtly stated, this information most likely refers obliquely to his experiment whereby he sent an impulse from his New York City lab up the Hudson to West Point, where he had travelled to receive the impulse, an experiment he conducted at this time. Gary Peterson Archives.

19. Phoebe Hearst to NT, 7/11/1896, NTM.
20. NT to William Randolph Hearst, 4/14/1919, NTM.
21. NT to William Randolph Hearst, 6/10/1921, NTM.
22. William Randolph Hearst to NT, 7/13/1921, NTM.

5. Wardenclyffe

1. "Cable Dispatch from London," *Buffalo Morning Express*, February 22, 1901 [Practical Application. Special to the *Buffalo Morning Express*, from *The New York Sun* correspondent, February 22, 1901, 1; *Electrical World and Engineer* 37 (1901): 364.]
2. NT, Can Radio Ignite Balloons? *Electrical Experimenter*, October 1919, p. 591.
3. Seifer, *Wizard*, 260–61.
4. Ibid., pp. 296–299. Throughout the spring and summer of 1901, as Tesla was working on plans to construct his laboratory and tower, although he settled on Stanford White as architect, Tesla was also working with the young competitor Titus de Bobula who wrote on August 28, 1901 requesting photographs of the site. Although de Bobula did not end up designing Wardenclyffe, thirty years later, de Bobula would indeed design a more modern looking Wardenclyffe-like tower, which in that instance resembled a Van de Graaff generator, NTM.
5. "Cloudborn Electric Wavelets to Encircle the Globe," NYT, March 27, 1904; cited in Seifer, *Wizard*, 289.
6. Theodore Stanton and Harriot Stanton Blatch, eds., *Elizabeth Cady Stanton as Revealed in Her Letters*. New York: Harper Brothers, 1922.
7. Theresa Krull, "Electrical Genius and His Struggles of Interest Today," *Indianapolis Star*, August 16, 1942, 5, BDA.
8. Ibid.
9. Telephone interview with Jane Alcorn, 2/5/2018.
10. "Ground Penetrating Radar Results, *The Tesla Files*, Prometheus Films, July–August 2017.

11. NT. The Magnifying Transmitter, *Electrical Experimenter,* June 1919, 176–178.

12. Ibid., 108–109; CSLI, 402–3.

13. The Art of Transmitting Electrical Energy Through the Natural Medium, N. Tesla, patent #787,412, filed 5/16/1900, published 4/18/1905.

14. Natalie Stiefel, "Nikola Tesla at Wardenclyffe," *Distant Sparks,* Summer 2005.

15. Ibid.; *The Echo,* July 1902.

16. Cited in Seifer, *Wizard,* 407–408; Nikola Tesla vs. George C. Boldt Jr., Suffolk County Supreme Court, April 1921, LAP.

17. Natalie Stiefel, op. cit., Summer 2005.

18. *Wizard,* pp. 407–408.

19. "Tesla's Latest: He Demonstrates How a Torpedo Boat Can Be Blown to Pieces," *Buffalo Evening News,* May 15, 1899.

20. Interview with Randy Hagerman by Joe Sikorski, *Invisible Threads,* 2021.

21. Gary Peterson, email to Seifer, 12/4/14.

22. Fierce Telecom website.

23. Carnetdevlo.org/wireless/loomis.

24. William Preece, "On the Transmission of Electrical Signals," *Electrical Engineer,* August 30, 1893, 209; cited in Seifer, *Wizard,* 108.

25. "Cloudborn Electric Wavelets to Encircle the Globe," NYT, March 27, 1904 (condensed); cited in Seifer, *Wizard,* 283.

26. NT 1916, 74–107.

27. Jovan Cvetic, email to M. Seifer, 1/18/15.

28. R. M. Winans, "Wireless Power" [interview with NT], *Buffalo Courier,* March 12, 1912, 4.

29. "World System of Wireless Transmission of Energy, 1927," in NT 1981, 83–86.

6. House of Morgan

1. L. S. Klepp, book review of *The House of Morgan* by Ron Chernow, *Entertainment Weekly,* March 9, 1990 (paragraph condensed).

2. NT to Thomas Fortune Ryan, 12/20/1905, NTM; *Wizard,* pp. 296–299.

3. NT to Henry Clay Frick, 12/28/1905, NTM; Henry Clay Frick papers, series correspondence. Frick Art Reference Library Archives.

4. Ibid., 2/13/1906.

5. NT to JPM, 2/17/1905, LOC.

6. RUJ to NT, 12/28/1897, LOC.

7. NT to Osgood S. Vili, 3/2/1901; in Branimir Jovanovic *Wireless: The Life, Work, and Doctrine of Nikola Tesla*, Belgrade: Vulcan Press, 2016, 130.

8. NT to JPM, 10/13/1904, LOC.

9. George Scherff to NT, 4/10/1906, LOC; T. C. Martin to NT, 12/24/1905, NTM; M. Seifer, "Nikola Tesla: Forgotten Inventor," in *Psychohistory: Persons and Communities*, edited by J. Dorinson and J. Atlas, New York: LIU, June 1983, 209–31; Marc Seifer, "Tesla: The Lost Wizard," *ITS Proceedings, 1984*, Colorado Springs, CO, 31–40.

10. NT to JPM, 02/02/1906, LOC.

11. Katharine Johnson to Robert Johnson, 9/21/1906, University of Delaware Archives.

12. NT to Anne Morgan, 3/31/1913, NTM.

13. "Tesla Will Destroy An Army a Minute," *New York Herald*, March 11, 1908.

14. Allan L. Benson, "Nikola Tesla, Dreamer. His Three-Day Ship to Europe and His Scheme to Split the Earth, *World Today*, February 1912, 1763–66.

15. Jean Strouse, *Morgan: American Financier*, New York: Random House, 1999, 631–32.

16. NT to JPM, Oct 15, 1906, LOC; in Seifer, *Wizard*, 323.

17. NT to JPM, 10/17/1904; in Seifer, *Wizard*, 314–15.

7. Remote-Control Robotics

1. NT 1919, 124. According to Elon Musk, his Tesla roadster should have "complete self-driving cars" by 2020. "On Monday, Musk said Tesla has a huge advantage over autonomous vehicle competitors because it gathers a massive amount of data in the real world. This quarter, he said, Tesla will have 500,000 vehicles on the road, each equipped with eight cameras, ultrasonic sensors and radar gathering data to help build the company's neural network. The network allows vehicles to recognize images, determine what objects are and figure out how to deal with them." Michael Liedtke and Tom Krisher, "Full Autonomy: Tesla Gears Up for Completely Self-Driving Cars, but Doubts Linger." *Providence Journal*, April 23, 2019, A12.

2. "Nikola Tesla, Methods of Controlling Automata at a Distance, in NT 2013, 206.

3. "Tesla Declares He Will Abolish War," NYHT, 11/8/1898, p. 6, Cameron Prince Archives.

4. Tesla Claims Invention: Japanese are using his dirigible torpedo with a new powerful explosive, *The Missoulian*, Missoula, Montana, 2/18/1904, p. 2.

5. NT 1900, 184–85; cited in Seifer, *Wizard*, 202–3.

6. In response to an email concerning how Tesla's telautomaton could "learn," Ken Corum responded, "The 'AND' gate is the basis for all digital electronics. . . . [Thus] the boat can learn because it is operated by what we now call 'states,' hence a state machine." Corum then gave as an example, "Step one, do X; Step two, do Y. Step three, if X > Y, then do Z. Each state is a step. Each state is the flip of a relay, or as things get more complex, series of relays [which Tesla had on his boat]. That control is called a program. . . . Data is nothing more than the relay ON or OFF (as 1 or a 0). Put 8 relays in a row and you have a byte of data. Have 1024 sets or 8 relays each, you now have 1K memory to store information. . . . Tesla's boat clearly demonstrates electrical automation for the very first time and in my opinion, is the first digital computer. . . . Today with digital electronics you can have billions and billions of relays. . . . The principles used today have never changed since Tesla." Ken Corum, email to M. Seifer, 3/11/2019.

7. NT 1900, 184–85.

8. Lloyd Espenschied, "Discussion of History of Modern Radio-Electronic Technology," *Proceedings of IRE* (1959): 1253–58.

9. J. H. Hammond Jr. and E. S. Purington, "Rebuttal to Espenschied," *Proceedings of IRE* (1959): 1258–66. Anderson, *Nikola Tesla: Guided Weapons*.

10. "Tesla as 'the Wizard' Expands Electricity as a Guest of the Commercial Club," *Chicago Tribune*, May 14, 1859, 1, BDA.

11. "Tesla Declares He Will Abolish War," NYHT, 11/8/1898, p. 6.

12. Abraham Menashe, *Dipping into Light*, Science Fiction/BBC, 1938.

13. Antonio Perez-Yuste, "Early Development of Wireless Remote Control: The Telekino of Torres-Quevedo," IEEE.org, January 2008.

14. "Tom Edison's Son Explodes Desk by Accident," NYT, May 3, 1898, 7:1.

15. Anderson, *Nikola Tesla: Guided Weapons.*
16. Seifer, *Wizard*, 194–95.
17. Willard L. Candee, *Electrical Review and Western Electrician*, November 25, 1911, 1059.
18. "Tesla's Latest Wonder: Magician of Electricity Says He Will Abolish War," *Reading Times*, PA, 11/10,1898, p. 2, BDA.
19. NT, "Torpedo Without a Crew," *Current Literature*, February 1899, 136–37.
20. M. Huart, "The Genius of Destruction," *Electrical Review*, December 7, 1898, 7; Seifer, *Wizard*, 196–97.
21. Mark Twain to NT, 11/17/1898, NTM; cited in Anderson, *Nikola Tesla, Guided Weapons*, 131.
22. NT, letter to Benjamin F. Miessner, author of *Radiodynamics: The Wireless Control of Torpedoes and Other Mechanisms*. New York: Van Nostrand, 1916, LOC.
23. JHH Jr. to NT, 2/16/1911, NTM; cited in Seifer, *Wizard*, 348–49.
24. NT to JHH Jr., 2/18/1911, NTM; in Seifer, *Wizard*, 349.
25. Lloyd Espenschied, "Discussion of History of Modern Radio-Electronic Technology," *Proceedings of IRE* (1959): 1254.
26. Hammond to Lowenstein, 5/11/1911; in Lloyd Espenschied. "Discussion of History of Modern Radio-Electronic Technology," *Proceedings of IRE* (1959): 1254.
27. Ibid.
28. Aitken, *Continuous Wave Technology.*
29. JHH Jr and E. S. Purington, "Rebuttal to Espenschied," *Proceedings of IRE* (1959): op. cit., 1259–60.
30. Miessner to JHH Jr., 2/6/1912, in "Rebuttal of Espenschied" by J. H. Hammond Jr. and E. S. Purington, *Proceedings of IRE* (1959): 1259.
31. Tesla-Fessenden U.S. Patent Interference Case, August 1902, pp. 87, 97–98.
32. Miessner to JHH Jr., 2/12/1912, op cit.
33. JHH Jr., patent #1,522,882, granted January 13, 1925.
34. Hedy Lamarr, patent #2,292,387 Secret Communication System; Melinda Wenner, "Hedy Lamarr: Not Just a Pretty Face," *Scientific American*, June 3, 2008.
35. Hammond Jr. and E. S. Purington, "Rebuttal to Espenschied," *Proceedings of IRE* (1959): JHH Jr. and E. S. Purington (1959): op. cit., 1262.
36. Fritz Lowenstein, patent #1,470,088, 10/9/1923.

37. Puharich, *Beyond Telepathy*; Seifer, *Wizard*, 354.

38. Nancy Rubin. *John Hays Hammond Jr.*, Gloucester, MA: The Hammond Museum, 1987.

39. Nikola Tesla, "Breaking Up Tornadoes," *Everyday Science and Mechanics*, December 1933; in NT 1984, 248–49.

8. Marconi and the Germans

1. Guglielmo Marconi, "Syntonic Wireless Telegraphy," *Electrical Review*, June 15, 1901, 755.

2. Adapted from Chapters 39–41 of *Wizard*.

3. Wikipedia.

4. "Mr. Tesla Before the Royal Institution, London," *Electrical Review*, March, 19, 1892, 57; in Martin, *Researches, Writings and Inventions*, 200.

5. NT, *New York World*, 4/13/30.

6. R. N. Vyvyan, cited in Marconi, *My Father Marconi*, 138.

7. NT 2008, 4/28/1910, 478.

8. By doubling that figure and dividing by the speed of light, he also calculated the resonant frequency of the Earth, which was 1/12th or 0.8484 of a second.

9. "Your Excellency," Tesla began writing from his Wardenclyffe laboratory, "Noting that [my] inventions . . . have been appropriated by others . . . boldly exploited to my detriment, I am compelled to protest humbly, yet emphatically, against the intended use of Your Excellency's name and authority in connection with the sending of a wireless message announced in the journals . . . [as] this would create an erroneous impression all over the world, which might permanently injure my material interests and temporarily detract from my scientific reputation as original investigator." NT to Teddy Roosevelt, 1/15/1903, NTM, quoted in Kruk, *Nikola Tesla: The Force Awakens*, 66–67.

10. NT to K. G. Frank, 10/12/1915, NTM. Tesla writes, "My discoveries in [AC transmission and motor design] are the foundation of an immense industry in Germany, but it is no benefit to myself . . . for such has been the uniform experience of American inventors. . . . Ask any. . . . Bell, Edison, Thomson, Bush, Weston and others, what pecuniary benefits they have received from Germany, and the answer will be none." NT, "Patents in Germany, Complaints of Ill Treatment of American Inventors," NYS, May 7, 1908.

11. Ambrose Fleming letter to R. N. Vyvyan, circa 1937, R & R Auction House.

12. Bernstein, *Three Degrees Above Zero*, 83.

13. Marconi, *My Father Marconi*, op. cit., 189–90.

14. KSP, in Seifer, *Wizard*, 361.

15. NYT, August 9, 1914.

16. Tesla-Fessenden, U.S. Patent Interference, Case, 1902, 87, 97–98, M. Seifer Archives; Seifer, *Wizard*, 280–81.

17. E. J. Wheeler, ed. "Rival Systems of Wireless Telegraphy and the Differences Between Them," *Current Opinion*, February 1914.

18. Seifer, *Wizard*, 361.

19. NYT, August 17, 1914.

20. K. G. Frank to NT, 10/7/1915, NTM. Although Tesla is asking Frank about the possibility of obtaining German patents, he already had patents in Germany that the Kaiser would not recognize. It should be noted that over his lifetime, Tesla had about 120 fundamental U.S. patents and nearly another 200 patents worldwide in such countries as Argentina, Australia, Austria, Belgium, France, Great Britain, Russia, India and Japan.

21. KSP, in Seifer, *Wizard*, 361.

22. Tuckerton Radio Station to J. Daniels, Secretary of the Navy, July 3, 1916, NA. Homag stands for, in English, High Frequency Machine Stock Company for Wireless Telegraphy.

23. R. M. Winans, "Wireless Power," interview with NT, *Buffalo Courier*, March 12, 1912, 4; NT to JJA, 3/22/1909, NTM.

24. Vincent Astor to NT, 5/15/1912, NTM.

25. NT to K. G. Frank, 2/16/1915, NTM.

26. "U.S. in Charge of Sayville," *Brooklyn Eagle*, July 9, 1915.

27. Gary Peterson, "The Application of EM Surface Waves to Wireless Energy Transfer," April 8, 2017.

28. K. L. Corum and J. F. Corum, "Bell Labs and the Radio Surface Wave Propagation Experiment, Texzon Technologies: Texas Symposium on Wireless Microwave Circuits and Systems, Baylor University, Waco TX, March 31, 2016"; Tesla 1916; Arnold Sommerfeld. "Uber die Ausbreitung der Wellen, in der drahtlosen Telegraphie," *Annalen der Physik* 28, no. 4 (March 16, 1909): 665–736; note cited by Anderson in Tesla 1916, 133.

29. Tesla explains in his 1916 deposition that Wardenclyffe was set up to reduce considerably the radiation aspect and increase the conductive properties by sending the EM energy through the Earth. "You want high potential currents, you want a great amount of vibratory energy; but you can graduate this vibratory energy. By proper design and choice of wave lengths, you can arrange it so that you get, for instance, 5% in these EM waves and 95% in the current that goes through the Earth. That is what I am doing. Or, you can get, as these radio men, 95% in the energy of EM waves and only 5% in the energy of the current. Then you are wondering why [they] do not get good results." NT 1916, 132–33; phone interview with Gary Peterson, 3/31/2020.

30. "Germans Treble Wireless Plant," NYT, April 23, 1915.

31. NT to F. W. Woolworth, 10/16/1912, NTM. In *The Tesla Files*, Prometheus Films, shown on the History Channel in June 2018, astrophysicist Dr. Travis Taylor speculated that when Tesla moved into the Hotel New Yorker, circa 1934–1943, he may have placed a laboratory in the top floor in a particular corner storage room, near where the telephone switchboard was located. Pipes in this corner room went all the way to ground, and also the Hotel New Yorker had its own generator separate from the grid, thus he would have had access to electric power. Taylor's speculation that Tesla, in a sense, created a Wardenclyffe wireless setup, has some merit, and in support of that, I pointed out that Tesla planned to do precisely this in the Woolworth building fifteen years earlier, but this segment was cut from the final show. A key point to keep in mind was that Tesla was hit by a taxicab in June of 1937 at the age of eighty-one, and was thus too incapacitated to do any real work after that point, that is, from 1937 until his death in 1943. Thus, if he did do wireless experiments using the hotel's height and electrical setup, it would have to have been during the years 1934 through the first half of 1937, when his interest seemed to be more involved in developing his particle beam weapon.

32. NT to RUJ, 12/24/1914, LOC.

33. NT to JPM Jr., 12/23/13, LOC.

34. R. P. Hobson to NT, 6/6/1902, NTM; cited in Anderson, *Nikola Tesla: Guided Weapons*, 134–35.

35. Lieutenant Commander A. R. Sulfidge to N. Tesla, 4/29/1896, NTM.

36. R. B. Bradford, chief of the Bureau of Engineering and Development, U.S. Navy, to NT, 10/23/1900, NTM.

37. NT to U.S. Navy, chief of Bureau of Equipment, 11/24/1900.

38. NT to J. D. Long, secretary of the Navy, 10/19/1900, NTM.

39. William Crawford to NT, 6/19/1900, NTM.

40. R. B. Bradford, chief of the Bureau of Engineering and Development, U.S. Navy, to NT, 12/17/1900.

9. The Great War

1. Edward H. Smith, "Tesla Describes Wireless Warfare of the Future," the *World* magazine, January 30, 1916.

2. "Denies German Ownership," NYT, August 17, 1914.

3. "Kintner v. Atlantic Communication Co., et al. Marconi Wireless Telegraph Co. of America v. Atlantic Communications, District Court, NY, March 4, 1921," legal.com/decision-1921.

4. K. G. Frank to NT, 9/17/1915, NTM.

5. NT to JPM Jr., 02/19/1915, LOC.

6. "Tesla Sues Marconi," NYT, August 4, 1915.

7. "Sayville Trebles Output," NYT, April 23, 1915. As stated in the text, Sayville's accomplishment was achieved by taking Tesla's advice, by increasing the ground connection and creating "a double tuned circuit which allowed the primary capacitance to be very large . . . what may have happened . . . is Tesla showed [the Germans] how to increase the primary capacitance and vary the mutual coupling from the primary to the secondary to match the break speed. Those adjustments could improve their output easily by a factor of 10" (email Ken Corum to M. Seifer, 1/18/2021).

Where the *New York Times* reported the creation of three five-hundred-foot towers, Joe Sikorski's research has suggested that "Telefunken erected sixteen small towers the first of which was 150 feet tall for short range messages with four taller towers to support the [five-hundred-foot-tall main] antenna of the 100 watt transmitter. The towers themselves were more of a support to what was called an 'umbrella antennae' system—which was a series of wires that connected to the tower masts and created a sort of web resembling an elaborate maypole. But every newspaper seemed to publish a different amount of towers." Sikorski confirmed nine smaller supporting towers (Joe Sikorski, email to M. Seifer, 4/1/2020), and various photographs confirm Sikorski's assessment. Tesla's deposition, written in 1916, however, emphasizes that he played a key role in getting Telefunken to reduce the radiation aspect and increase dramatically the ground connection, and this was certainly confirmed by the reports that Sayville had in rapid fashion tripled their power. The importance of this event in supporting the viability of Tesla's theories cannot be understated. See NT 1916, 133.

8. FDR, assistant secretary of the Navy, re: Tesla priority in wireless, September 14, 1916, NA.

9. "Prof. Pupin Now Claims Wireless His Invention," *Los Angeles Examiner*, May 13, 1915; Seifer, *Wizard*, 371–73.

10. Marconi v. Atlantic Communications, 1915; Kintner v. Atlantic Communication Co., et al. Marconi Wireless Telegraph Co. of America v. Atlantic Communications, District Court, NY, March 4, 1921. See also Marconi Wireless v. Kilbourne and Clark, Decided May 3, 1920, ravellaw.com. Kintner had taken over Fessenden's company, and Tesla had established priority over Fessenden in court in 1902, see Seifer, *Wizard*, 280–82; Marconi at the Key: Inventor, with Judge Veeder, Visits Wireless Station, NYT, 5/2//1915; *Invisible Threads* by Joe Sikorski, 2021.

11. NT, "The Disturbing Influence of Solar Radiation on the Wireless Transmission of Energy." *Electrical Review and Western Electrician*, July 6, 1912; and NT 1984, 123.

12. Stephenson, Mr. Nikola Tesla and the Electric Light of the Future. Scientific American, 3/30/1895, pp. 16408–09; A Way to Harness Free Electricity Discovered by Nikola Tesla, *The World Sunday Magazine*, 3/8/1896.

13. NT 1916, 29–30, 105; NT to JPM, 12/26/1901, NTM.

14. Anderson, "John Stone Stone," cited in Seifer, *Wizard*, 372–73.

15. Marconi Wireless Telegraphy Company of America, plaintiff-appellant, vs. Kilbourne and Clark Manufacturing Company, defendant, appellant, United States Circuit Court of Appeals, 9th Circuit, Brooklyn, NY, March 10, 1916.

16. Lessing, *Man of High Fidelity*, 42–43.

17. Seifer, *Wizard*, 373–74; Engineering and Technology History Wiki, ethw.org.

10. The Fifth Column

1. NT to K. G. Frank, 5/21/1916, NTM.

2. K. G. Frank to NT, 5/16/1916, NTM.

3. Blum, *Dark Invasion*, 70.

4. Ibid., 146; Kaiser Sends First Wireless to Wilson, NYT, 1/29/1914; Wilson's Word to Kaiser, NYT, 6/21/1914.
5. Jones and Hollister, *German Secret Service.*
6. Ibid.; Return of Alien Property, U.S. Congress, Committee of Ways and Means, Statement by Senator Sutherland, Alien Property Custodian, November 23, 1926; a heavyweight whose life reads like a German version of a Le Carré novel, von Papen would later survive the Nuremberg Trials, live through the 1950s and 1960s, and pass away at age 85, in 1969.
7. Jones and Hollister, *German Secret Service*, 49.
8. Wikipedia.
9. Jones and Hollister, *German Secret Service*, op. cit., 49.
10. "Captain Boy-Ed," letter to the editor, NYT, July 9, 1915, 10.
11. *New York Times* Current History series, vol. 17, *The European War.*
12. Ibid., 113–14.
13. "JP Morgan Shot by Man Who Set the Capitol Bomb," NYT, July 3, 1915; NT to JPM Jr., July 1915, LOC.
14. NT, "Wireless Controls German Air Torpedoes," NYT, July 10, 1915.
15. NT, "Science and Discovery Are the Great Forces which Will Lead to the Consummation of the War," NYS, December 30, 1914; Edward H. Smith, "Tesla Describes Wireless Warfare of the Future," *The World Magazine*, January 30, 1916, 5–6.
16. Charles Apgar, The Amateur Radio Station Which Aided Uncle Sam, EE, 11/1915, pp. 337–338; *Invisible Theads* by Joe Sikorski, 2021.
17. NT to George Scherff, 12/25/1917, LOC.
18. RUJ to NT, 3/1916, LOC.
19. K. G. Frank to NT, 11/29/1915, NTM, German translation by Google.
20. NT to K. G. Frank, 12/1/1915, NTM, German translation by Devin Keithley.
21. Karl Ferdinand Braun to Tesla, 7/11/1915, NTM, cited in Kruk, *Nikola Tesla: The Force Awakens*, 12.
22. Tesla to Karl Ferdinand Braun, 9/11/1915, NTM, cited in Kruk, *Nikola Tesla: The Force Awakens*, 13.
23. W. H. Bragg, June 11, 1931, in NT 1961, cited in Kruk, *Nikola Tesla: The Force Awakens*, 8–9.
24. NT to U.S. Lighthouse Board, 9/27/1899, NA.
25. NT to RUJ, 11/10/1915, BLCU.

26. Seifer, *Wizard*, 380.

27. Wikipedia.

28. "Tesla No Money," *New York World*, March 16, 1916.

29. NT to K. G. Frank, 12/17/1915, NTM.

30. NT 1919, cited in Seifer, *Wizard*, 382.

31. NT to Edison, 2/9/1912, NTM; Edison to NT, 2/23/1912, NTM.

32. NT to Edison, 6/22/1914, NTM.

33. NT to Edison, 12/16/1914, NTM.

34. Edison to NT, 12/16/1914.

35. NT, Edison Medal Speech, 1917.

36. J. B. Smiley to F. Hutchins, 7/16/17, LAP.

37. NT to Waldorf-Astoria, 7/12/1917, LAP.

38. "Tesla's New Device Like Bolts of Thor," NYT, December 8, 1915.

39. Joseph Alsop, "Beam to Kill Army at 200 Miles, Tesla's Claim on 78th Birthday," *NYHT*, July 11, 1934, pp. 1, 15, in NT, 1981, p. 112.

40. NT to William Crawford, 4/28/1913, NTM.

41. NT to JPM Jr., 4/8/16, LOC.

42. "Spies on Ship Movements," *NYT*, 2/17/1917.

43. Mitchell Palmer, *Report of all Proceedings Under the Trading with the Enemy Act During 1918*. United States Alien Property Custodian Publication, 1919.

44. Blum, 2014.

45. "19 More Taken as German Spies," *NYT*, 4/8/1917.

46. "Reason for Seizing Wireless," *NYT*, 2/9/1917.

47. "Zenneck Arrested," *The Evening News*, 7/6/1917; "More Aliens Seized," *New York Herald Tribune*, July 7, 1917.

48. "German Attaches Must Quit US," *Brooklyn Eagle*, July 7, 1917.

49. James Corum to M. Seifer, Philadelphia Tesla Conference, July 2010.

50. *New York Times* Current History series, vol. 17, *The European War.*

51. K. G. Frank to NT, 7/11/1935, NTM

52. NT to K. G. Frank, 11/13/1935, NTM.

53. Wikipedia.

54. Attorney Crammond Kennedy for Homag to Secretary of the Navy Joseph Daniels, 7/3/1916; Joseph Daniels to Crammond Kennedy, 7/14/1916 and 3/28/1917, NA.

55. FDR to Secretary of Commerce re: Tesla's priority over Marconi, 9/14/1916, NA.

56. Seifer, *Wizard*, pp. 390–394.

57. The story of ice skaters atop Wardenclyffe adapted from Jane Alcorn interviews with old-time Shoreham residents, Tesla Museum at Wardenclyffe; names and incidental details fictionalized.

58. George Scherff to NT, 08/20/1917, LOC.

59. "Destruction of Tesla's Tower at Shoreham, LI Hints of Spies," NYS, August 5, 1917.

60. NT to E. M. Herr, 10/19/1920, LOC.

61. E. M. Herr to NT, 11/16/1920, LOC.

62. George Westinghouse Corp to NT, 11/28/21, LOC.

63. NT to GW Corp, 11/30/1921, LOC.

11. Tesla's Mysterious 1931 Pierce-Arrow

1. Frederick M. Kerby, "Nikola Tesla Tells How We May Fly 8 Miles High at 1000 mph." *Reconstruction*, July 1919; TFCbooks.com.

2. NT. "Nikola Tesla's View of the Future of Motive Power." *Manufacturer's Record*, December 29, 1904, Prometheus Films Archives.

3. Interview with Francis Fitzgerald of the Niagara Power Commission, Buffalo, New York, by Ralph Bergstresser, 1945, FBI papers, J. Kinney Archives.

4. Gary Peterson website.

5. Heinz Jerbens to M. Seifer, 11/15/2003.

12. Telephotography

1. Arthur Korn to NT, 1931 in NT 1960.

2. COLI, 497–99.

3. NT, "Developments in Practice and Art of Telephotography, *Electrical Review*, December 11, 1920, in NT 1956, A-94–A97; Seifer, *Wizard*, 184.

4. The photophone, Wikipedia; Zoom conversation, Marc Seifer with Vint Cerf (and others) at Tesla Science Center at Wardenclyffe Honors Vint Cerf, 10/30/2020. During the discussion which at first centered around whether or not Marconi had direct contact with Tesla, which was a question that Vint asked and I commented on, Vint mentioned the importance of Hedy Lamarr's frequency skipping invention. He also discussed Bell's invention of transmitting multiple phone conversations along the same phone line, circa 1880, and I commented that Bell also invented the ability to transmit voice via

light beams which became a forerunner for fiber optics. When I asked
him what his greatest surprise was after developing the Internet, Vint
went over the history of the invention beginning about 1969 and then
outlined three major developments: in particular, the world wide web,
which was invented by Berners-Lee, circa 1989; the introduction of
powerful graphics to this interface; and Steve Jobs' development of
the smart phone, including a motion picture camera and so on, which
evolved in the early 2000s.

5. NT, The Wonder World To Be Created By Electricity, *Manufacturer's Record*, 9/9/1915, in NT, 1956, p. A-182.
6. Carol Bird, "Tremendous New Power Soon To Be Unleashed, Nikola Tesla, Starting His 78th Year, Works on Revolutionary Power Project and Also Is Completing Process for Photographing Thought," *Kansas City Journal-Post*, September 10, 1933.
7. "MIT Media Lab: Where Tomorrow's Technology is Born," *60 Minutes* segment, CBS TV, April 22, 2018, discussing work of Arnav Kapur.

13. The Day Tesla Died

1. O'Neill, *Prodigal Genius*, 381–82 (passage greatly condensed).
2. "General Andrew McNaughton and the Canadians, *Time*, cover story, August 10, 1942, 30–32.
3. FDR Memorandum, 1/2/1943; Marvin H. McIntyre, reply, White House Memorandum, 1/5/1943, FDR Presidential Library; Prometheus Films Archives.
4. FDR Memorandum to Eleanor Roosevelt, 1/9/1943, NTM cited in Dmitry Kruk, 2020, p. 52.
5. V. Bush to FDR, 12/16/1942 in Zachary, p. 206.
6. Bloyce Fitzgerald's service record and accompanying material, Herb Vest Archives, Dallas, TX, email to M. Seifer, 10/2/2020.
7. F.E. Conroy Memorandum to "Director," FBI files, 10/17/1945, FOIA.
8. Office of the chief medical examiner, coroner's report concerning Nikola Tesla, Hotel New Yorker, January 7, 1943, 10:30 P.M. Dr. Palmer's notes read, "White adult male. . . . Found dead by maid Alice M [illegible]. Sudden death while asleep. . . . Nothing susp." Prometheus Films Archives.
9. Tesla, "Prolific Inventor Dies," NYT, January 8, 1943, 19:1.

10. *Tesla's Death Ray: A Murder Declassified*, Discovery Channel, January 2018, segment 3.

11. "2,000 Are Present at Tesla's Funeral," NYT, January 13, 1943, 24:2,3.

12. "Nikola Tesla Eulogy," written by Louis Adamic, read by Mayor Fiorello LaGuardia over radio station WYNC, Sunday, January 10, 1943; Ernst Lawrence's and James Franck's Western Union telegrams, Louis Adamic papers, Yale University, c/o Charles de la Roche Archives.

13. W. R. Hearst to Raul Walsh, 9/15/1942, R&R Auction House.

14. Talbot, *Devil's Chessboard*, 25.

15. NT to JPM Jr., 11/29/1934, FBI Archives, FOIA.

16. Wikipedia. Speculations on the web that Foxworth was murdered because he had special knowledge about Tesla's death ray strike this author as completely absurd.

17. Percy Foxworth, FBI memorandum, 1/12/1943, FBI Archives.

18. *The Tesla Files*, Prometheus Films, History Channel, May and June 2018, segments 2 and 5; rumor stems from speculation in Adam Trombley Lecture, 1988, International Tesla Society, Colorado Springs, Colorado, as reported by Tim Eaton who attended this lecture, in email to M. Seifer, 2/17/2017.

19. "Dr. Tesla Gives Home to an Errant Pigeon," NYT, 1935; The leg band translation refers to the International Federation of Racing Pigeons, with this pigeon, number 1283, born in 1934. "N.U." might stand for the owner.

20. Phone interview by M. Seifer with Ralph Bergstresser, circa 1989; Michael Riversong interview with Ralph Bergstresser, ITS Conference, 1988. "Aerial Defense Death Beam Offered to U.S. by Tesla," *Baltimore Sun*, June 12, 1940; William Laurence, "Death Ray for Planes," NYT, September 22, 1940, 2:3; "Proposing the Death Ray for Defense," *Philadelphia Inquirer*, October 20, 1940.

21. William Laurence, "Vast Power Source in Atomic Energy Opened by Science," NYT, May 5, 1940.

22. OAP Tesla file, 11/21/1940, OAP Archives.

23. David Bazelon, OAP to Wright Field, 10/24/1947, OAP Archives.

24. "Proposing the Death Ray for Defense," *Philadelphia Inquirer*, October 20, 1940.

25. W. H. Ballou, "Tesla's Death-Dealing Aero to End War Along Border," *Washington Post*, January 24, 1916, Professional Historical Newspapers.

26. "Tesla's New Device Like Bolts of Thor," NYT, December 8, 1915; Seifer, *Wizard*, 387–88.

27. J. Edgar Hoover memorandum, 1/9/1943, FBI files; memorandum to J. Edgar Hoover re: Tesla's particle beam weapon and William Laurence's September 24, 1940 NYT article on same, FBI files.

28. FDR note, 1/2/1943, NA, Prometheus Films Archives.

29. F. E. Cornels memorandum to J. Edgar Hoover, FBI files, 10/17/1945.

30. Culbert, *Information Control and Propaganda*, CISUPA.com.

31. Fenn, *At the Dragon's Gate*, 16–17.

32. O'Neill, *Prodigal Genius*, 290–310. As a point of interest, the Manhattan Storage Warehouse was designed by McKim, Mead & White.

33. A perusal of Tesla's extensive correspondence with Kirsch from the Tesla Museum Archives reveals nothing of interest along those lines. Most of the correspondence between Tesla and Kirsch and Tesla's secretary, Mrs. Skerrit, occurred in 1913 when Kirsch was sent to Montreal to set up a contract with Miller Brothers and Sons to manufacture and market Tesla's turbine pumps, Canadian patent #135,174. While filming *The Tesla Files*, traveling with Rob O'Brien, a writer for Prometheus Films, and Jason Stapleton to the Riverhead Courthouse on Long Island, we located a mention of blueprints on the construction of Wardenclyffe in Tesla's files, but no blueprints were forthcoming. Although unable to locate them in Stanford White's papers, blueprints of Tesla's tower can be found at the Tesla Museum.

34. Leland Anderson would talk Agnes out of auctioning off the Tesla correspondence to her parents so that the papers could be donated, en masse, to the Butler Library at Columbia University.

35. Nancy Cataldi and Carl Ballenas, *Maple Grove Cemetery*, Charleston, SC: Arcadia Publishing, 2006, 99.

36. Ralph Bergstresser, "List of Person's Associated with Nikola Tesla," 1945, Joe Kinney Archives, FBI files.

37. FBI memorandum, 3/29/1943, Dover, Delaware.

38. Bloyce Fitzgerald to NT, 3/9/1939, NTM.

39. Ibid.

40. Ibid.

41. Bloyce Fitzgerald to NT, 12/20/1942, NTM.

42. Ibid.

43. F. E. Cornels memorandum to Percy Foxworth, assistant director of the FBI, New York office, January 9, 1943, FBI records.

44. Rhodes, *Making of the Atom Bomb*, 352–53, 412.

45. Bloyce Fitzgerald would end up in a rather ignoble situation, relocating out West, where he raised as much as fifty thousand dollars in Arizona and California from various investors who believed he had developed "an ultrasonic gadget that could be used to kill or repel insects, locate schools of fish for fisherman, cut the cost of making beer and test vibrations in jet airplanes." Unfortunately, none of these claims panned out, and "Fitzgerald was arrested in Phoenix for passing scores of bogus checks. Extradited to Los Angeles, Fitzgerald stood trial in California in the fading months of 1952. Apparently the scholarly-appearing defendant . . . who was also billed in the press as an alleged swindler and con-man, avoided jail time because his investors refused to believe they had been victims." Fitzgerald was only thirty-six years old when these charges were filed. Unfortunately, his writing of bad checks continued, and Fitzgerald was arrested a decade later in Los Angeles, "accused also of absconding with an expensive late model car . . . which earlier had been reported as stolen in Los Angeles." This 1964 *Nevada Sate Journal* article went on to say, "Fitzgerald is said to be an electronics expert with many contacts in defense and aerospace industries." *Arizona Republic*, November 19, 1952, 7, Prometheus Films Archives; "Troubles Mount for Fitzgerald, *Arizona Daily Star*, November 19, 1952, 22, BDA; "Accused Swindler Returned to Stand Trial in California," *Tucson Daily Citizen*, November 20, 1952, 6, BDA; "Swindle Suspect to Be Tried for Promoting Electronic Device," *Arizona Daily Star*, November 21, 1952, 2, BDA; "Theft Suspect Given to LA Police," *Arizona Republic*, November 23, 1952, BDA; "Bad Check Suspect Nabbed Here," *Nevada State Journal*, December 8, 1964, 12, BDA.

46. Edward A. Tann, FBI memorandum to D. Ladd, 1/12/1943, FBI Archives.

47. J. Edgar Hoover, FBI memorandum, 1/21/1943, FBI Archives.

48. T. J. Donegan, FBI Memorandum, 1/11/1943, FBI Archives; Wikipedia.

49. Branimir Jovanovic, Wireless: The Life, Work & Doctrine of Nikola Tesla. Belgrade: Vulcan Press, 2016, p.160.

50. Seifer, *Wizard*, 438.

51. NT to G. S. Viereck, 12/20/1934, Christies Auction House, BDA.

52. Nikola Tesla, as told to George Sylvester Viereck, "A Machine to End War," *Liberty Magazine*, February 9, 1935, 5–7.

53. Seifer, *Wizard*, 438–39.

54. George Sylvester Viereck memoranda by J. Edgar Hoover and FBI associates, 10/28/1941, 1/8/1942, 6/2/1943, 2/3/1943, and 4/22/1947, FBI Archives, FOIA. [are these memos that Viereck wrote *to* Hoover and associates? If not, why are they *his* memos?]

55. Branimir Jovanovic, *Wireless: The Life, Work, and Doctrine of Nikola Tesla*. Belgrade: Vulcan Press, 2016, 160.

56. Seifer, *Wizard*, 427–28.

57. NT to KJ, 11/3/1898, BLCU.

58. NT to JPM, 4/22/1904 and 4/1/1904, LOC.

59. Wikipedia.

60. Jacob Schiff to NT, 8/19/1906, NTM.

61. "U.S. Official Lauds Work of Three Jews in Development of Communications: Fritz Lowenstein, Arthur Korn, and Emile Berliner," *Wisconsin Jewish Chronicle*, April 7, 1939, 20, BDA.

62. NT to Simon Guggenheim, 10/16/1932, NTM.

63. Seifer, *Wizard*, 428.

64. NT to Carl Laemmle, 7/15/1937, Corbis Archives.

14. The Trump Report

1. Trump Report, 1/30/1943, OAP Archives, FOIA.

2. Rhodes, *Making of the Atomic Bomb*, 259–60.

3. Ibid., 269–70.

4. Ibid., 635; Kenneth Davis, *FDR: The War President: 1940–1943*.

5. Isaacson, *Einstein: His Life and Universe*, 481; Clark, *Einstein: The Life and Times,* 549–50; Vannevar Bush to Frank Aydelotte, 12/30/1941.

6. Christgau. *Enemies: WWII Internment*, 65–69.

7. OAP Archives, FOIA.

8. Vannevar Bush to NT, 7/1/1931, in NT 1961, NTM.

9. *Encyclopedia Britannica*; Kenneth Davis, *FDR: The War President: 1940–1943*.

10. FDR, letter to Vannevar Bush, 11/17/1944, White House Archives.

11. James Conant to FDR, 4/25/1941; *The Tesla Files*, Prometheus Films Archives.

12. General Andrew McNaughton to NT, 7/5/1938, NTM.

13. C. J. Mackenzie to General Andrew McNaughton, 12/6/1940, in *The Mackenzie-McNaughton Wartime Letters,* edited by Mel Thistle, 56–57, Toronto: University of Toronto Press, 1975.

14. Ibid., 7/14/1942, 112–113.

15. "Yankee Scientist: Vannevar Bush," *Time,* cover story, April 3, 1944, 52–56.

16. MIT, biographical sheets on John G. Trump, MJS.

17. Zachary, *Endless Frontier,* 128.

18. Seifer, *Wizard,* 452–53.

19. George, *Surreptitious Entry.*

20. "Mr. George of Naval Intelligence once had a professional locksmith open the safe previously opened by Kosanović at the Hotel New Yorker as well as [open] several other locked trunks . . . in the presence of all the other men." John Newington to Walter Gorsuch, 2/15/1943, OAP Archives, FOIA.

21. Bloyce Fitzgerald, letter to NT, 12/20/1942, NTM.

22. Seifer, *Wizard,* 459; Colonel T. B. Holliday, letter to James Murphy of the OSS, September 6, 1945; James Murphy of the OSS to Colonel T. B. Holliday, September 21, 1945, NA; Kevin Leonard, researcher, *The Tesla Files,* Prometheus Films Archives.

23. Myrna Oliver, "Laurence Craigie, First U.S. Military Jet Pilot Dies," *Los Angeles Times,* February 28, 1994.

24. L.C. Craigie, letter to FBI, 10/19/1945, FBI Archives.

25. LAP.

26. Seifer, *Wizard,* 391.

27. Douglas, *Radio Manufacturers of the 1920's.*

28. NT to S. M. Kitner, Westinghouse Corporation, 4/7/1934; *The Tesla Files,* Prometheus Films Archives.

29. Branimir Jovanovic, *Wireless: The Life, Work, and Doctrine of Nikola Tesla,* Belgrade: Vulcan Press, 2016, 139.

30. Johnson, *Remembered Yesterdays,* 542–43.

31. R. U. Johnson to Bainbridge Colby, 12/6/1932, MJS.

32. R. U. Johnson to Bainbridge Colby, 5/28/1931, MJS.

33. NT to Frank Vanderlip, 2/9/1935, NTM, cited in Branimir Jovanovic, *Wireless: The Life, Work, and Doctrine of Nikola Tesla,* Belgrade: Vulcan Press, 2016, 161.

34. Wikipedia.

35. Irving Jurow to M. Seifer, 7/5/1993, MJS.

36. NT to Andrew Robertson, Westinghouse Corporation, 5/22/1941, LOC.

37. NT to Carl, 3/2/1942, Tim Eaton Archives.

38. Andrew Robertson. *About George Westinghouse and the Polyphase Electric Current*. New York: Newcomen Society, 1939, 28; Seifer, *Wizard*, 430–31.

39. Trump Report, 1/30/1943, OAP Archives.

40. "Nikola Tesla: The New Art of Projecting Concentrated Non-dispersive Energy Through Natural Media, 1937," in *ITS Proceedings, 1984*, 144–50.

41. Homer Jones, chief, Division of Investigation and Research, NDRC report to James Markham, deputy Alien Property Custodian, 1/30/1943, OAP Archives.

42. Zachary, *Endless Frontier*, 134.

43. Frank B. Jewett, letter, January 1943, cited in *Electrical Engineering* 62, no. 8 (August 1943).

44. J. Edgar Hoover to Vannevar Bush, 5/13/1943; Bush to Hoover, 5/18/1943, cited in Zachary, *Endless Frontier*, 159–60.

45. Clarence A. Robinson Jr., "Debate Seen on Charged-Particle Work," boxed argument within the article "Soviets Push for Beam Weapon," *Aviation Week and Space Technology*, May 2, 1977, 17.

46. Zachary, *Endless Frontier*, 179.

47. Vannevar Bush and James Conant. "Guided Missiles and Techniques: Training of Pigeons, Tech Report, Div. 5," December 1946, NDRC.

15. The Russian Connection

1. Trump Report, January 30, 1943, OAP Archives, FOIA.

2. Evgeny Zhirnov, "No Armor Able to Protect Against a Damaging Effect," *Kommersant* (the Russian Daily), June 18, 2012, Kommersant. Ru.doc.1946; NT to Tolokonsky, undated circa July 1934, NTM.

3. G. P. Brailo from Amtorg to NT, 8/4/1934; NT to A. A. Vartanian, 12/1934, NTM.

4. Agreement between NT and Amtorg Trading Corporation, representatives of the USSR, NTM; Prometheus Films archives.

5. Wikipedia.

6. NT, Brief Statement Concerning the Present Status of the Work Conducted Since April 20, As Per Agreement, 1935.

7. Evgeny Zhirnov, "No Armor Able to Protect Against a Damaging Effect," *Kommersant* (the Russian Daily), June 18, 2012; M. Galaktionov to K. Voroshilov, re: Tesla; declassified Soviet Union communiqué, July 26, 1935; Charles de la Roche Archives, email communiqué, 11/9/2018.

8. Wikipedia.

9. Ibid.

10. Evgeny Zhirnov, op cit., "No Armor Able to Protect Against a Damaging Effect," *Kommersant* (the Russian Daily), June 18, 2012.

11. Kruk, *Nikola Tesla: The Force Awakens*; Branimir Jovanovic, *Wireless: The Life and Work of Nikola Tesla*, Belgrade: Vulcan Press, 2016, 160–61.

12. NT to A. A. Vartanian, 11/8/1935, NTM.

13. Seifer, *Wizard*, 454–55.

14. Clarence A. Robinson Jr., "Soviets Push for Beam Weapon," *Aviation Week and Space Technology*, May 2, 1977, 16–23.

15. Ibid.

16. Andrija Puharich presented the paper at the 1984 International Tesla Society Meeting in Colorado Springs, Colorado, and the paper was published in those proceedings. According to Margaret Cheney, intelligence officers associated with DARPA (which was formed in the 1950s, well after Tesla's death) had access to microfilm copies of probably all of Tesla's notebooks, including those compiled in Colorado and Wardenclyffe and also, naturally, every detail Tesla put down concerning his particle beam weapon. After the Tesla Museum published Tesla's Colorado Springs notebook, officers at DARPA compared the published book with their own copies and found them to be essentially the same. Cheney's source was an unnamed intelligence officer. Cheney: *Tesla: Man Out Of Time*, 309–10.

17. *Tesla's Death Ray: A Murder Declassified*, Discovery Channel, January 2018, segment 3.

18. Private conversation with former DARPA contractor, 8/18/2020; Scalar Waves Conference review most likely of a Tom Bearden lecture, 12/22/1987, declassified yet still redacted Confidential/Noforn document, CIA-RDP96-00792R000500240001-6.

19. *Encyclopedia Britannica*, History of Radar.

20. Seifer, *Wizard*, 376; NT, "Science and Discovery Are the Great Forces Which Will Lead to the Consummation of the War," NYS, December 20, 1914, in Tesla 1956, A162–A171.

21. "New Yankee Tricks to Circumvent the U-Boat," *The Fort Wayne Journal Gazette*, Sunday, August 18, 1917, 18; Prometheus Films Archives.

22. H. Winfield Secor, "Tesla's Views on Electricity and the War: Exclusive Interview with Nikola Tesla," *Electrical Experimenter*, August 1917, 220–21, 270.

23. Ibid., 270.

24. Zachary, *Endless Frontier*, 165–68, 203.

25. "Yankee Scientist: Vannevar Bush," *Time*, cover story, April 3, 1944, 52.

26. NT, patent #787,412, 1956, 334 (quote condensed).

27. Seifer, *Wizard*, 474, 476; "Seafarer ELF Submarine Command and Control Communications Systems Records, MSS-249," nmu.edu.

16. Negotiations with the British Empire

1. H. Grindell Matthews, "The Death Power of Diabolical Rays," NYT, May 21, 1924, 1:2, 3:4,5.

2. Seifer, *Wizard*, 427. "Then there is Dr. Antonio Longoria who says that he destroyed a death ray machine which he invented in 1933 because it was too dangerous. Of this machine, Albert Burns, president of the Inventors Congress of 1934 said that he had seen it kill pigeons, rabbits, dogs and cats at considerable distances. Now, Dr. Longoria says that he is willing to re-assemble his apparatus in the event that the United States is subjected to an unwarranted attack." From "Proposing the Death Ray for Defense," *Philadelphia Inquirer*, October 20, 1940. Longoria also claimed that he could "bring down pigeons on the wing, at ranges up to four miles." From "To Specific," *Time*, October 23, 1939, 11–12.

3. "When Marconi Tested Nikola Tesla's 'Ray of Death,'" February 28, 2019, Saper-L.com; "Scientists and Their Real-Life Death Rays," Grunge.com.

4. Helen Welshimer. "Dr. Tesla Visions the End of Aircraft in War," *Everyday Week Magazine*, October 21, 1934, 3.

5. NT. "On Light and High Frequency Phenomena," *Electrical Engineer*, March 8, 1893, 248–49; Seifer, *Wizard*, 88.

6. Branimir Jovanovic, *Wireless: The Life and Work of Nikola Tesla*, Belgrade: Vulcan Press, 2016, 165.

7. NT to British War Office, 8/28/1936, NTM.

8. British War Office to NT, 9/14/1936, NTM.

9. H. L. Lewis, British War Office, to NT, 10/27/1936, NTM.

10. Wikipedia.

11. General Andrew McNaughton, *Life*, cover story, December 18, 1939, 9; Wikipedia.

12. General Andrew McNaughton, letter to C. J. Mackenzie, 8/6/1942, in Mel Thistle, ed. *The Mackenzie-McNaughton Wartime Letters*, Toronto: University of Toronto Press, 1975, 115–16; General Andrew McNaughton, *Time*, cover story, August 10, 1942, 31–32.

13. General Andrew McNaughton to NT, 7/6/1937, NTM.

14. Mel Thistle, ed. *The Mackenzie-McNaughton Wartime Letters*, Toronto: University of Toronto Press, 1975, 153, 159.

15. NT to General Andrew McNaughton, 7/18/1937, NTM.

16. NT to General Andrew McNaughton, 9/14/1937, NTM.

17. "Invisible Dust Curtain to Halt War Planes, Says Nikola Tesla," *Popular Mechanics*, July 11, 1934, 693.

18. NT to General Andrew McNaughton, 9/14/1937, NTM; James J. Hill to NT, 8/22/1896, NTM. In this letter, Hill invites Tesla to meet Hermann Thofehrn, who invented a "very compact and powerful little electric locomotive for use in mines . . . [that] can pass through a tunnel 4 feet high by 3 feet wide," "The Thofehrn Electric Mining Locomotive," *Engineering and Mining Journal*, January 17, 1897.

19. John G. Trump, Trump Report, Exhibit E, OAP Archives.

20. NT to General Andrew McNaughton, op. cit., 9/14/1937, NTM.

21. General Andrew McNaughton to NT, 9/16/1937.

22. NT to director of mechanization, British War Office, 2/8/1938, NTM via Joe Kinney Archives.

23. NT to General Andrew McNaughton, 2/10/1938, NTM.

24. General Andrew McNaughton to NT, 2/12/1938, NTM.

25. NT to director of mechanization, British War Office, 4/5/1938, NTM

26. Anderson, *Nikola Tesla: Lecture. Before the New York Academy of Sciences*, 4/6/1897, Breckenridge, CO: 21st Century Books, 1944; NT, The Hurtful Actions of Lenard and roentgen Tubes, Letter to Ed, *Electrical Review*, May 1, 1897.

27. "Tesla as 'the Wizard' Expands Electricity as a Guest of the Commercial Club," *Chicago Tribune*, May 14, 1899, p. 1, BDA.

28. Leland Anderson, talk at ITS, 1988. NT, The new Art of Projecting Concentrated Non-dispersive Energy Through the Natural Media, 1937, in ITS Proceedings, 1984, p. 145.

29. Interview with Nancy Czito, Julius Czito's daughter-in-law, Washington, DC, 1984, cited in Seifer, *Wizard*, 456.

30. Richard Rhodes, "Clashing Colleagues," *University of Chicago Magazine*, Fall 2017, 24 (discusses Enrico Fermi and Leo Szilard).

31. General Andrew McNaughton to NT, 7/5/1938, NTM. At this point, Tesla had a change in attitude and this suggests that the meeting with Dr. Rose took place and that he may have also received compensation.

32. NT to the undersecretary of state, British War Office, 10/27/1938.

33. Ibid.

34. NT to Konstantine Fotic, 11/9/1937, translated by B. Dragasevic; D. M. Stanojevic, 11/1/1941, Royal Yugoslav Consulate General Document, Fotic Papers, Hoover Institute, Stanford, CA.

35. NT to Director of Mechanization, The War Office, London, 11/7/1937, Fotic Papers.

36. NT to Fotic, 11/9/37, translated by B. Dragasevic, Fotic Papers.

37. Ibid., 10/28/1937.

38. NT to Simo Vrlinich, President, Serb National Federation, Pittsburgh, PA, 6/11/1941, Fotic Papers.

39. Zachary, *Endless Frontier*, 160–61.

40. Ibid., 158–59; the boa constrictor quote from Kaempffert is cited in "Tesla, a Bizarre Genius, Regains an Aura of Greatness" by William Broad, NYT, August 28, 1984.

41. Mel Thistle, ed. *The Mackenzie-McNaughton Wartime Letters,* Toronto: University of Toronto Press, 1975, 56, 57, 112, 215–16; General Andrew McNaughton to C. J. Mackenzie, 8/6/1942; FDR White House memo, 1/2/1943, NA.

42. *Tesla's Death Ray: A Murder Declassified*, Discovery Channel, January 2018.

43. NT. The New Art of Projecting Non-dispersive Energy Through the Natural Media, 1935, *ITS Proceedings* 1984, pp. 144–150.

44. M. Seifer discussions with Travis Taylor at Hotel New Yorker, October 2017.

45. David Sharp, "Navy Brings Out Futuristic Guns," *Providence (RI) Journal*, February 18, 2014.

46. Ibid.

17. The Birth of the New Physics

1. A. Einstein, soundtrack to the film *Atomic Physics*, J. A. Rawls Ltd., 1998.

2. Isaac Newton, in Koestler, *Sleepwalkers*, 503.

3. Oliver Nichelson, "Tesla's Self-Sustaining Electric Generator and the Ether," *ITS Proceedings, 1984*, 67; see also Seifer, *Transcending the Speed of Light*, 61–99.

4. Ibid.

5. Ouspensky, *New Model of the Universe*, 356.

6. Dingle, forward to *Duration and Simultaneity* by Henry Bergson, 1965, XX.

7. James Clerk Maxwell, *The Scientific Papers of James Maxwell*, edited by W. D. Niven, Cambridge: Cambridge University Press, 1890, 775.

8. L. Williams, *Origin of Field Theory*, New York: Random House, 1966. 135.

9. Ibid., 61.

10. Ibid., 62.

11. Mikhail Shapkin, ether theory manuscript allegedly attributed to Nikola Tesla, Web 2002. Tesla's actual quote is as follows: "We have made sure by experiment . . . that light propagates with the same velocity irrespective of the character of the source. Such constancy of velocity can only be explained by assuming that it is dependent solely on the physical properties of the medium, especially density and elastic force," "An Inventor's Seasoned Ideas, Nikola Tesla Pointing to 'Grievous Errors' of the Past," NYT, April 8, 1934.

12. G. Lombardi, "The Michaelson-Morley Experiment," drphysics.com, 1997.

13. "Experiments in Alternating Currents with Very High Frequency," in NT 1894, 149.

14. Seifer, *Transcending the Speed of Light*, 121.

15. Fritjof Capra. *The Tao of Physics*, Berkeley, CA: Shambhala Press, 1975, p. 168.

16. Lincoln Barnett. *The Universe and Dr. Einstein*, NY: Time Inc. Book Division, 1948/62, p. 60.

17. Clark, *Einstein: The Life & Times*, 78.

18. Hermann Minkowski, in Clark, *Einstein: The Life & Times*, 123.

19. Clark, *Einstein: The Life & Times*, 124.

20. Charles Musès, "Hyperstages of Meaning," subchapter in Musès and Young, *Consciousness and Reality*, 130.

21. George Gamow, *Thirty Years That Shook Physics*, Garden City, NY: Doubleday, 1966, 46.

22. Ibid., 124.

23. E. Gora, letter to the editor, re: Dirac, *Journal of Occult Studies*, 1978, 207–8.

24. NT, "On Light and Other High Frequency Phenomena," *Journal of the Franklin Institute*, August 1893, 91.

25. Sharla Stewart, "How to Catch a Higgs," *University of Chicago Magazine*, April 2001, 21.

26. Guy Wilkinson, "Measuring Beauty (Towards a Higher Theory of Physics), in *Extreme Physics*, Scientific American Publications, 2019, 7.

27. Seifer, *Transcending the Speed of Light*, 84–85, 119–20, 313; Ouspensky, *New Model of the Universe*, 358. The idea that "light has weight" was established by Professor Lebedev of Moscow: "Light when falling on bodies produces a mechanical pressure on them."

28. Peter Weiss, "Jiggling the Cosmic Ooze: A New Blueprint for All the Universe's Mass and Energy May Be Just Around the Corner." *Science News* 159 (March 10, 2001): 152–54.

29. Jeffrey Kluger, "The Cathedral of Science," *Time*, July 23, 2012, 33–35.

30. Krause. *CERN: How We Found the Higgs Boson*.

31. Peter Weiss, "Jiggling the Cosmic Ooze," *Science News* 159 (March 10, 2001): 152–54.

32. Jeffrey Kluger, "Interactive Panorama: Step Inside the Large Hadron Collider," *Time*, July 12, 2012, 35.

33. Ouspensky, *New Model of the Universe*, 354.

34. Ibid., 394–96.

35. Ibid.

18. Tesla's Dynamic Theory of Gravity

1. Robert Millikan to NT, 1931, in NT 1960.

2. Tesla, May 20, 1891, in NT 1893, 145.

3. Ibid., 148.

4. Ibid., 149.

5. Rutherford to NT, in NT 1960.

6. "Nikola Tesla, Mr. Tesla's Vision," NYT, April 21, 1908.

7. NT to Corinne Robinson, 03/27/1899, Harvard University Archives.

8. Walter Russell to Royal Lee, 3/24/1954, Toby Grotz Archives.

9. Walter Russell to NT, 7/24/1934 and 3/24/1939, NTM.

10. Tesla Gets Scott Award For Electrical Inventions, NYHT, 10/6/1934.

11. "Death-Ray Machine Described," NYS, July 11, 1934, in NT 1981, 108–9.

12. "Tesla at 78, Bares New 'Death-Beam,'" NYT, July 11, 1934, 18:1, in NT 1981, 109.

13. Joseph Alsop, "Beam To Kill Army at 200 Miles, Tesla's Claim on 78th Birthday," *New York Herald-Tribune*, July 11, 1934, 1, 15, in NT 1981, 110–14.

14. Most of it, however, does appear in my book *Transcending the Speed of Light*, which was published in 2008, but again, I did not use the precise term of Tesla's, "dynamic theory of gravity," pp. 116–118, 229–231, 284–299. In a slightly different unpublished rendition of the article "Sending Messages to the Planets Predicted by Dr. Tesla on His Birthday," NYT, July 11, 1937, Tesla wrote, "It is absolutely impossible to convert mass into energy. It would be different if there were forces in nature capable of imparting to a mass infinite velocity. Then the product of zero mass with the square of infinite velocity would represent infinite energy. But we know that there are no such forces and the idea that mass is convertible into energy is rank nonsense" NTM: MNT, CDXXIX, 67A. Tesla is apparently talking about Einstein's famous equation $E = mc^2$, and if that is the case, this is a very interesting observation. What really is the speed of light squared in a physical sense? If we take a length "a" equal to a line, then a^2 is a plane, yet a plane, which is a higher dimension than a line, is made up of an infinite number of lines. In the same sense, a cube, a^3 is infinity to a plane, a^2. Seen in this light c^2, if considered a higher dimension, would be infinite to the speed of light, c, and would also thus be a "force" that Tesla speculates does not exist. Thus, when Tesla says "if there were forces in nature capable of imparting to a mass infinite velocity," c^2 certainly fits that bill, and if that were the case then Tesla actually states the premise herein which explains the monumental amount of energy trapped inside of mass. The secret is in understanding what c^2 means in a physical sense.

15. Nikola Tesla, circa 1937, Kenneth M. Swezey Papers, Smithsonian Institute, 1–2.

16. NT 1919, *Electrical Experimenter,* cited in Seifer, *Wizard,* 121.

17. Wikipedia.

18. Halexandria.org.

19. NT to Corinne Robinson, 03/06/1899, Harvard University Archives.

20. Marriage invitation of Corinne Douglas to Joseph Wright Alsop, 11/4/1909, NTM.

21. NT, unpublished 1936 manuscript simply titled "Nikola Tesla: New York," Kenneth M. Swezey Papers, Smithsonian Institution. In thinking about this idea of this invisible ether providing energy to existing matter, it occurred to me that humans faced a similar situation in the late 1700s when Joseph Priestly discovered oxygen. Just as all animals need to breathe in invisible air in order to survive, a highly similar process is suggested by Tesla's dynamic theory of gravity, whereby physical matter requires a constant input of ether for the very same reason.

22. Albert Einstein to Hendrik Lorentz in 1916, in Isaacson, *Einstein: His Life and Universe,* 318.

23. Albert Einstein, in Clark, *Einstein: The Life and Times,* 78.

24. NT 1891, NT 1893, 148.

25. Gamow, *Thirty Years That Shook Physics,* Garden City, NY: Doubleday, 1966, 119–20.

26. Ibid., 129–30.

27. Isaacson, 2007, 319.

28. NT, "Pioneer Radio Engineer Gives View on Power. Tesla Says Wireless Waves Are not Electromagnetic, but Sound Waves in Nature. Holds Space not Cured. Predicts Power Transmission to Other Planets." *New York Herald Tribune,* September 11, 1932.

29. E. Lerner, "Magnetic Whirlwinds," *Science Digest,* July 1985.

30. Tesla 1915.

31. NT, "Man's Greatest Achievement," *Milwaukee Sentinel,* July 13, 1930.

32. Seifer, *Wizard,* pp. 202–203; NT, 6/1900, pp. 184–185.

33. NT, "Mr. Tesla's Vision. How the Electrician's Lamp of Aladdin May Construct New Worlds," NYT, April 21, 1908.

34. Bearden, *Energy from the Vacuum,* xx.

35. NT, unpublished 1936 manuscript simply titled "Nikola Tesla: New York," Kenneth M. Swezey Papers, Smithsonian Institution.

36. NT, "Radio Power Will Revolutionize the World," *Modern Mechanix and Inventor,* July 1934, 40–42; 117–19.

37. NT, "Promises to Transmit a Force," NYT, July 11, 1935, in NT 1981, 128–30. It should be noted that Tesla never met Einstein. There is a famous 1922 photograph of Einstein standing next to Charles Steinmetz with a man who resembles Tesla standing directly behind the two of them. A myth has existed for many years that this man is Tesla, but actually his name is Jon Carson, and he was an employee of AT&T. Other than the one letter Einstein wrote congratulating Tesla on his 75th birthday, there is no correspondence between them.

38. Lewis E. Rubin, OAP Archives, Intercustodial and Property Branch, U.S. Government, November 6, 1951.

39. Lew Price and B. Herbert Gibson, *Is There a Dynamic Ether: A New Reality for 21st-Century Physics*, promedia.net/users/greenbo, 2007; Ron Hatch, "A New Theory of Gravity: Overcoming Problems with General Relativity," Physics Essays, March 2007, researchgate. net; Vencislav Bujić, "Magnetic Vortex, Hyper-Ionization Device," linux-hots.org/magvid; Warren York, "Scalar Technology," teslatech. info, 2007; David Wilcox, "Convergence VIII: Scientific Proof of the Nature of a Multi-Dimensional Harmonic Universe," ascenion200. com, 2000; Seifer, *Transcending the Speed of Light*, 116.

40. Mike Hodges, email to M. Seifer, 3/27/2017, re: his new theory of everything, medium.com, 9/20/2017.

41. Isaacson, 2007, pp. 318–319.

42. Hannah Devlin, "Beam Me Up, Scotty! Scientists Teleport Photons 300 Miles Into Space," July 12, 2017, theguardian.com.

43. Ibid.

44. Travis Taylor, personal interview with M. Seifer, Wardenclyffe, New York, August 15, 2017.

45. Max Planck, *"Das Wesen der Materie [The nature of matter]"* (a 1944 speech in Florence, Italy).

46. Seifer, *Transcending the Speed of Light*, 116.

19. Final Thoughts

1. NT to KJ, 12/22/1895, BLCU.

2. "Nikola Tesla Tells How He'd Defend Ethiopia Against Italian Invasion," *Detroit Times*, September 22, 1935, 3.

3. D. L. Petkovic, "A Visit to Nikola Tesla," *Politika* XXIV, no. 6824.
4. William Terbo, reminiscences.
5. Mrkich. *Nikola Tesla: The European Years*, 7–12.
6. Arthur Compton to NT, 7/8/1931, in NT 1960.
7. O'Neill, *Prodigal Genius*, 304.
8. NT, "Fragments of Olympian Gossip." This is the title of a poem Tesla wrote circa 1928 for his friend George Sylvester Viereck, on the one hand poking fun at Albert Einstein and the new paradigm arising in quantum physics, yet on the other hand suggesting the Serbian wizard's ability to "listen on my cosmic phone," cited in "Tesla: Life and Legacy," PBS.org.
9. NT, "Man's Greatest Achievement," *Milwaukee Sentinel*, Sunday, July 13, 1930.

Acknowledgments

I first learned about Tesla in 1976, the same year that I began my life as a professional journalist. Since I was also studying parapsychology at that time, I had stepped into a very bizarre world where such things as telepathy, remote viewing, psychokinesis, UFOs, and life after death became plausible possibilities. In that sense, I had entered a minefield because it became very difficult to discern where reality ended and fantasy or wishful thinking began. But all that changed after I read John O'Neill's biography on Nikola Tesla titled *Prodigal Genius* and Tesla's book of his lectures, articles, and patents. Yes, there certainly was a great deal of mystery and possible fantasy associated with Tesla, but one thing that was not fantasy were his patents, and thus the concrete, provable actuality of his accomplishments. I could see that Tesla was indeed the inventor of what we could call the AC hydroelectric power system, and at the same time, I could also see the numerous attempts to obscure his contributions and thereby relegate him to nonperson status.

I became fully committed to learning everything I could about his life, spending six years in the archives of several universities writing a doctoral dissertation on why his name disappeared from the history books and another dozen years honing a definitive biography.

Because of this work, not only did I gain an encyclopedic understanding of the details of Tesla's life, but I also met amazing Tesla experts who helped me attain an even greater understanding of the full scope of his thoughts, theories, and achievements. In a way, this work culminated with me starring in the five-part History Channel miniseries *The Tesla Files*, which gave me access to another level of understanding that would have been impossible to glean in any other way. I was already working on a sequel to *Wizard: The Life and Times of Nikola Tesla, Biography of a Genius* when the show began, but as the show progressed, my understanding of Tesla's life expanded, not only because of the help I received from Kevin Burns, the producer from Prometheus Films who made all this possible, and his production staff, but also because of a number of other people who contacted me from all corners of the globe to share additional esoteric information. Simply put, I was placed in a unique situation that gave me access to heretofore unknown or unpublished knowledge about Tesla's life, details that could not have been obtained in any other way. *Wizard at War* is also shaped extensively by the use of primary sources, including numerous never before published letters to and from Tesla; information from his private notebooks from Colorado Springs and Wardenclyffe, from the Tesla Museum in Belgrade, the National Archives, the Library of Congress, and many university libraries, particularly the Butler Library at Columbia University, the

357

University of Rhode Island, Harvard University, and Brown University; and information obtained from the FBI and the Office of Alien Property that was released to me through the Freedom of Information Act.

The list of people who helped me shape this book is long. And if I miss mentioning anybody, I apologize profusely. I can say, however, with certainty, that this book would not have been possible without the help I received from the following individuals: Kevin Burns for conceiving of *The Tesla Files* and for hiring and believing in me; numerous people from the show, including producers Joe Lessard, Scott Hartford, and David Silver, director Scott Rettberg, writer Rob O'Brien, on-screen buddies Jason Stapleton and the absolutely brilliant rocket scientist Travis Taylor; Scott Stanley; Hunter Bartholomew; and Rebecca Banks for uncovering a memorandum by FDR on his interest in Nikola Tesla. Also New Yorker Hotel manager Joe Kinney, for providing me with an amazing letter Tesla wrote to the British War Office.

Other key individuals without whom this book in its present form would simply not have been possible include my screenplay writing partner and ultimate Tesla geek, Tim Eaton; people at Wardenclyffe, including Barbara Daddino, who, like Tim, sent me numerous key articles; Jane Alcorn, for sharing some amazing anecdotes and inviting me to speak at several conferences; Velimir Abramović for his wonderful description of Tesla's laboratory and meeting with Swami Vivekananda; Diane Draga Dragasevic for translating the letters between Tesla and Ambassador Fotic; and Lynn Sevigny, whose illustrations of Tesla's particle beam weapon and Wardenclyffe tower, which can be found in *Wizard*, helped me in many ways. Numerous Tesla experts also helped immeasurably, in particular Branimir Jovanovic, head of the Tesla Museum in Belgrade, who has been my friend and colleague in this quest for over thirty years, and his predecessors, Alexander Marincic for first opening the door of the Tesla Museum in 1986 and inviting me to speak at several European conferences, and Vladimir Jelenković for gifting me several very expensive Tesla tomes; also, the museum's research assistants, including Milica Kesler, Alexander Ivkovic, and Bratislav Stojiljkovic, who asked me to help decipher the handwriting of Marion Crawford.

In America, numerous Tesla experts also came to my aid in a variety of ways, including Ljubo Vujovic, who early on encouraged me; Gary Peterson, Jovan Cvetic, and Jim and Ken Corum, for helping me understand how Wardenclyffe works and how Tesla's remote-controlled robot had within its construction a binary system that lay at the basis of the first computers; Michael Krause, Eric Dollard, Dmitry Kruk, Chris Cooper, and Toby Grotz, for inviting me to my first Tesla conference in 1984; Tibor Hrs Pandur for research assistance and correcting some of my math; journalists William Broad, John H. Wasik, Phil Cozzolino and Scott Smith; Cameron Prince for taping a key lecture of mine on Tesla and the God particle and designing the highly regarded Tesla Universe website; Stanley Chang for providing a rare Tesla photo; Charles de laRoche for helping me obtain

Russian documents; Ernst Willem, Nemanja Jevremovic, Herb Vest, and Vasilj Petrovic for talking me into going to the 2006 unveiling of the new Tesla statue on the Canadian side of Niagara Falls; Les Drysdale for creating that inspiring statue; Tom Valone for inviting me to speak in Washington; Steve Ward for explaining how the Tesla coil works; Steve Elswick for inviting me to speak at a half dozen international Tesla conferences held in Colorado Springs; Nenad Stankovic for publishing quite a number of my articles in his *Tesla Magazine*, and Irving Jurow, who was on-site working for the Office of Alien Property the very day Tesla died.

Other key individuals without whom the book would not have been possible include several who are no longer with us: Tesla's grandnephew, William Terbo, who wrote the introduction to my first Tesla book; Elmer Gertz, who met Tesla at Viereck's apartment; Ralph Bergstresser, the last person to see Tesla alive and the one most responsible for forwarding Tesla's top secret paper on his particle beam weapon to Andrija Puharich, who released it to the world and who helped me on other aspects of the book; Robert Golka, for constructing two gigantic Tesla towers and for stopping by my house on numerous occasions and sharing his wealth of knowledge; Bill Wysock; John Ratzlaff; and Leland Anderson, the godfather of Tesla researchers. Other key individuals who helped include Tom Bearden, Daniel Snyder, Ryan Cochrane, Nicholas Lonchar, David Hatcher Childress, Marina Schwabic, Goran Lazovich, Nancy Czito, Karen Bouchard, archivist from Brown University, Sheila Mason from the Miller Center, Dara Baker from the Franklin D. Roosevelt Presidential Library, filmmakers Dave Grubin and his *Tesla* film for American Experience, Donna Davies and her crew for Ruby Tree Films and Joe Sikorsky, for his movies *Tower to the People* and *Invisible Threads* and for creating a rather wonderful 1.5 minute promo for this book!; George Noory from the *Coast to Coast* radio show, and Howard Smukler for starting me on this entire quest by giving me the O'Neill biography in 1976.

Also Stanley Krippner, my mentor for my doctoral dissertation, which was a 714-page treatise on why Tesla's name disappeared from the history books; Alan Wilson, the editor who first published *Wizard* and thereby changed my life; typists Allyn Pazienza and Pat Mullaney; my literary agent, John White; my sister, Meri, and her husband, John Keithley; their son, Devin Keithley; my brother, Bruce Seifer; Uri Geller; Nelson DeMille; J. T. Walsh and Sandy Neuschatz. At Kensington Publishing House, I would like very much to thank the entire staff, particularly CEO Steve Zacharius, publisher Lynn Cully, associate publisher Jackie Dinas, Vida Engstrand, the marketing director, copy editor Jeff Lindholm, production editors Arthur Maisel and Sherry Wasserman, and especially Michaela Hamilton, editor-in-chief, for believing in me and backing the project. I would also like to thank my wife, Lois Pazienza, who has had to live with Tesla these last forty-plus years!

INDEX